D1117021

MENAGERIE

MENAGERIE

The History of Exotic Animals in England
1100 – 1837

Caroline Grigson

OXFORD
UNIVERSITY PRESS

OXFORD
UNIVERSITY PRESS

Great Clarendon Street, Oxford, OX2 6DP,
United Kingdom

Oxford University Press is a department of the University of Oxford.
It furthers the University's objective of excellence in research, scholarship,
and education by publishing worldwide. Oxford is a registered trade mark of
Oxford University Press in the UK and in certain other countries

First Edition published in 2016
Impression: 1

Published in the United States of America by Oxford University Press
198 Madison Avenue, New York, NY 10016, United States of America

British Library Cataloguing in Publication Data
Data available

Library of Congress Control Number: 2015941386

ISBN 978–0–19–871470–5

Printed in Great Britain by
Clays Ltd, St Ives plc

For Joe Banks

ACKNOWLEDGEMENTS

Many people have helped and encouraged me during the long gestation period of this book. I would particularly like to thank Arthur MacGregor, Ray Desmond, Adrian Lister, Marco Masseti, Ian Rolfe, Sally Festing, Sheila Hamilton-Dyer, Florence Pieters, Katlijne Van der Stighelen, Christopher Plumb, Isobel Armstrong, and Laurence Cook, as well as Rosemary Baird for information about Goodwood and for facilitating a memorable visit to the remains of the menagerie there, Iain Bain for sending me copies of Thomas Bewick's correspondence, Louise Martin for supporting my honorary membership of the staff of the UCL Institute of Archaeology, Roberto Portela Miguez for supporting me as a scientific visitor to the Natural History Museum in London, Paul Cooper and Hellen Pethers of the Natural History Museum Library, Gina Douglas at the Linnean Society, Sam Alberti and Sarah Pearson of the Hunterian Museum at the Royal College of Surgeons, and my son Joe Banks for his unflagging encouragement and interest when presented with the each new 'astounding' discovery. My friend and colleague Juliet Clutton-Brock died as this book was in press, I owe her an immense debt of gratitude for many discussions of fact and interpretation and for writing the Foreword.

I have also received much help from the often anonymous librarians at the British Library, Cambridge Public Library, Liverpool Public Library, the Westminster Archive Centre in London, the National Archives, the London Metropolitan Archive, the London Library, the Science Library at UCL, the Institute of Archaeology library at UCL, and the London Library. And, of course, I need to express my gratitude to my agent, Georgia Glover, at David Higham for contacting Oxford University Press on my behalf, to the two anonymous reviewers for recommending the book for publication, and to Luciana O'Flaherty, Matthew Cotton, and other staff members at the Press for their exemplary efficiency and enthusiasm, as well as Hilary Walford for her meticulous copy-editing. Finally it would be ungracious not to acknowledge how much the research for the book was facilitated by internet access to the Rhino Resource Centre, the British Newspaper Archive, the Biodiversity Heritage Library, and Google.

CONTENTS

LIST OF PLATES

LIST OF FIGURES

FOREWORD

Why are human beings so fascinated and often enraptured by other species of animals? It is a question that is not easy to answer, but maybe it is because the essence of humanness is centred in boundless curiosity combined with the instinct to nurture and a need to exert personal power. The keeping of dangerous yet strange and unusual animals in captivity has been recorded in pictures and writings for thousands of years—from the bas reliefs in the British Museum we can see how lions were let out of cages to be hunted by the kings of ancient Assyria, a 'sport' that continues today with the canned hunting of lions in South Africa. And from Diodorus Siculus we learn how hunters caught a snake, 30 cubits long, and brought it to the Egyptian King Ptolemy Philadelphus (285–46 BC) for his collection of wild and exotic animals. From the time of this Greek historian (80–20 BC) there has been an outpouring of literature about the keeping of wild animals in captivity and in 'menageries'—a word from the French, not used in English until 1712.

Today, there are hundreds of zoos and zoological gardens worldwide and thousands of books, articles, films, and television programmes about the wild animals held in them, while there is also a plethora of books about their history. But this book about the history of menageries in England is different, not so much because of the great age of this country's collections of wild animals, including the great cats, elephants and rhinos, but because of the extraordinarily detailed information that is known about the individual animals, their names, their owners, their keepers, and their travels. This knowledge has been hugely increased with the recent digitization of books and particularly of local newspapers, which has made much fascinating and valuable but previously obscure information easily available to historians. And who better to tease out these facts than Caroline Grigson whose professional life has been spent as an archaeozoologist and a curator at the Royal College of Surgeons of England amidst John Hunter's great eighteenth-century collection of wild animal preparations.

This book is very much more than a history of British menageries, it will intrigue and delight the reader with an infinite number of enticing details of which the following description of how a young orang-utan travelled to the cold north is a touching example:

> It usually slept at the masthead wrapped in a sail, but off the Cape of Good Hope it began to suffer from the cold, especially early in the morning, when it would descend from the mast, shuddering with cold, and running to any of its friends would climb into their arms, clasping them closely to warm itself. In Batavia it had eaten mainly fruit, but on board ship it ate all kinds of meat, especially raw, and was fond of bread, although it always preferred fruit if available. It preferred tea and coffee to water and would even drink wine, sometimes stealing the Captain's brandy bottle.

Juliet Clutton-Brock (16 September 1933–21 September 2015)

And here are Beares, wolfes sett,
Apes, owles, marmoset,

(from the Chester Play of The Deluge)

I

The Normans to the Tudors

1.1. In the Beginning

Britain's most famous and long-standing collection of exotic animals, the royal menagerie at the Tower of London, was probably founded in 1204, when King John brought three ship- or crate-loads of wild beasts to England from Normandy. However, seventy-five years earlier, at Woodstock in Oxfordshire:

> King Henry the First enclosed the park at *Wudestoc* with a wall, though not for *deer*, but for all foreign *wild beasts*, such as *lions*, *leopards*, *camels* and *linxes*, which he procured abroad of other Princes.[1]

Henry's menagerie also included hyaenas, a porcupine, and a rare owl. It was said that, whenever he travelled about the kingdom or went to war, he took some of the animals with him as a sign of his royal status.[2]

Most of the animals that arrived at the Tower were gifts to the reigning monarch from other European kings, a lion as the 'King of the Beasts' being particularly appropriate. Other spectacular animals were sometimes incarcerated there, including three leopards sent to Henry III in 1235 by Frederick II, the Holy Roman Emperor; a 'white' bear from the King of Norway in 1251, which was allowed to swim in the Thames; and four years later an African elephant, 'a beast most strange and wonderful to the English people' (Figure 1.1), a gift from Louis IX of France.[3]

The Tower was not the only place where exotic animals were kept in Britain. Peafowl—native to India and Shri Lanka—might decorate barnyards or gardens. Very realistic representations of peacocks appear on Romano-British mosaics, and their bones have been excavated from two Roman archaeological sites,[4] but more birds probably arrived from Europe with the Normans. The earliest images of peacocks that look as though they were based on living birds date to the early fourteenth century, as, for example, in the

Figure 1.1. Henry III's elephant at the Tower of London in 1255 with its keeper Henricus de Flor (from Matthew Paris, *Chronica Maiora*, ii, MS 16, fo. iir).

Queen Mary Psalter and the *Sherborne Missal.*[5] As well as being admired for their appearance, peacocks were a luxury food, to be served on the most impressive occasions, such as the peacock feast laid on for Edward III by Robert Braunch, the mayor of Kings Lynn.

It was not until the fourteenth century that parrots, or more usually parakeets, became familiar animals in England, both being referred to as popinjays by Chaucer and other early writers. Images of parakeets are common decorations on illuminated manuscripts. The earliest depiction is a poor illustration in the Bible decorated by William of Devon in about 1250, followed by two in the *Holkham Bible* and the *Queen Mary Psalter* and one in an East Anglian *Apocalypse with Figures*, which shows a green bird with a divided tail, a common condition in caged parrots. These, and many subsequent images, show that they were all ring-neck (rose-ring) parakeets, originating in India or Africa. Whether they could have survived the long journey from India is open to doubt, so they may have been one of the African subspecies. Another possibility, involving a much shorter journey, is that the parakeets of medieval Europe were derived from the feral population of rose-rings in Egypt, which may have been there as early as the Hellenistic period (fourth–first centuries BC), to be distributed by Italian merchants. Escaped Indian rose-rings have become feral in many parts of the world, and have recently spread from south-east England as far north as Manchester.

A few monkeys were imported at a surprisingly early date; there are two archaeological finds of their bones—one from the pre-Roman Navan fort in Northern Ireland and the other from a Roman cess pit in Dunstable.[6] In the thirteenth century they were occasionally kept as pets by the wealthy, even by nuns—in 1284 the Abbess of Romsey Abbey was reprimanded by the Archbishop, John Peckham, for keeping 'monkeys, or a number of small dogs in her own chamber' and stinting the food for her nuns by diverting 'rosted flesh or milk and wastel-bread' to her pets.[7] By the fourteenth century monkeys were fairly familiar animals in Britain, commonly depicted in the decorations of English psalters, where they are usually engaged satirically in human pursuits—for example, in the *Macclesfield Psalter*, an ape-doctor treats a sick bear. Most of the images show ordinary monkeys with long tails, probably green monkeys or patas monkeys from Africa, but Barbary apes from north-west Africa were used by Chaucer as derogatory metaphors in the speech of his storytellers in the *Canterbury Tales*. In the *Reeve's Tale* a miller is described as having a skull 'as bald as any ape's head', and in the *Parson's Tale* short-cut smocks or jackets are objected to because they expose a man's private parts and buttocks 'like the hinder parts of a she-ape in the full of the moon'.

Most exotic animals brought to Britain in medieval times would have derived from the Mediterranean coast of North Africa, or been brought by the Islamic trans-Saharan trade, with slaves, ivory, and gold, from the southern Sudan, Ghana, and Mali, to northern Morocco and Algeria.[8] Others may have been brought along the spice routes, part maritime part overland, that connected Mediterranean ports, such as Antioch and Alexandria, with Arabia, Persia, and the great entrepôt of southern Asia, Calicut (Kozhikode) on the Malabar Coast of India.[9] They would have been obtained from Arab middlemen by the Genoese and Venetian merchants who brought their trade goods across Europe to the great northern fairs of Calais, Flanders, and Brabant, to exchange for woollen cloth and other goods for export to the east.[10] Some animals were probably brought across Europe by travelling showmen—the traditional Italian organ-grinder with his performing monkey may have had a long history.

Towards the end of the century the Christian world began to enlarge. Convinced that it was possible to reach India by sea around the coast of Africa, the Portuguese Prince Henry the Navigator sent ships further and further south, until by 1485, the year in which the first Tudor monarch, Henry VII, came to the throne of England, the Portuguese had reached as far south as Cape Cross. It would not be long before the search for the sea route to the east opened the way, not only to India and the East Indies, but across the Atlantic to the New World.

1.2. The Age of Exploration

The last years of the fifteenth century saw the beginning of the Age of Exploration, when European navigators discovered new lands, new continents, new plants, new animals, and unheard of riches, spices, and gold and silver. The Portuguese soon established settlements on the Atlantic coast of Africa, enabling them to break the trans-Saharan monopoly of the Arabs, and facilitating, among other things, the transport of animals from tropical West Africa to Lisbon.

In their quest for a sea route to the East Indies, other early navigators sailed west across the Atlantic, and found instead the West Indies, the great continents of North and South America, and the isthmus of Central America. In 1492 Christopher Columbus discovered the Bahamas and various West Indian islands, which he claimed for Spain, and on his triumphant return in 1493 he rode into Barcelona with a cavalcade of officers and servants, as well as six Indian natives (Caribs?) dressed in fine clothes with gold bracelets on their wrists and ankles. They carried pearls, gold masks, strange fruits, and brilliantly coloured Amazonian parrots from Cuba as presents for King Ferdinand and Queen Isabella.[11] When Vicente Pinzón sailed to the coast of Brazil in 1499, he acquired a female opossum carrying several young in her pouch, and on his return to Spain a year later he presented this extraordinary marsupial, alive, to King Ferdinand, less three of her young that had died on the voyage.[12] A year later the Portuguese reached Brazil and soon settled along its eastern seaboard, to the east of the demarcation line established by papal treaty that divided the colonial worlds between Spain to the west and Portugal to the east.

Meanwhile, other Portuguese navigators were attempting to establish an eastern sea route to the East Indies. In 1497 Vasco da Gama's caravels rounded the Cape of Good Hope, sailed north along the east coast of Africa, and then crossed the Arabian Sea to Calicut, thereby opening up the seaway to India, and in 1510 another Portuguese explorer, Afonso de Albuquerque, took Goa, where he established the first European settlement in India, followed a year later by the capture of Malacca in Malaya. As a result the Portuguese were able to challenge the previously undisputed trading rights of the Arabs, the Zamorin sailors of Calicut, and the Italians, and to establish a monopoly over the lucrative spice trade from India and the East Indies, which was to last for over a century, until they were superseded by the Dutch.

Gama's royal sponsor King Manuel I made his court in Lisbon a centre for chivalry, art, and science, with several menageries in his royal parks. From Africa he received baboons, elephants, gazelles, antelopes, lions, and a host of birds, including grey parrots; monkeys and macaws came from Brazil, and, from India, birds, rare 'Indian dogs' (perhaps domestic dogs, possibly the red dog *Cuon* from Java), trained cheetahs and elephants, including the famous elephant, Hanno, in 1511, followed four years later by an Indian rhinoceros, both sent from Cochin by Albuquerque. It was this animal on which the iconography of rhinos was based for many years, including Dürer's famous drawing and engraving. Manuel used his animals to reinforce his image as a magnificent and powerful sovereign—he would ride through the city in a cavalcade led by the rhino, followed by five Indian elephants, and a cheetah carried on a Persian horse.[13] Manuel also bought and sold exotic animals and for a short time controlled the international market in monkeys and parrots; many were exported by the powerful German bankers, the Fuggers, via their trading post in Lisbon.[14]

The importation of animals and of course many other more lucrative goods from the west, that is from Central America, Venezuela, Columbia, Peru, and the West Indies, was largely in the hands of the Spanish, while the Portuguese controlled trade with Brazil, and, until the beginning of the seventeenth century, the priceless spice trade from Africa, India, and the Far East. The English had only cod from Newfoundland, although by means of frequent piratical raids on Spanish and Portuguese caravels they often acquired massive cargoes of treasure, as well as African slaves for sale in the New World, a practice soon encouraged by Queen Elizabeth I.

Much of Britain's commercial and colonial success was due to the efficiency of the trading companies established in Tudor times, the most famous being the English East India Company founded in 1600, but two earlier companies must also have facilitated the importation of animals. The Muscovy Company was established in 1555 to trade with Russia, and, in 1581, the Levant Company, whose ships came home each year bringing silk, cotton, coffee, indigo, and oak-apple galls from their 'factories' in Smyrna, Istanbul, Ankara, and Aleppo and whose members were referred to as 'Turkey merchants'. Even more important was the acquisition in 1588 of England's first toehold in West Africa, when English merchants purchased the Gambia from the Portuguese and set up the Guinea Company to traffic in slaves, gold, and 'elephants' teeth' (ivory).

After the successful completion of two voyages to Indonesia in the closing years of the sixteenth century, on which the Dutch acquired huge quantities of pepper, nutmeg, mace, cloves, and cinnamon—not to mention a live cassowary and a parrot, 'which they sold for 170 livers, tho' 'twas none of the finest'—they were determined to break the Portuguese monopoly of the spice trade and in 1602 set up the Dutch East India Company (*Vereenigde Oostindische Compagnie*, or VOC).[15]

No doubt many of the animals brought back by the Spanish, Portuguese, and Dutch ended up in the royal and princely courts of Europe, but the only exotic animals apart from horses to arrive in Britain in any quantity in the sixteenth century were three useful domestic birds—guinea fowl from West Africa, turkeys from the Americas, and Muscovy ducks, which originated in South America, but probably came to Britain via West Africa. Guinea pigs were introduced into Europe by the Spanish soon after the conquest of Peru in 1532.[16] Their presence in England in Elizabethan times is confirmed by a nearly complete skeleton dating from about 1574, excavated at Hill Hall Manor in Essex,[17] as well as a portrait of three richly attired children, painted in about 1580, which includes a young girl holding a pet guinea pig.[18] The much repeated assertion that one of Queen Elizabeth I's favourite pets was a guinea pig is almost certainly untrue; perhaps the original reference was to Queen Elisabeth of Spain.[19] The coincidence of the terms guinea fowl and guinea pig is best explained by the derivation of both from Guinea in Africa or Guiana (now usually spelt Guyana) in South America, from the Portuguese term for far-off lands, *guiné*. The other possibility is that they were brought to England on the trading ships known as the guineamen.

Queen Elizabeth was the intended recipient of a most unusual animal, an Indian elephant, a gift from Henri IV of France, but there is no record that it ever crossed the Channel. It had been sent to Henri in about 1591 and disembarked at Dieppe; however, the following year, having decided that the elephant's upkeep was proving too expensive, Henri travelled to Dieppe to arrange to have it transported across the Channel to England.[20] The story simply stops there. Had the elephant ever arrived at the Tower, its presence would have been noted by visitors such as Frederick, Duke of Württemberg, in 1592, who recounted:

> In this tower also, but in separate small houses made of wood, are kept six lions and lionesses, two of them upwards of a hundred years old. Not far from these is also a lean, ugly wolf, which is the only one in England; on this account it is kept by the Queen—and indeed there are no others in the whole Kingdom, if we except Scotland, where there are a great number.[21]

During Queen Elizabeth's last illness the old lioness that bore her name pined away and died, so that the soothsayers were ready with their forecasts that Her Majesty's end was also near.[22]

The menagerie at the Tower continued to house lions, which were sometimes used for baiting—a practice that upset Henry VII. John Caius in his *Treatise on English Dogges* wrote:

> Henry VII (a Prince both politique and warlike) commanded all such dogges (how many soever there were in number) should be hanged, being deeply displeased, and conceaving great disdaine, that an yll favoured rascall curre should with such violent villany, assault Lyon, King of all beastes.[23]

In 1561 Caius saw lions at the Tower that were so tame they would kiss their keepers, and three other much fiercer, smaller felids from Africa, which he identified as a lynx, and a male and female *Uncia*. He sent drawings and descriptions of them to Conrad Gesner, the great Swiss naturalist, who incorporated them into a volume of his *Historiæ Animalium*. In modern parlance these animals should be two snow leopards and a lynx, but neither species occurs in Africa—from Caius's descriptions, it seems more likely that these animals were actually a pair of leopards and a young caracal (Figure 1.2).[24]

Dancing bears and the occasional monkey on horseback had probably been taken round the country by travelling showmen for centuries, being so commonplace as to be rarely recorded, and as we have seen many people acquired rose-ring parakeets, but it was not until the beginning of the seventeenth century that other exotic animals arrived in any quantity. It follows that those that were imported before then were the playthings of the very rich, the families of kings or of 'nobles and other great men'. The animals provided a relatively harmless way to show off wealth, prestige, and foreign connections, but presumably they also provided enjoyment for their new owners.

Figure 1.2. One of the leopards (*Uncia*) and the caracal (*linx*) seen by John Caius at the Tower of London in 1561 (top); his own North African squirrel and *Zibeth or sivet-cat* (bottom) (from Edward Topsell, *The Historie of Foure-footed Beastes ... collected out of all the volumes of C. Gesner* (1607)).

Henry VIII was not above owning a pet himself. John Ray, the seventeenth-century naturalist, recounted a story about Henry's 'intelligent parrot' kept at Hampton Court. It was not the usual rose-ringed parakeet, but the larger, rarer, and more expensive African grey, favoured for its ability to speak.

> A story which Gesner saith was told him by a certain friend, of a parrot, which fell out of Henry VIII, his Palace at Westminster into the river of Thames that runs by, and then very seasonably remembering the words it had often heard some whether in danger or in just use, cried out amain, A Boat, A Boat, for twenty pound. A certain experienced boatman made thither presently, took up the bird and restored it to the King, to whom he knew it belonged, hoping for as great a reward as the bird had promised. The King agreed with the boatman that he should have as the Bird being asked anew should say: And the Bird answers, Give the Knave a Groat.[25]

In about 1545 Henry VIII commissioned a family portrait to commemorate the reinstatement of his daughters, Princess Mary and Princess Elizabeth, as heirs to the throne in succession to Prince Edward.[26] Two arches afford tantalizing glimpses of the heraldic garden at Whitehall Palace; standing under one is Will Somers, the King's Fool, on whose shoulder is perched a small brown monkey, protected from the cold by a pink, hooded dress. Will's head is bent forwards, perhaps to allow the animal to groom his scalp, a poignant illustration of the intimate relationship between the jester and his most important prop, the animal that made a monkey of whomever Will was parodying.

A member of the suite of the Lord Admiral of France wrote from London to Lady Lisle, the wife of the Governor of Calais, in 1534, thanking her for her hospitality in Calais and sending her a thank-you present:

> Further, there hath been brought to him out of France certain small beasts which are come from Brazil . . . and the said beasts are called two marmosets, the smaller ones; and the larger is a long-tailed monkey, which is a pretty beast and gentle. And you must understand that the said beasts eat only apples and little nuts, or almonds, and you should instruct those who have charge of them that they give them only milk to drink, but it should be a little warmed. The larger beast should be kept near the fire, and the two other small ones should always be hung up for the night close to the chimney in their boite de nuit, but during the day one may keep them caged out of doors. I send you the said three beasts by a merchant of Rouen, who is this bearer and a man of substance . . .[27]

A year later Lady Lisle made so bold as to send a monkey to Anne Boleyn. The gift was not a success, as John Hussee wrote to Lady Lisle on 21 June 1535: 'And as to touching your monkey, of a truth, madam, the Queen loveth no such beasts nor can scarce abide the sight of them . . .'.[28]

Although there are few specific records of pet monkeys in Shakespeare's works, they were clearly well known to him as pets with a reputation for lecherousness and as show animals—jack-an-apes travelling with performing bears. Tubal in the *Merchant of Venice*, referring to one of Antonio's creditors, taunts: 'one of them showed me a ring that he had of your daughter for a monkey', to which Shylock responds, 'Out upon her! Thou torturest me, Tubal: it was my turquoise; I had it of Leah when I was a bachelor: I would not have given it for a wilderness of monkeys.' In *Henry IV*, Part 2 Falstaff declares: 'he was the very genius of famine; yet lecherous as a monkey, and the whores called him mandrake . . . ', while King Henry in *Henry V* muses: 'Or if I might buffet for my love or bound my horse for her favours, I could lay on like a butcher and sit like a jack-an-apes . . . '.

A famous painting, probably executed in 1568, shows the family of William Brooke, Lord Cobham of Cobham Hall in Kent, seated around a dining table; the picture seems to celebrate his Lordship's pride, not only in his family—six healthy children at a time when many children died in early childhood—but also in their pets.[29] His twin girls were indulged with a bird so rare and special that it was allowed to walk around the dining table—a yellow-shouldered Amazon blue-fronted parrot. Their younger sister holds another rare animal from South America, a marmoset, while perched on the hand of their elder brother Maximilian is a pet goldfinch, its leg attached by a fine thread to a small perch hung with tiny bells. Goldfinches are native birds, commonly imprisoned as cage birds until the practice was prevented by law in the nineteenth century.

Lord Cobham's enjoyment of his family and their pets was surely equalled by his pride in his celebrated garden: 'A rare garden there, in which no varietie of strange flowers and trees do want, which praise or price may obtaine from the furthest part of Europe or from other strange Countries . . . '.[30] No doubt 'the other strange Countries' included the source of his family's pets, South America. The marmoset and the Amazon parrot were probably brought from Brazil by the Portuguese and obtained by Brookes from Lisbon or perhaps Antwerp, the great entrepôt of Europe, for in the 1560s as many as 500 ships from Danish, English, Hanseatic, German, Italian, Portuguese, and Spanish ports sailed in and out of its harbour every day. For much of the sixteenth century the Antwerp branch of the Fugger Bank received large shipments of parrots, monkeys, and wild cats from India, Africa, and South America, which were kept in cages in a large garden next to their offices, awaiting purchasers.[31]

In 1596 Sir Robert Cecil, Queen Elizabeth's Secretary of State, received a 'parakito' from Sir John Gilbert, accompanied with instructions:

> He will eat all kinds of meat, and nothing will hurt him except it be very salt. If you put him on the table at meal time he will make choice of his meat. He must be kept very warm and after he hath filled himself he will set in a gentlewoman's ruff all the day. In the afternoon he will eat bread or oatmeal groats, drink water or claret wine: every night he is put in the cage and covered warm.[32]

In the middle of the sixteenth century the rich were beginning to embellish their gardens with plants from overseas and sometimes with elaborate aviaries housing both native and exotic birds. In 1586 William Harrison had over 100 varieties of plant, 'If therefore my little plot . . . be so well furnished, what shall we think of Hampton Court, Nonsuch, Tibaults [Theobalds], Cobham garden, and sundry others appertaining to divers citizens of London?'[33] Clearly there were many opportunities to acquire exotic plants, and presumably small caged birds as well. The most famous aviaries were at Kenilworth Castle, Theobalds, and Nonsuch Palace.

At Kenilworth in 1575 Robert Dudley, Earl of Leicester, entertained Queen Elizabeth and her splendid entourage on her summer progress for eighteen days of feasting, dancing, fireworks, masques, waterborne singers, deer-hunting, and the obligatory bear-baiting, all described by Robert Laneham in a famous letter to a friend. A magical new garden had been built by Leicester to replace the medieval one—against the outer wall of the castle was a shady terrace for viewing the four knot gardens, each surrounding a 15-foot high pyramid surmounted by a huge porphyry orb. In the centre was an immense white marble fountain, topped with a ragged staff and with elaborate statues—Neptune and his steeds, Thetis and her dolphins, Triton with fish, Proteus with his 'sea buls', as well as carvings of whirlpools, whales, and sea fish. The water cascaded into a pool stocked with carp, tench, bream, perch, and eels. The garden also had 'obelisks, spheares, and white beares all of stone', 'too fine arbers redolent by sweet tress and floourz', 'fair alleyz green by grass', all 'Respirant' with 'fragrant earbs and floourz, in form, cooler and quantities so deliciously variant: and frute Trees bedecked with their Applz, Peares and ripe Cherryez'. Backing on to the terrace was a large aviary: 'a square cage, sumptuous and beautifull ... of a rare form ...'. It was 20 foot high, 30 foot wide, and 14 foot broad; surmounting the breast-high earth walls were large windows, arched at the top and separated from one another by columns, every part beautified with 'great Diamons, Emerauds, Rubyes and Sapphyres' and 'garnished' with gold. The windows were covered with a fine mesh to confine the birds, and holes had been cut into the wall inside for roosting and nesting. The birds could perch on holly trees planted at each end. There were fine vessels for drinking water and 'diversities of meates ... and sundry graines'. A skilled and diligent Keeper cared for the 'lively Burds, English, French, Spanish, Canarian, and (I am deceaved if I saw not sum) African. Whearby, whither it becam more delightsum in chaunge of tunez and armony too the eare: or els in differens of coollerz, kindez & propertyez too the ey ...'.[34] No doubt the English birds were songbirds captured in the surrounding countryside, and the African birds must have been colourful parrots and their relatives. The identity of the French and Spanish birds is a mystery—perhaps exotic birds for which the French or Spanish acted as middlemen—and the Canarian birds must have been some of the newly arrived, sweet-singing, and colourful 'canarie-birds'.[35]

Wild canaries are, or rather were, native of the Canary Islands, as well as Madeira and the Azores. During the fifteenth century the islands were taken over and plundered by the Spanish, who maintained a strict monopoly over the export of canaries to Europe, allowing only males to be taken. By the middle of the sixteenth century this was a highly profitable business, with thousands of wild males exported annually, being valued for their exotic origin and most of all for their sweet singing.[36] The earliest record of canaries in England is the payment of 10*s*. in 1539 to 'Mr Raynoldes servant, for bringing a cage of canary birds' noted in the household accounts of Henry VIII.[37] Gesner, in the first edition of the volume on birds in his *Historiæ animalium* published in 1555, states that, because canaries were brought from afar and highly valued for their song, few people apart from the rich and noble could afford to keep them. He also quotes a poem, which includes the lines:

> This herbe-greene birde of sparrows quantitie
> More pleasinge tunes from little throte out throwes . . .[38]

The earliest menagerie in England apart from that at the Tower seems to have been in Oxford. It is said that in among the flower gardens, vegetables, orchards, dovecots, and fish ponds of Magdalen College was a small collection of animals. Few details are known, except that in 1486 peacocks, swans, and hares were acquired (destined for the table?), and in 1504 Henry VII gave the college a female bear. The claim that the college presented some marmosets to Henry in 1495–6 is highly unlikely, considering that South America, the home of marmosets, had not yet been discovered; however, the term 'marmoset' in early use denoted any small monkey.[39]

Another early menagerie was kept by Sir Thomas More in the garden of his house, the Old Barge in the shadow of St Paul's Cathedral. In a famous letter, probably written in 1517 a few months after his last visit to England, the great humanist Erasmus wrote admiringly of More: 'One of his amusements is in observing the forms, characters and instincts of different animals. Accordingly, there is scarcely any kind of bird that he does not keep about his residence, and the same of other animals not quite so common, as monkeys, foxes, ferrets, weasels, and the like.'[40]

When More and his wife Dame Alice moved with their family to their new house in Chelsea, an inventory of their household goods included 'a 'gret cage fir birds' valued at least at £10. This aviary is said to have contained exotic birds, presumably parrots, but, as far as one can tell, the only exotic mammal in his household was a monkey, or perhaps successive monkeys. Both husband and wife were inordinately fond of dogs and monkeys, and when a monkey was recovering from an injury it was allowed to move around the house without its chain.[41] Holbein painted a portrait of More and his family in their Chelsea home in 1526. Although the painting itself no longer exists, a full-size copy by Rowland Lockey at Nostell Priory shows a small spaniel nestling at More's feet, and, sitting on the floor beside Dame Alice, a small brown monkey (Plate 1).[42] Other copies of the painting survive, but none shows the monkey. However, a sketch by Holbein depicts a monkey with a long tail, partly secreted in Dame Alice's gown. The monkey is holding the ends of a chain hanging loosely from a collar, suggesting that it was unrestrained, at least for the time being.[43] After More's canonization in 1935, the sculptor Eric Gill was commissioned to carve an image of the saint for Westminster Cathedral, and the resultant relief included the saint's pet monkey clinging to his robes. When Gill died in 1940, the carving was still in his studio, complete with monkey, but by the time it was erected on the altar of the Chapel of St George and the English Martyrs in Westminster Cathedral the monkey had been removed on the orders of Cardinal Griffin.[44] Presumably the monkey had been considered improper or unsuitably frivolous for a saint and erased; others might regard its erasure as an act of humourless vandalism.

Like Erasmus and Gesner, Dr John Caius—the Cambridge don, who refounded Gonville College (now Gonville and Caius College) in 1558—was a humanist scholar, dedicated to a revival of classical learning in reaction to medieval philosophy. The movement benefited enormously from the invention in the previous century of printing with movable type, which allowed a relatively wide circulation of books, particularly humanist editions of the classics, and the creation of large libraries. In zoology the most revered of the classical authors were Aristotle, newly translated into Latin from the Greek, as well as Pliny and Columella, but many other authors were quoted in what were basically compendia of previous knowledge. In this Age of Discovery the letters and published journals written by explorers and travellers often included accounts of 'strange and wondrous beastes' encountered in the newly discovered parts of the world; maps too were often decorated with pictures of the animals. Some of this new information as well as the observations and theories of his friends and colleagues was incorporated by Gesner into his magnificent *Historiæ Animalium.* Although the more bizarre conjectures of earlier authors were sometimes incorporated without question, a healthy scepticism based on the actual observation of animals began to creep in. Although often used to explain the works of God, this new rationalism was to underpin the great scientific advances of the emerging sciences of mathematics, chemistry, physics, medicine, and, of course, biology.[45]

Caius sent descriptions in Latin and images of several exotic animals that he had seen in England—some of which he owned—to Gesner, who incorporated them into those volumes of the *Historiæ Animalium* that appeared during and after 1554. Notes on these animals were also included in an essay by Caius, *De Rariorum Animalium atque Stirpium Historia*, dedicated to Gesner and printed in 1570 without illustrations. Gesner's texts were translated into English by Edward Topsell in his *Historie of Foure-Footed Beastes* (1607) complete with copies of most of Gesner's illustrations. Topsell's work was much criticized on account of its mistakes in translation and its lack of originality—an early version of 'dumming-down'; nevertheless it made both authors' work available to a much wider public.

Most of the animals that Caius described came from Mauritania. This was not the country that bears that name today, but northwest Africa—those parts of Algeria and Morocco that border the Mediterranean, extending west through the Straits of Gibraltar to the Atlantic coast of Morocco (the Barbary Coast), and through the Atlas Mountains towards the Sahara. The animals included a striped squirrel—probably a ground squirrel, whose behaviour Caius had studied in some detail[46]—a very shaggy dog, '*Canis Getulus*',[47] two North African sheep,[48] and a 'zibett', probably a civet cat[49] (Figure 1.2). He wrote that another bovid, which he acquired in 1561, 'seemed in confinement to be very gentle, full of play, and frolicksome, like a goat'; he called it *Tragelaphus seu Hircocervpo* because its front half resembled that of a goat and its rear half a deer, but he also referred to it correctly as a Barbary sheep.[50] Another live bovid, which he called *Bucula Cervina*, or *Moschelaphum*, since it resembled both a deer and a young heifer, seems to have been a bulbul or northern hartebeest, extinct since the nineteenth century.[51] Caius also described and figured an elk from Norway, which he did not recognize as such because it did not conform to Caesar's insistence in the *Gallic Wars* that the elk of the northern forests had to sleep upright because it had no knee joints, a claim dismissed as nonsense by the more realistic Gesner. After arriving in England with its keeper, the elk gave birth to a fawn. Gesner named it *Hippelaphus*, as he and Caius both thought its front half resembled a deer and its rear half a horse.[52]

Caius owned a small green monkey, *Cercopitheco Vario*, from West Africa or the Cape Verde Islands, a tiny marmoset from South America, and a chameleon, and notes that he had seen a white bear and a black fox, both imported from Muscovy (Russia).[53] He saw many kinds of parrots in England, including an African grey, a large and brilliantly coloured scarlet macaw, which he said came from Brazil—probably one of the first to be imported into Britain—and some smaller green birds, probably parakeets. He wrote that the parrots came from Spanish Isle (Hispaniola), Egypt, and India. Egypt as a source of parrots is of particular interest, as the possibility that ring-necked parakeets in Britain were taken from feral populations in Egypt has already been mentioned.[54]

Although collections of birds in aviaries were integral to several grand Tudor gardens, most people in England seem to have been interested in exotic mammals only if they were useful, edible, or could be used as pets, baited in cruel sports, or induced to perform tricks. Even the desire for novelties, which was already apparent to Caius in 1570, extended mainly to dogs, pets, and horses. It was in his *English Dogges* that he wrote:

> for we English men are marvailous greedy gaping gluttons after novelties, and covetous coruorauntes of things that be seldom, rare, straunge, and hard to get.[55]

The desire for animals that were rare, strange, and hard to get was to be indulged in to a much greater extent in the next century, by two of the Stuart monarchs, James I and Charles II, who both had notable menageries in St James's Park in London, but it was not until the early eighteenth century that other 'great men' and the occasional lady began to set up menageries for enjoyment, and, to a much lesser extent, scientific study.

2

The Stuarts, 1603–1688

2.1. James I to the Interregnum

With Queen Elizabeth's death on 24 March 1603, the Tudor dynasty came to an end, to be succeeded by the Stuarts in the person of her distant relative James VI of Scotland. On receiving the news of her demise, he travelled down from Edinburgh to take up the crown of England as James I. Prior to his triumphal procession through London to Westminster he was lodging in the Tower; on 'being told of the Lyons he asked of their being & how they came thither, for that in England there were bred no such fierce beastes', and was told that, although lions did not breed here, English mastiffs were as courageous as lions. He ordered Edward Alleyne the 'Master of the Royal Bears and Bulls and Mastiff Dogs' to fetch 'three of the fellest dogs in the Garden'—that is, from the Paris Garden on the south bank of the Thames—in order to test the courage of an English mastiff against a lion. The experiment was made in the presence of the King, the Queen (Anne of Denmark), and their eldest son, Prince Henry. The mastiffs were let in one at a time; each of them furiously attacked the lion, who eventually declined the combat, but not before all three dogs had been injured, two so badly that they died within a few days, although the third recovered. Henry commanded Alleyn to bring it to St James's Palace, 'and charged him to keep the dog and make much of him, saying, hee that had fought with the King of beasts, should never after fight with an inferiour creature'.[1]

In the spring of 1605 the King ordered that part of the moat on the west side of the Lion Tower should be filled in to form an area where the lions could be baited; doors were cut into the wall of the Tower so that they could be driven out into the open air. When the King visited on 3 June, the lions were so reluctant to leave their dens that they had to be forced out with flaming torches, and 'when they were come downe into the walk, they were both amazed, and stood looking about them, and gazing up into the ayre'. The King seated with his retinue on a viewing platform ordered Ralph Gill, the Keeper of the Lions, to throw two racks of mutton down to them followed by several cockerels, all of which they devoured. Then a lamb was let down on a rope, but the lions merely sniffed at it, so it was 'very softly' drawn up again, unscathed. Having disappointed the King, the lions were returned to their den in disgrace, and another fiercer lion was brought out to be baited by three mastiffs; it 'spoiled them all'.[2]

In June 1610 the King and Queen, accompanied by Prince Henry and Princess Elizabeth, and with 'diuers great Lords, and manie others, came to Tower to see a triall of the Lyons single valour, against a great fierce Beare, which had kild a child, that was negligently left in the Bear-house'. The King, hoping for a grand battle, ordered the bear to be forced into the lions' den, but the lions only cowered in the corners. Two weeks later the bear was given to the mastiffs, probably at the Paris Garden, to be baited to death, 'and unto the mother of the murthered child was given twenty pounds out of the money which the people gave to see the bear kil'd'.[3]

Procreation among the lions was a rare event—if any cubs were born, they were anxiously cherished. When the lioness named Elizabeth whelped on 5 August 1604, the King was so 'greatly entertained' that he issued orders to Sir William Waad, the Lieutenant of the Tower, to feed the cub carefully and keep it warm. After the birth the lioness carried the cub around in her mouth; the next day the keeper managed to retrieve it from her, but it died. It was then disembowelled and embalmed for presentation to the King. Elizabeth gave birth to a second cub only six months later, but it also died, despite, or perhaps because of, being taken from its mother to be reared by hand.[4] The birth of a third cub is recorded on 24 October 1605[5] and yet another in September 1607, when Sir William wrote a rather anxious letter to Robert Cecil (by then Earl of Salisbury, Secretary of State and James's spymaster):

> Last night there is born a fine young male lion whelp of the former lions, Henry and Anne. Mr Gill the keeper, in regard of the infection[6] round about him, is in the country. I have directed care to be used to protect the little one, as is fit, if the whelping of it so late in the year, and the cold coming on, do not hurt it before it get strength. Both the lions keep together with the little whelp with that care as is very tender and full of love.[7]

There are few records of additions to the Tower menagerie during James I's reign, but when Justus Zinzerling visited in about 1610 he saw not only the lions, but also leopards, an eagle, and a wolf.[8]

At the time of James's accession, St James's Park, which lay to the west of Westminster Palace and south of St James's Palace, was swampy and occasionally flooded by the Tyburn stream, which ran through it.[9] In 1605 James ordered Sir Thomas Knyvett, the Warden of the Mint, to construct fountains, walks, waterworks, and houses for the reindeer, red deer, ducks, and foreign fowl, and to employ two men to care for them.[10] Captain Newport brought two young crocodiles and a 'wild boar' (probably a peccary) from Hispaniola (the island now divided into Haiti and the Dominican Republic),[11] and in 1608 William Walker, the Keeper of Fowl, was responsible for the care of some ostriches and wild boar that had been brought to the Park 'out of France'.[12]

Salisbury was active in the procurement of exotic animals for the King; in 1607, not content with the monkeys and marmosets for which he had paid £50, he wrote a letter to his cousin, Sir John Ogle, in The Hague, asking him to 'employ some merchant that is your friend to hearken amongst those that deal with Indian [i.e. foreign] commodities, whether they have any of the small sort of monkeys . . . '. He enclosed a drawing showing the type of monkey required. It is significant that Ogle was in Holland and therefore had access to merchants of the Dutch East India Company. A few weeks later Ogle wrote back:

> I send your Lordship by this bearer, Henry Seager, 2 such monkeys as you desired. I doubt not but they will be to your liking, for they are the only ones and the best at this time to be gotten. I thought it best to send them by him because he helped me to them, and knows best to bring them safely thither. He tells me he is the man that brought over the last into England. At the spring I am given to understand there will be more choice. I will then see if I can fit you with some other to your liking, to dispose to some such friends as shall be desirous of them . . .[13]

Salisbury was also instrumental in awakening the English East India Company to the idea of importing live animals. The fledgling company had established its first 'factory' in Asia in 1603, not in India, but in the great port city of Bantam on the island of Java. In 1607 an application was made to the company for permission to send out a man in one of their ships to bring home 'paratts, munkies, and marmasitts' for Salisbury; the company replied that, 'as the man may die and his things miscarrie', it would be better if the company's servants saw to the collection and care of the animals. So the commanders of the Third Voyage were instructed to bring back 'straunge beasts and fowles that you esteeme rare and delightfull'.[14] 'In some of these isles aforesaid [the East Indies] are sundry beastes and birds of strange shape and qualitie, as in Banda, from whence Captayn William Keeling, being general of the Third Voyage, brought hither a fowle called a cassuare, as big as a Turkie, and it hath no wings . . .'.[15] Keeling arrived back in England on 10 May 1610, the 'cassuare' was unloaded, and Salisbury presented it to the King. It was the first ever seen in England, a truly strange and exotic gift. Cassowaries are flightless birds about 5 feet in height, with blue-black feathers and horny protuberances on their heads; bright red and blue wattles ornament their bare necks and five or six stout quills reduced to bare shafts hang from their much-reduced wings. James's 'East India bird' was even reputed to eat hot coals.[16] A 'very fine white parrot with a yellow crest' (that is, a sulphur-crested cockatoo), native to Indonesia and perhaps acquired in the market at Bantam, also arrived, probably in the same shipment.[17] The instructions to the company were repeated in each of the following four years, the animals being intended for presentation to the King or 'any of the noble lords that are our honorable friends'.[18]

Another cassowary was sent to King James by Maurice, Prince of Orange; it had been in Holland for over seven years before arriving in England, where it lived for another five. According to James's physician William Harvey, it was kept in a cage adjacent to one that held a pair of ostriches, 'the Cassoware oftentimes over-heard them in the act of coition . . . She unexpectedly conceived egges (stirred up, as I suppose, by a certain sympathy . . .)'; Harvey dissected the unfortunate bird and found a second 'corrupt' egg in her 'uterus'.[19] Parts of one of James's cassowaries were preserved for posterity in Tradescant's museum in Lambeth: 'A legge and claw of the Cassawary or Emeu that dyed at S. James's.'[20]

Salisbury's help was also solicited by the Earl of Southampton in the hope that the King's interest in his menagerie might be harnessed to further the interest of the newly founded Virginia Company in North America. In December 1609 Southampton reported to Salisbury that 'The King is eager to have one of the Virginia Squirrels, that are said to fly'; several flying squirrels duly arrived, accompanied by the first marsupial recorded in England—an opossum, for whose upkeep in 1611 Walker was allowed 5*s*. a month.[21]

In March 1611 the King's lover Robert Carr was created Viscount Rochester and appointed 'Keeper of the Palace of Westminster'; one of his duties was 'to keep and preserve wild beasts and fowl in St James's Park and Garden and Spring Garden'. His account dated 18 July 1611 'for keeping the fowle and Beastes att St James Parke and Gardens' lists the various underkeepers and the sums allowed to each for feeding the animals.[22] The Mint account for the same year includes the sum of £55 15*s*. 8*d*. for 'Sondrye persons for work and changes in the park of St James and Sprynge Garden there, viz: for meate for Indian beastes, cranes, puettes, hernes, guynea-hennes, duckes, turtle doves, seagulles, pheasauntes, busterdes, shovellers, the tame falcone, red deare, beaver, barbarye sheep and others . . . '.[23]

'Indian beastes' means foreign beasts, 'hernes' were herons, 'guynea-hennes' guinea fowl. Barbary sheep seem likely to have been domestic sheep from North Africa rather than wild Barbary sheep. 'Puettes' are a mystery. An aviary for James's birds was built on the southern boundary of the park along the track subsequently named Birdcage Walk.

Many other people presented exotic animals to the King. In 1613 Giovanni Perundini was rewarded with £76 13*s*. 4*d*. for bringing a tiger (more probably a leopard)' a gift from the Duke of Savoy, along with a lioness.[24] The 'lion and certain other beasts' were consigned to the Paris Garden to be baited. As well as furs, jewels, and many other luxurious gifts that Tsar Mikhail Fyodorovich sent James from Muscovy in 1617, there were animals—pet sables in jewelled collars, white hawks, and gyrfalcons.[25] Sir Thomas Roe, on his return from his three-year embassy at Agra in 1619, presented James with gifts from the Great Moghul, 'two antelopes and a straunge and beautifull kind of red deare', as well as a rich tent, rare carpets, umbrellas, and 'such like trinkets'.[26] The antelopes were probably blackbuck, and the deer either sambars (rusas) or barasinghas (swamp deer).

The Muscovy Company sent the ship *God-Speed* on an expedition to Cherie (= Bear) Island halfway between the North Cape and Svalbard. It returned to London on 20 August 1608, loaded with walrus tusks and carrying a young, live walrus:

> we brought our living Morse [walrus] to the Court, where the King and many honourable personages beheld it with admiration for the

> strangenesse of the same, the like whereof had never before been seene alive in England. Not long after it fell sicke and died. As the beast in shape is very strange, so is it of strange docilitie and apt to be taught ...[27]

Two polar bear cubs were captured on another expedition to the Arctic sponsored by Sir Francis Cherie, a backer of the Muscovy Company, to buy fish from the Lapps and Russians. Two ships, the *Lionesse* and the *Paul*, Jonas Poole Master, left Harwich on 15 April 1609. Having acquired the necessary fish, they sailed due north to Cherie Island, where Poole recorded on 30 May:

> We slew 26 seals, and espied three white bears. We went aboard for shot and powder, and coming to the ice again, we found a she-bear and two young ones: Master Thomas Welden shot and killed her. After she was slain, we got the young ones, and brought them home, into England, where they are alive in Paris Garden.

The cubs were placed in the care of the 'Master[s] of the Royal Game of Bears, Bulls and Mastiff Dogs', Philip Henslowe and Edward Alleyn, who were awarded 2*s*. a day for the maintenance of the two white bears, a lion, and other beasts.[28] The cubs appeared on stage in a revised version of the anonymous play *Mucedorus*, performed before James I at Whitehall on Shrove Tuesday 1610, advantage having been taken of their presence in London to add several new scenes involving them, which the audience found hilarious. On New Year's Day 1611, when Ben Jonson's *The Masque of Oberon* was performed in the Banqueting Hall at Whitehall, Prince Henry, in the title role, rode on to the stage in a chariot drawn by the two white bear cubs.

James I also kept exotic animals, hidden away from the public gaze, at Theobalds in Hertfordshire. Having acquired the house and park in 1607 from Salisbury in exchange for the Tudor mansion at Hatfield,[29] he enlarged and enclosed the estate, and made various improvements mostly related to his obsession with hunting and hawking, but he also kept some exotic animals there. In 1611 the ambassador from the Duke of Savoy brought a tame ounce or leopard to Theobalds, which disgraced itself by attacking a pet red deer fawn, nursed by a woman 'entertained for the purpose'.[30] James may also have received some deer of a most unusual species—large, ponderous European elks. In July 1612 Sir James Murray was ordered to reward 'the servants of the Marquis of

Brandenburg for transporting and conveying hither a present of elks from the said Marquis'; more elks followed in 1624, brought from 'Count Palatine of Weare' (Frederick V, the husband of James's daughter Elizabeth).[31]

In 1622, when negotiations were underway for James's second son, the future Charles I, to visit Spain with a view to obtaining the hand of the Spanish Infanta, Count Gondomar, the Spanish ambassador to England, wrote to James from Madrid, including in the letter the promise that he would send him two camels and a pair of donkeys for propagation at Theobalds.[32] James's 'two sweet boys' George Villiers (then merely *Earl* of Buckingham) and Charles travelled out to Spain, incognito, in a hare-brained scheme that came to nothing.[33] While the boys were in Spain, the elderly James, though desperately ill with nephritis, was as keen as ever to enlarge his menagerie and wrote to Buckingham enquiring not only about the promised camels and donkeys, but also about elephants.[34] Buckingham replied from Madrid, compliant in the matter of obtaining animals, but with a veiled threat:

> Sir, —Foure asses I have sent. Tow hees and tow shees. Five camels, tow hees, tow shees with a young one; and one ellefant which is worth your seeing. These I have impudently begged for you . . . My Lord Bristow [the English ambassador at Madrid] sayeth he will send you more camels . . . And I will lay waite for all the rare coler burds that can be hard of. But if you do not send your babie jewels enough, I'le stop all other presents. Therefore louke to it.[35]

The elephant and camels reached London in July: 'Going through London at night, they could not pass unseen and the clamour and the outcry raised by some street loiterers at their ponderous gait and ungainly step brought sleepers from their beds in every district through which they passed.' They were accommodated temporarily in the King's Mews. James sent instructions from Theobalds that the elephant was to be well fed and nobody was to be allowed to view it. Similarly the camels were to be exercised daily in St James's Park and 'every precaution taken to screen them from the vulgar gaze', before being sent on to Theobalds, where a large camel house was being built to accommodate them. Secretary Conway wrote to Sir Francis Cottingham conveying the rather tetchy query from the King as to when the Spanish keepers of the elephant and camels were going to relinquish their charges to 'other [presumably local] hands'.[36] Compromise appears to have been reached, as the elephant was allowed four keepers—two Spaniards and two Englishmen. The Spaniards insisted that from September to April the elephant must not be allowed to drink water, but must be given a gallon of wine a day. As the elephant cost £275 12*s*. a year to maintain and as £150 was presented to the Spanish official who brought it, the King's secretary Sir Richard Weston opined that the Lord Treasurer would not approve of presents that cost the King as much as a garrison to maintain. It is not clear whether or not the elephant followed the camels to Theobalds; perhaps it died of alcohol poisoning.[37] James I died at Theobalds in 1625; how long any of his beloved animals survived is unknown, but some of those in St James's Park may still have been alive in the early years of the reign of Charles I.

It was not only King James who enjoyed animals. Queen Anne arranged for a large aviary to be built for her at Greenwich Palace. Anne, the daughter of Frederick II of Denmark and Norway, was married to James in November 1589 in Oslo, when she was only 14 years old, and crowned Queen at Holyrood Abbey in Edinburgh the following May. After James's accession, she made her home in London at Somerset House in the Strand and employed the great French garden architect and hydrological engineer Salomon De Caus to work for her there, as well as at Hampton Court[38] and Greenwich Palace. He built an Italianate garden at Greenwich with an elaborate grotto, designed to appeal to all the senses; its arcades were encrusted with shells, mother-of-pearl, flowers, grasses, and sweet-smelling herbs. It housed dozens of birds, prevented from flying away by wire gratings.[39] Here too, at Greenwich, she employed Inigo Jones, to build the first Palladian House in Britain, the Queen's House, which still stands, although it was unfinished when she died in 1619. There seems to be no record of the species of bird in Anne's aviary, but we can assume that, as well as indigenous singing birds caught in the English countryside, she would have had exotic birds including 'Virginia nightingales' (red cardinals) imported from the newly founded colony of Virginia, and canaries, which at that date could be bought only at vast expense; she may also have been the owner of the ostrich that Sir Thomas Browne saw at Greenwich when he was a schoolboy.[40]

De Caus had been brought to court in 1608 by James to teach perspective drawing to Prince Henry and his sister Elizabeth, and two years later Henry had appointed him 'Architect' for his palace at Richmond.[41] One of the structures planned by De Caus was an artificial mountain, two or three storeys high, planted with sizeable fir trees, with a talking statue of Memnon on the top, and a vast cavernous aviary within,[42] but a few months before the Prince's death in 1612 the entire project was abandoned, probably for lack of funds, which, considering the scale of the enterprise, is unsurprising.[43]

Princess Elizabeth seems to have shared her parent's enthusiasm for exotic animals. When she was 16 and living in Kew, where Lord and Lady Harrington set up an establishment for her, the entries in her private accounts for 1612–13 include such items as 'Paid for strewing herbs, and cotton to make beds for her grace's monkeys, 3*s.* 3*d.*' and 'Paid to a joiner for mending her grace's parrot cages'.[44] In a portrait painted when she was a child, Elizabeth has a macaw on one shoulder, a parrot on the other, a little lovebird on her hand, and a monkey and dog at her feet.[45] On New Year's Day 1613 she received many valuable gifts, among them a parrot given by Sir Thomas Roe—a personal friend of both Elizabeth and Prince Henry; a servant was immediately dispatched to buy 'canarie seedes' for its dinner.[46] Elizabeth's parrot was probably a South American species, for in 1610 Henry had sent Roe 'upon a discovery to the West Indies'. He had sailed 200 miles up the Amazon, then unknown to British explorers, making numerous excursions by canoe up its tributaries; he explored the coast from the mouth of the Amazon to the Orinoco, returning to England in July 1611 via Trinidad. Roe's friendship with Elizabeth lasted for many years, and after her marriage to Frederick the Elector Palatine in Westminster Abbey in 1613, when the couple were both aged only 16, Roe was one of the large train of noblemen who accompanied them to Heidelberg.[47]

Elizabeth took at least one of her monkeys with her to Heidelberg, and Sir Dudley Carleton sent her two more that he had acquired in Italy. One of Elizabeth's ladies wrote a thank-you letter:

> Her highness is very well, and takes great delight in those fine monkeys you sent hither, which came very well, and are now grown so proud as they will come at nobody but her highness, who hath them in her bed every morning; and the little prince, he is so fond of them as he says he desires nothing but such a monkey of his own. They be as envious as they be pretty, for the old one of that kind, which her highness had when your lordship was here, will not be acquainted (with) his countrymen, by no means; they do make very good sport and her highness very merry; you have sent nothing would a been more pleasing.[48]

Elizabeth was soon able to renew her acquaintanceship with her former drawing master Salomon de Caus, for Frederick employed him to design the magnificent *Hortus Palatinus* at Heidelberg. The huge multi-terraced garden, built on a previously barren hillside, included orchards of fruit trees (imported entire from England), blooming parterres, fountains, and, crowning the whole, an entrance arch, said to have been constructed in a single night as a surprise for Elizabeth. Let into a wall close by is the inscription 'Frederick V to Elizabeth, his dearest wife, AD 1615'. Her beloved animals were housed in an aviary, a monkey house, and a menagerie. Frederick later became the short-lived King of Bohemia—the 'Winter King'—and Elizabeth his sad 'Winter Queen'.

In his essay *On Gardens* published in 1625, Francis Bacon began 'GOD Almightie first Planted a Garden...'. After discussing the layout and planting of the ideal garden, he advised: 'For Aviaries, I like them not, except they be of that Largenesse, as they may be Turffed, and have Living Plants, and Bushes, set in them; That the Birds may have more Scope, and Naturall Neastling, and that no Foulnesse appeare, in the Fluore of the Aviary.' This advice was for the ordering of a royal or 'Prince-like' garden intended to create a version of paradise on earth; in less princely establishments birds were to be more confined. Bacon suggests the provision of a square garden enclosed with high hedges trained over arches and over each arch a small turret surmounting a space 'enough to receive a Cage of Birds'.[49] In Bacon's *New Atlantis* one of the elements of his 'Modell and Description of a Colledge' was a menagerie. Although the animals in the parks and enclosures within 'Solomon's College' could be enjoyed for their appearance or rarity, their main purpose was to be scientific. They were to be used in experiments designed to shed light on human anatomy, physiology, and pathology, and might be experimented upon to create new, more perfect species.[50] The *New Atlantis* went through numerous editions and was influential in the founding of the Royal Society in Charles II's time, though, as far as one can tell, none of Bacon's followers actually instituted a menagerie, nor was there one in Gresham College, where the early meetings of the society were held.

James I died in 1625 and was succeeded by his second son, Charles I. There is very little information about exotic animals in England during the twenty-three years of Charles's reign—the country was in turmoil for much of that time and he had rather more urgent matters to preoccupy him. However, shortly after his accession a lioness whelped in the Tower, which was taken as a good omen—'which some take as a presage that all things are like to succeed as in the former time'[51]—and he gave audience to a chiaus (messenger) from Algiers, who brought a present of 'tigers' (probably leopards) and lions that were sent to the Tower.[52]

In 1635 Robert Gill the Keeper petitioned the King, reminding him that his father and grandfather before him had served Charles, James I, and Queen Elizabeth, as 'Keeper of his Majesty's lions and leopards in the Tower of London', and, although King James had forbidden anyone to take lions or leopards around the country to show them for gain, Thomas Ward had been showing a lion in Oxford, at Sturbridge Fair, and in other places. Ward had since parted with it to Martin Brocas and John Watson, who were doing the same, despite the fact that 'the lion had grown so fierce that he almost killed a child, and bitten his keeper so that he lay eight weeks of the sore . . .'.[53]

Some animals were still kept in St James's Park during the early years of Charles I's reign, including the cassowary 'whose fine green channelled egg' was acquired by Thomas Browne.[54] In 1625 a warrant was 'made to the cofferers of the household to pay to John Walker "keeper of the King's House and yard within the Park of St James", divers sums of money for keeping and breeding fowl there', and on 31 January the following year Philip Earl of Montgomery was allotted £72 5*s.* 10*d.* yearly for life for keeping the Spring Garden and St James's Park, and the beasts and fowl there, with extra payments for repairs, labourers, and weeders.[55] Whether such appointments lasted for the remainder of Charles's reign or through the ensuing Civil War and Interregnum is uncertain; it seems that it was not until the reign of James I's grandson Charles II that St James's Park was again used to accommodate exotic animals.

Shortly after his accession Charles married 16-year-old Henrietta Maria, the youngest daughter of Henri IV of France. She had been brought up in a gay, informal household, surrounded by animals. Her father had enlarged and refurbished the chateau at Fontainebleau and had employed Claud Mollet to improve the gardens. He added a great aviary holding ostriches, herons, seagulls, peacocks, shelducks, pheasants, partridges, and cormorants (gifts from James I) trained to fish in the fishponds. And, in the usual bloodthirsty fashion of the time, as well as the aviary, there were facilities for staging big cat fights.[56] Frivolous and gay, Henrietta Maria was well known for her love of animals, so she was probably delighted with the cage of birds brought to her in 1631 by Captain Weddell, in one of the East India Company's ships.[57] Her household included a train of dwarfs, black servants, monkeys, and dogs of many sizes, and, like her father, when she travelled on the royal progresses, she took a selection of animals with her.[58] Her portrait with her favourite dwarf, Sir Jeffrey Hudson, painted by Van Dyck in 1633, includes a pet monkey named Pug,[59] a blue monkey from Africa. She gave this valuable animal to Sir Jeffrey, and it became his constant companion.

In 1636 John Tradescant the elder installed one of the earliest orange gardens in England for Henrietta Maria at Oatlands, where he had been appointed six years earlier as 'Keeper of his Majesty's Gardens, Vines, and Silkworms', a position that linked his name with that of the 'Rose and Lilly Queen', as she is commemorated on the Tradescants' grave at the church of St Mary-at-Lambeth (now the Garden Museum),[60] uniting the English rose with the French Fleur de Lys. But Henrietta Maria's favourite home was Wimbledon Manor in Surrey, where she was able to indulge her love of birds and of gardens. Here, no doubt remembering the garden at Fontainebleau, she employed Claud Mollet's son, André, to improve the gardens. He oversaw the planting of 'great and large borders of rosemary and white lavender', built a large orangery with forty-two orange trees, 'one lemmon tree bearing grate and very large lemons', and planted many other exotic trees.[61] There were several gardens with mazes, wildernesses, knots, alleys, a 'muskmilion ground' for melons, and, near the house, an elaborate aviary:

> Three open turrets very well wrought for the sitting and perching of birds, and also having standing in it one very fair and handsome fountain, with three cisterns of lead belonging to it, gilded, which, when they flow and fall into the cisterns, make a pleasant noise. The turrets, fountain and little court are all covered with strong iron wire and lie directly under the window of the two rooms of the said Manor House called the Balcony Room ... This bird-cage is a great ornament both to the House and Garden.[62]

Although exotic animals especially monkeys and parrots were kept as household pets, few if any private people owned *collections* of animals; one exception may have been the Duke of Buckingham. In 1625, after he had been made Lord of the Admiralty, he set about improving his estate at New Hall in Essex, employing Tradescant the elder as his gardener; he was interested in exotic animals as well as plants. Tradescant wrote to the Secretary of the Navy asking him on Buckingham's behalf to deal with

> Al Marchants from all Place But Especially the Virgine & Bremewde & New Foundland ... that when they [go] into those parts they will take care to furnishe His Grace with All manner of Beastes & fowels & Birds alive or if not, with heads, horns, beaks, claws, skins, feathers, slips or seeds, plants, trees, or shrubs. Also from Guinea or Benin or Senegal ...

Whether or not the hoped for birds arrived, Buckingham did not have long to enjoy them, for he was assassinated in 1628.[63]

Other rich men were interested in exotic animals. In 1636 Sir Edmund Verney, a Royalist soldier and MP, wrote to his son:

> A merchant of London wrote to a factor of his beyoand sea, desired him by the next shipp to send him 2 or 3 Apes; he forgot the r, and then it was 203 Apes. His factor has sent him fower scoare and sayes hee shall have the rest by the next shipp, conceiving the merchant had sent for tow hundred and three apes; if yor self or frends will buy any to breede on, you could never have had such chance as now.[64]

In about 1638, when Sir Hamon L'Estrange was walking in London, he

> sawe the picture of a strange fowle hang out upon a [illegible] and my selfe, with one or two more then in company, went in to see it. It was kept in a chamber, and was a great fowle, somewhat bigger than the largest turkey-cock, and so legged and footed, but shorter and thicker, and of a more erect shape, coulourd before like the breast of yong cock fesan, and on the back of dunn or deare coulour. The keeper called it a Dodo.[65]

The dodo was probably the one stuffed by Tradescant, listed in the catalogue of his collection as 'Dodar from the island of Mauritius; it is not able to flie being so big'. His collection ended up in the Ashmolean Museum in Oxford—parts of the dodo's skull and a foot, blackened by age, survive and are now in the Oxford Museum of Zoology.[66]

Another relict that has survived is the tortoise given to William Laud in about 1628, when he was Bishop of London. At first it was kept in his garden at Fulham Palace, but five years later, when he became Archbishop of Canterbury, it was moved to Lambeth Palace. It lived on through the Civil War, the Interregnum, the reigns of Charles II, James II, William III, Queen Anne, and Georges I and II, quietly munching cabbages in the vast garden of the palace, until 1753, when, according to one account, it was killed by the frost when a gardener, for a small bet, dug it up during its winter hibernation.[67] The tortoise's damaged shell still exists, housed in a glass case in the library at Lambeth Palace.

The Civil War that began in 1642 did not end with the execution of Charles I in 1649; it dragged on until 1651, with the establishment of the Commonwealth, and then, two years later, its leader Oliver Cromwell became virtual dictator of the country as Lord Protector. Hyde Park and various other royal parks and mansions were sold for the benefit of the Commonwealth. Although the House of Commons ordered that St James's Park should be spared, it was neglected, and many of the trees were cut down by the citizens for fuel. After Cromwell had taken up his abode in Whitehall Palace, he was often seen walking in St James's Park, and his wife is said to have kept cows there, and to have built a dairy in Whitehall. The park was not open to the public at that time, but ladies and gentlemen connected with Cromwell's Court were permitted to promenade there.[68]

Few animals apart from horses were imported during the Interregnum, and in 1657 the only animals left in the Tower were six lions.[69] However, as livestock had been successfully transported across the Atlantic to New England by the colonists, Cromwell requested that a young moose and a deer fawn should be sent to England. In response, in 1655 Roger Williams, the governor of the plantation at Providence in New England, sent him a young deer as a present. It arrived at Falmouth on board the *Winsby*, before being transhipped to the *Reserve* for transport to London.[70] It was probably the first North American deer to be seen in Britain, but the species was not recorded.

Most pleasures, especially expensive pleasures like keeping exotic animals, were, of course, frowned upon during the grim years of the Civil War and Interregnum, and places of entertainment such as theatres and bear pits were closed down. However, exotic animals such as a pair of ostriches seen by Mr Clarke in London were occasionally shown; more information comes, not surprisingly, from John Evelyn, who was keeping out of the public eye in his garden at Sayes Court in Deptford, in south-east London. In 1654 he wrote in his diary: 'I saw a tame *Lion* play familiarly with a *Lamb*: The Lion was a huge beast, I thrust my hand into his mouth, & found his tongue rough, like a *Catt's* ...'. In 1657 his diary entry for 18 June read:

> I saw at *Greenewich* a sort of Catt brought from the *East Indies*, shaped & snouted much like the *Egyptian Ratoone*, in the body like a *Monkey*, & so footed: the eares & taile like a Catt, onely the taile much longer, & the Skin curiously ringed, with black & white: With this taile, it wound up its body like a Serpent, & so got up into trees, & with it, would also wrap its whole body round; It was of a wolly haire as a lamb, exceedingly nimble, & yet gentle, & purr'd as dos the Cat.[71]

It is uncertain what animal he was describing—possibly a ring-tailed lemur from the island of Madagascar.

After Oliver Cromwell's death in 1658 and the appointment of his son as Lord Protector, the country was plunged into political chaos. The situation was resolved when General Monck smoothed the way for the restoration of the monarchy. On 25 May 1660, when Charles II arrived at Dover from exile in Holland, Monck was on the beach to welcome him.

2.2. Charles II and James II

As pleasure returned to London after the Restoration of the Monarchy in 1660, Charles II set about renovating St James's Park as a place of enjoyment where he could parade with his dogs, a mistress, and a train of courtiers. John Evelyn's friend Adrian May was employed to oversee the construction of a formal garden in the French style, to supervise the gardeners brought over from France, and to set up enclosures and vivaries for animals.[72] Some of the ponds were converted into a canal, nearly 1,000 yards long, and others in the south-east corner were converted into a network of small canals around small islands resembling a Dutch decoy. Four long avenues of trees were planted, two flanking the canal, one along the north side (the Mall) and the fourth along the south side (Birdcage Walk).

Having spent some of his teenage years with his exiled mother Henrietta Maria at her court at Saint-Germain near Paris and at Fontainebleau, where, as we have seen, his grandfather Henri IV had kept birds, Charles was keen to embellish the park with animals. The East India Company, anxious to woo royal favour, sent an order to Surat in India for rare beasts and birds. They were hoping to acquire a rhinoceros, but the only animals to arrive were some spotted deer (chital). The King, however, was so pleased with them that he ordered that a pair of the 'handsomest deer' be sent by every ship.[73]

James I's aviaries along Birdcage Walk were extended, and a pheasantry was built where Marlborough House now stands.[74] By July 1661 John and William Walker had been reappointed 'Keepers of the King's house and yard in St. James's', and the park had been restocked with pheasants, guinea fowl, partridges, tame pigeons, rabbits, monkeys, parrots, and other birds, and an allowance made of *2s.* for food for the 'Cossawarway' (cassowary).[75] In November 'two greate and one small antelops, two pellicans, and two noorees [parrot in Malay] or Maccasser parrots' (Sulawesi hanging parrots) were dispatched by the Company's representative in Madras, 'wishing they may come live home, and in such a case, that Your Worships may make of them a royall present to the King's Majestie'.

Three envoys from Tsar Alexis of Russia arrived in London in 1662. The King ordered that they were to be received and accommodated in great state. Samuel Pepys was one of a huge crowd of citizens who watched their procession through the City of London, guarded by trained bands of the City and King's Lifeguards, and accompanied by most of the wealthier citizens of London attired in black velvet coats and gold chains. A month later the envoys were granted an audience with the King and Queen in Whitehall and presented them with 'rich furs, hawkes, carpets, cloths of tissue, and sea-horse teeth [walrus tusks]', but also, surprisingly, some pelicans.[76]

The traveller Peter Mundy visited the park in 1663 and wrote the most complete account of the animals there:

> Of beasts there are several sorts, Vz., an elke, of whose skinne buffe [leather] is made, fallow deere, also Indiann antelops, a kind of deere, allsoe another sort which I have seene in India of a yellowish colour with many white spots in ranckes; a small kind of goates from Guinea, etts. Aboundance of fowl, and of several sorts, viz., both of land and water, viz., cranes, storks, shovelars, pelicans, ets.... Peacocks, peahens, a white raven flying to and from the parcke, a

> hen with a brood of partridges, doubtlesse of her own hatching. Of waterfowl aboundance, as outlandish geese, duckes, etts of several shapes, collours and sizes. Among the rest, claegeese [barnacle geese?]... a great number of our ducks and mallard, widgeon and teal, pewetts etts., which swimme and fly to and fro, frequenting the several ponds. They have little hutts or cabins of boards fitted for them to breed... At a place near St Jameses house I saw a cassawarwa, a strange fowle somewhat lesser than an estridge, the body about four foote high. There was also a shee bustard...[77]

A French visitor in May noted in addition some sheep from the Cape Verde Islands (a Portuguese possession), 'which are much smaller than ours, and in which the coat is short and fawn-coloured like that of deer'.[78] Charles ordered Sir William St Ravy to take two of the pelicans, some Indian [foreign] ducks, and large number of 'extraordinary animals', as well as eighty deer, to his cousin Louis XIV's fledgling menagerie at Versailles, then under construction—a menagerie that would soon become the largest, grandest, most comprehensive in the whole world.[79]

Evelyn wrote in February 1665:

> I went to St *Ja: Park*... and examined the Throate of the *Onocratylus*, or *Pelecan*... a Melancholy water foule, brought from *Astracan* by the Russian Ambassador... also 2 Balearian *Cranes*, one of which having had one of his leggs broken and cut off above the knee, had a wodden or boxen leg & thigh, with a joynt for the <knee> so accurately made, that the poore creature could walke with it & use it as well as if it had been natural: it was made by a souldier: The Parke was at this time stored with infinite flocks of severall sorts of ordinary & extraordinary Wild foule, breeding about the Decoy, which for being neere so greate a Citty, & among such a concourse of Soildier, Guards & people, is very divert: There were also Deere of severall countries, Wite, spotted like Leopards, Antelope: An Elke, Red deeres, Robucks, Staggs. Guinny Goates; Arabian sheepe &c:...[80]

In 1664 the court of the East India Company 'directed Captaine Prowd to repair aboard the severall shipps and enquire what rarities of birds, beasts, or other curiosities there are aboard fitt to present His Majestie'. Among many subsequent orders was one sent to Madras in 1669, requesting eight spotted deer and, 'if there bee any small parretts, about the bigness of sparrows, which are called noories, or any other of the like nature which are rarities and pleasant to the eye or eare, send us some of them'.[81] By the time the naturalists John Ray and Francis Willughby visited the park, probably between 1666 and 1672 (the year of Willughby's death), some raptors had been added to collection—a golden eagle, a vulture, a crested Indian falcon, 'brought out of the East-Indies', and two eagle owls.[82]

In a letter to Surat in 1676, an East India Company official wrote: 'His Majestie desires no more cassawarrens'. Instead he requested more deer, especially females, a crowned crane, an 'East India goose', and any wild fowl that were different in colour from those in England. Eight deer were sent in response, as well as two sarus cranes, two antelopes (probably blackbuck), and an elk (perhaps a sambar or a barasingha, as elks *sensu stricto* are not native to India).[83] The King's interest in breeding deer may have been prompted by his love of field sports. Perhaps for the same reason he attempted to acclimatize red-legged partridges imported from France or the not very exotic islands of Jersey or Guernsey. The birds were released around Windsor and Richmond, but died out after a few years. It was many years before such an attempt was to be successful.[84]

Charles married the Portuguese princess Catherine of Braganza in 1662; her vast dowry included Tangier on the north-west tip of Morocco (Barbary) and Bombay on the western coast of India. Bombay was immediately leased to the East India Company for a mere 10 guineas a year, but Tangier remained an English possession until 1684. Sultan Moulay Ismail, the 'barbarous Emperor of Morocco', sent an ambassador to London, who arrived early in January 1682 hoping to bargain for the ownership of Tangier. Among the gifts he brought for the King were two lions, named Charles and Catherine, which were sent to the Tower, and thirty ostriches; when Charles saw the ostriches, he laughed; they were placed in St James's Park, and his courtiers were invited to help themselves to them.

Edward Browne,[85] one of the King's physicians, took one of the ostriches to his house in Salisbury Close off Fleet Street, but it died suddenly during the night of 3–4 February, probably from the cold. He promptly set about dissecting it and sent detailed post-mortem reports to his father, Sir Thomas Browne, the distinguished scientist, antiquarian, and physician. Edward was planning to give a lecture on the ostrich, probably to a meeting at the Royal Society, and his father advised him to tone down the text: 'The King or gentlemen will bee litle taken with the anatomie of it ... butt are like to take more notice of some other things which may bee sayd upon the animal, and which they vnderstand. Have a care of yourself this sharpe wether.' Among much else, Edward's dissection revealed that the ostrich was a young male with an intestine 20 yards long; he promised additional detail on the skeleton 'when the bones are cleane'. One of the many questions that had bothered Sir Thomas over the years was the claim that ostriches could digest stones and other hard objects. While his ostrich was still alive, Edward had decided to investigate the matter by experiment and had made it swallow a piece of iron weighing 2½ ounces; he was not surprised to find that, when the bird's first stomach was opened, the iron was completely unaltered.[86]

In his *Elysium Britannicum*,[87] John Evelyn set out his prescriptions for the princely garden; although he was much impressed by some of the enclosures he had seen on the Continent with their contingents of 'exotik creatures', and suggested 'lesser enclosures to be contrived in some remoter part of our Elysium with partitions & accommodations suitable to their natures', the only quadrupeds he notes as suitable for England are, disappointingly, squirrels and tortoises, which he recommends for their 'rare meate'. However, he has much to say about birds and aviaries; the reason is clear: birds provided music: 'We would therefore so plan our Aviarie in some part of the Garden, as the singing of the birds might resound even to the house, and be the oftner visited.'

Evelyn suggests the provision of a volary for larger birds such as cranes, bustards, and peacocks, and, for smaller birds: 'An Aviary of 60 foote long, 15 broade & 30 high will be sufficient to hold 500 smale Birds together with a competent number of Turtles [doves], Quailes, Partridge & Pheasant' and

> Cages of 20 & 30 foote in length will hold birds enough to make the Welkin ring with their musique ... Those who keep curious Birds, Nightingalls & such as appear in the Summer onely, must add a stove to attemper the aire in winter, with curtains made of oiled canvas ... we would rather recommend the keeping of such Birds in cages apart, & within the mansion house ...

He advised that the aviary should be planted with hemp, canary seed, millet, oats, holly, and hawthorn to provide seeds and berries as food for the birds, which were mostly indigenous, but also included 'the Canary Bird, Virginian Nightingall , & some others'. He goes into a great deal of detail regarding the positioning and construction of the aviary, with its perches and nest holes, and the provision of suitable food and water; in his own words: 'Thus we have don with Aviaries in which we have bin the more accurate and prolix because we have ever esteemed it amongst the Sweetest varieties & ornaments of our Elysium, in which we have taken wonderful delight... '. Evelyn built an aviary in his own garden at Sayes Court in Deptford, where, in 1656, the Marquis of Argyll 'since executed', mistook the turtle doves for owls.

In his *Systema Horti-culturae: Or, the Art of Gardening*, first published in 1677, John Worlidge wrote: 'One of the pleasures that may be esteemed belonging to a Garden, is an Aviary, which must be near your house, that you may take some delight in it there, as well as in your Garden.'[88] Although it is probable that many of the gardens of the rich were provided with aviaries, there are few references to them; this may suggest that they were too common to mention, or, contrariwise, that they were actually rare. The Duke of Lauderdale had several aviaries in his famous gardens at Ham House, near London, which Evelyn described in his diary in 1678—'the house furnish'd like a great Prince's; the Parterres, Flower Gardens, Orangeries, Groves, Avenues, Courts, Statues, Perspectives, Fountains, Aviaries... '[89]—and, as we shall see, Sir Thomas Browne had an aviary at his home in Norwich.

One of the most important institutions in the development of science was the Royal Society. It owes its origin to a dozen men who met on 28 November 1660 at Gresham College in London and decided to form 'a Colledge for the Promoting of Physico-Mathematicall Experimentall Learning'; by the time it received its second royal charter from Charles II in 1663, its title had been changed to 'The Royal Society of London for Improving Natural Knowledge'. Among its earliest and most distinguished fellows were Christopher Wren, Robert Hooke, Robert Boyle, Elias Ashmole, Evelyn, Pepys, and Francis Willurgby, soon to be joined by John Ray and Edward Browne. The papers they presented at their weekly meetings were published in the *Philosophical Transactions of the Royal Society of London*, a journal of international standing still in production.

As we have seen, much of the information on James I's birds comes from Willughby and Ray. Ray's *Ornithology of Francis*

Willughby was the first definitive book on ornithology, containing an extraordinary amount of information on the birds they had seen, both dead and alive, in Britain and on their travels in Europe. The species were arranged in systematic order, and many references to earlier scholars such as Gesner, Aldrovandus, and Carolus Clusius were included. Sir Thomas Browne provided them with descriptions and pictures of various birds; he is also credited with coining the word 'incubation'.[90] Although, like John Caius a hundred years before, Browne mined classical references, he relied more on the evidence of his own eyes and carried out numerous experiments to test his ideas; both he and his son Edward dissected large numbers of animals, and both had an extraordinary knowledge of comparative anatomy. Browne was also a great collector whose 'library, museum, aviary and botanical garden . . . were thought by Fellows of the Royal Society well worthy of a long pilgrimage'.[91] It is not known what birds were kept in his aviary, apart from an owl and an eagle, but for a while he kept an ostrich with a voracious appetite. He complained that, 'when it first came into my garden, it soon ate up all the gillyflowers, tulip-leaves, and fed greedily upon what was green, as lettuce, endive, sorrell; it would feed on oats, barley, peas and beans; swallow onions; eat sheep's lights and liver . . . '.[92]

There were few, if any, menageries in the sense of collections of exotic animals during the seventeenth century, apart from those owned by James I and Charles II, although some people acquired pet monkeys or a few cage birds. Some of the first exotic birds to arrive were red cardinals from eastern North America. Evelyn's mention of 'Virginian nightingales' (as they were known) seems to be the earliest, if not very definite, record in England, although one individual was recorded in Italy at the end of the fifteenth century.[93] In 1673 Willurgby saw some tiny colourful red avadavats probably from India, in a cage 'in the house of a certain citizen of London',[94] and in the following year Henry Oldenburgh wrote to Ray to tell him that Lord Mordaunt had 'Barbadoes Turtles that are not bigger than Larks' at his house at Parsons Green and that his Lordship was willing to permit any artist 'to take a draught'.[95] In another letter later that year Oldenburgh describes Indian fantail pigeons, brought by the East India Company:

> I lately saw 2 or 3 Sorts of East-Indian Birds, brought thence with the last Return-Ships; very fine Creatures: And they were, 1. A curious speckled Indian Hen. 2. Some East-Indian Pigeons, delicately shaped.

3. Some very small Birds, with short scarlet beaks, and curiously speckled feathers, etc. These if we could learn their names, and something of their Nature and Quality, were very well worth, in my Opinion, to be taken into your Book. I hear they are shortly to be brought from Wapping... to Tower-Hill; and if they be so, we may get a Draught of them, if you think fit, for the Engraver... The pigeons were Suratta pigeons, sprightly, with extraordinarily broad Tails, which they spread out almost Peacock like.[96]

Pepys's diary contains several references to parrots owned by his acquaintances, with no indication of species, but, according to Ray, the 'green parrot having the ridge of the wing red... is the most common with us', so they were probably male eclectus parrots from Indonesia, still favourite pets today.[97] Pepys saw a more unusual bird in April 1664: 'In the Duke's chamber there is a bird, given him by Mr Pierce the surgeon, comes from the East Indys—black the greatest part, with finest collar of white about the neck. But talks many things and neyes like the horse and other things, the best almost that ever I heard bird in my life.' This seems to be the earliest record of a hill mynah in England, although the 'collar' is usually yellow.[98]

When Pepys's employer Lord Sandwich was in Lisbon in February 1662 struggling with problems over the payment of Catherine of Braganza's dowry, Captain William Hill, commander of the *Augustine*, sailed back to England, bringing presents from Sandwich to his wife, among them a civet cat, a parrot, and some apes (probably Barbary apes from Gibraltar or Morocco).[99] In May, when Sandwich himself returned home, more presents arrived on board the expedition's flagship the *Royal James*, not only 'many birds and other pretty noveltys' and a fine Barbary horse for himself, but also a 'little Turke and a Negro', intended as pageboys for two of Sandwich's daughters.[100]

In May 1665 Pepys and his wife saw 'a fine rarity: of fishes kept in a glass of water, that will live so for ever; and finely marked they are, being foreign', belonging to Lady Penn, the wife of the Navy Commissioner Sir William Penn, their near neighbours in Seething Lane. Being 'finely marked', these may have been paradise fish from China or south-east Asia rather than goldfish, which the naturalist Thomas Pennant believed to have been first introduced in about 1691.[101]

Pepys owned several exotic animals himself. Returning home late one night in January 1661, he was furious to find that his pet monkey had got loose; he caught it, and, to his later shame, battered it almost to death.[102] A week later he was sent two cages of canaries by a 'stout seaman' Captain Rooth.[103] In 1664 he recorded giving his eagle to his distant cousin: 'we were heartily glad to be rid of her, she fouling our house mightily . . . '.[104] It is difficult to imagine a tame eagle in a crowded London house; perhaps it was a smaller bird of prey. Another improbable pet was a tame lion cub sent to Pepys by Mr Martin, the consul at Algiers, 'a gift he had taken the boldness to send as the only rarity the place offered'. Pepys wrote to thank him, assuring him that the cub was comfortably installed, 'as tame as you sent him and as good company', at Derby House in Canon Row, the house he had moved into in 1674 shortly after his appointment as Secretary to the Admiralty.[105]

In August 1661 Pepys wrote:

> By and by we are called to Sir W. Battens to see the strange creature that Captain Holmes hath brought with him from Guinny; it is a great baboone, but so much like a man in most things, that (though they say it is a Species of them) yet I cannot believe but that it is a monster got of a man and she-baboone. I do believe it already understands much english; and I am of the mind that it might be taught to speak or make signs.[106]

Holmes had just returned from an expedition to West Africa, so this could have been a chimpanzee, or possibly a gorilla, and therefore the first recorded importation of a living great ape to Britain.

Several large strange animals were brought from afar during Charles's reign, intended, not for him, but as commercial ventures, to be shown or sold in pubs or at fairs such as Bartholomew Fair in the City of London. A young female rhinoceros arrived on board an East Indiaman, the *Herbert*, in August 1684 from Machilipatnam on the Coromandel Coast north of Madras where the company had a factory (Figure 2.1). She was valued at £2,000, but, having failed to sell her, the owners exhibited her at the Belle Sauvage Inn at the foot of Ludgate Hill every day from 9 a.m. to 8 p.m.; the proprietor was said to have charged 1*s*. for a peek and 2*s*. for a ride, making a tidy profit of £15 a day.[107]

The ever curious Evelyn went to see her in October and wrote an engagingly graphic description in his diary,

> The Rhinocerous (or Unicorne) being the first that I suppose was ever brought into England . . . resembled a huge enormous swine . . . but what was the most wonderfull, was the extraordinary bulke and

Figure 2.1. England's first rhinoceros, 1684 (*The Exact Draught of that Famous Beast the Rhinoscerus*, anonymous engraving 1684: Glasgow University Library, Sp.Coll. Hunterian Av. 1.17 item fol 35).

> Circumference of her body, which . . . could not be lesse than 20 foote in compasse: she had a set of most dreadfull teeth, which were extraordinarily broad, & deepe in her Throate, she was led by a ring in her nose like a Buffalo, but the horne upon it was but newly Sprowting . . . in my opinion nothing was so extravagant as the Skin of the beast, which hung downe on her haunches, both behind and before her knees, loose like so much Coach leather . . . these lappets of stiff skin, began to be studdied with impenetrable Scales, like a Target of coate of mail, loricated like Armor . . . [When she lay down] she appeared like a great Coach overthrowne, for she was much of that bulke, yet would rise as nimbly as ever I saw an horse . . . to what stature she may arive if she live long, I cannot tell, but if she grow proportionable to her present age, she will be a Mountaine: They fed her with Hay, & Oats, & gave her bread.[108]

Sadly England's first rhinoceros did not last long. A newsletter dated 28 September 1686 reported: 'Last weeke died that wonderfull creature the Rhynocerus; the several proprietors having

Ensured £1200 on her life the Ensurers are catched for much money.'[109]

On the same day that Evelyn saw the rhinoceros (22 October 1684), he

> went to see a living Crocodile, brought from some of the W: Indian Islands, in every respect resembling the Egyptian crocodile, it was not yet fully 2 yards from head to taile, very curiously scaled & beset with impenetrable studds of a hard, boney substance ... they kept the beast or Serpent in a longish Tub of warme Water, & fed him with flesh &c.

Several Indian elephants were brought to London in Charles II's reign. The first in 1675, a calf, was intended for Lord Berkeley, a director of the East India Company.[110] It was transported on an East Indiaman from Bantam in Java in the care of two native attendants and landed at Whitefriars. It was soon put up for auction—'The Lord Berkeley's elephant (who is 5 foot and 4 inches high) is to be sold by the candle at the East India House, sett up at £1,000 and to advance £20 every bidding' (Figure 2.2).[111] In the event it went for £2,000 (or £1,600 according to Robert Hooke, who paid 3*s.* to see it) and was first exhibited at the Rising Star in St John's Court, near Clerkenwell Green, and then at 'the White Horse Inn over against Salisbury Court in Fleet Street'.[112] On 8 February 1676 permission was granted to Lord Berkeley by the Company to send the two 'blacks' 'that came over with the elephant' back to Bantam by the next ships.[113]

Berkeley's elephant may have been the same young animal that Robert Plot saw in Oxford in 1676 and may have survived for at least three years, for in July 1679 Thomas Browne wrote to his son: 'I heare that there are 2 elephants in London, when you have opportunity and leasure to see one, observe this well'; the 'vulgar error' that Browne was anxious to debunk was the oft repeated assertion that elephants could not bend their legs. It may also have been the animal shown at Bartholomew Fair on 1 September that year where Hooke and Sir Christopher Wren 'saw Elephant wave colours, shoot a gun, bend and kneel, carry a castle and a man, etc.'.[114]

There was a third elephant in the British Isles at this time, not in England, but in Ireland. It was shown by a Mr Wilkins in a 'booth' near the Custom House in Dublin, but, at 3 a.m. on 17 June 1681, the booth caught fire and the elephant was incinerated. It was

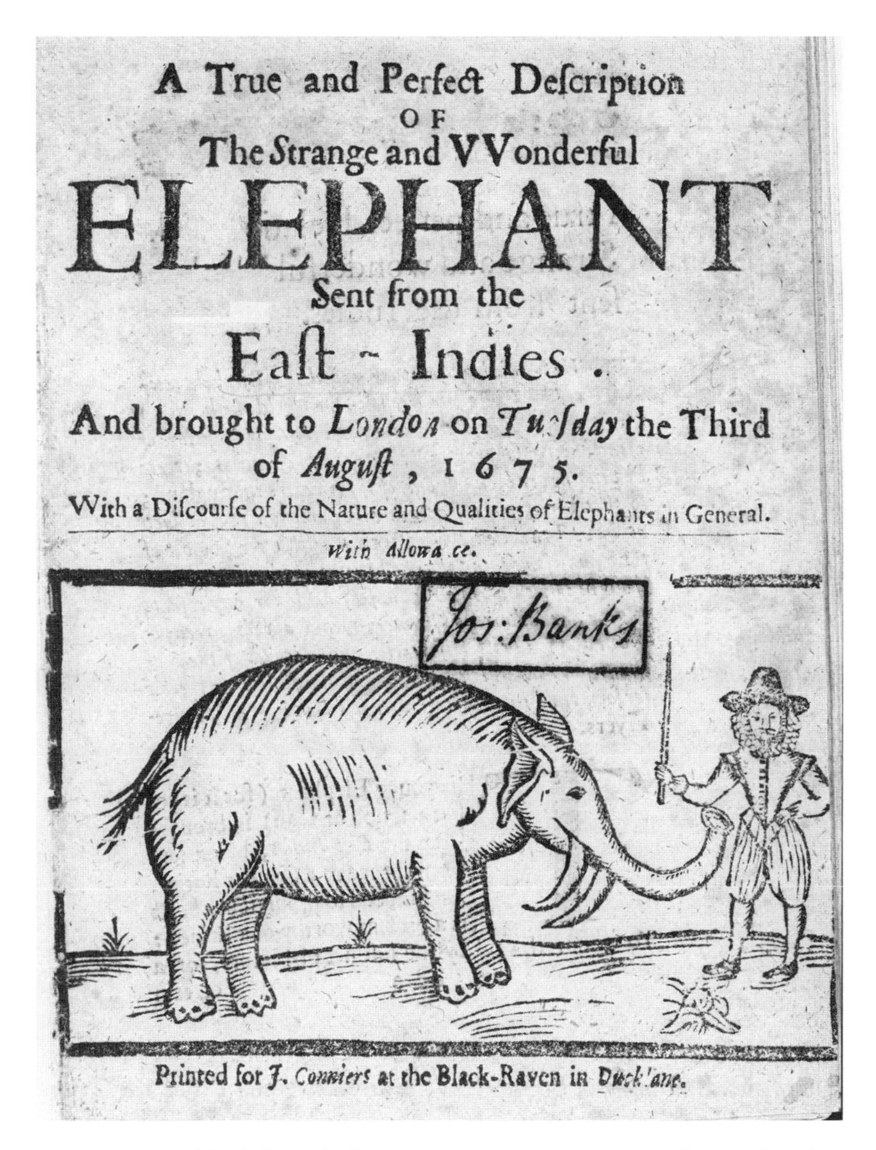

A True and Perfect Deſcription
OF
The *Strange* and VVonderful
ELEPHANT
Sent from the
Eaſt-Indies.
And brought to *London* on *Tueſday* the Third
of *Auguſt*, 1675.
With a Diſcourſe of the Nature and Qualities of Elephants in General.
With Allowance.

Jos: Banks

Printed for *J. Conniers* at the Black-Raven in *Duck-Lane*.

Figure 2.2. Lord Berkeley's elephant 1675 (anonymous page, *A True and Perfect Description OF The Strange and VVonderful ELEPHANT*...: BL, B.424.(2.)).

dissected by candlelight under the watchful eye of Dr Allan Mullen, who described his observations in a booklet published the following year, adding that its flesh tasted 'like that of lean Beef season'd with Salt of Hartshorn'. Wilkins was able to compensate for his loss by putting the mounted skeleton and some of the elephant's parts on display, reassured by the fact that the skeleton did not require food.[115]

A broadsheet in the British Library purports to be 'A True and Perfect Description of the Strange and Wonderful SHEE-ELEPHANT sent from the INDIES, which arrived at LONDON, August 1. 1683',[116] but unfortunately the 'shee elephant' is depicted with a penis, and the remainder of the text deals only with elephants in general. Yet another elephant, a young male, was offered for sale a year later, 'about six foot high, of a large breed, and now arriv'd in the George from India, stands to be sold at Mr Thomas Collis's Brewer in Lime-House'.

In his *Micrographia* published in 1665 Hooke mentioned that he had handled a caribou, 'a Greenland Deer, which being brought alive to London, I had the opportunity of viewing; its hair was so exceeding thick, long and soft, that I could hardly with my hand, grasp or take hold of his skin, and it seem'd so exceeding warm, as I had never met with any before'.[117]

Performing monkeys were sometimes on show; in September 1660 Evelyn saw

> in Southwark at St Margarites faire ... *Monkyes* & Apes daunce, & do other feates of activity on the high-rope, to admiration: They were galantly clad *alamode*, went upright, saluted the Company, bowing, & pulling-off their hatts: They saluted one another with as good grace as if instructed by a Dauncing Master. They turned heales over head, with a bucket of Eggs in it, without breaking any: also with *Candles* (lighted) in their hands, & on their head, without extinguishing them, & with vessells of water, without spilling a drop.[118]

Pepys saw similar shows on various occasions at Bartholomew Fair, 'but such dirty sport that I was not pleased with it'.[119] Other animals were sometimes displayed at the fair; in 1672 Hooke 'saw India Catt, Japan Peacock, Porcupine, Upapa (hoopoe), Vultur, Great Owl, 3 Casswaris' and in 1677 paid 2*d*. to see a 'tigre' (probably a leopard).[120]

As before, most of the animals in the Tower were lions, but a leopard was noted in 1704 as having been in the Tower 'ever since K. Charles the Second's Time', and there were also a few birds—two 'Swedish Owls', one of 'a great Bigness, called Hopkins',[121] and three golden eagles were seen by Ray and Willughby, probably in the late 1660s, when they visited what they referred to as the

Royal Theriotrophium.[122] In 1662 Pepys took the children of Sir Thomas Crewe and of Lord and Lady Sandwich 'to the Tower and showed them the lions and all that there was to be shown and so took them to my house and there made much of them; and so saw them back to my Lady's—Sir Th. Crewes children being as pretty and the best behaved that ever I saw of their age'.[123] On 4 February 1685 one of the lions died, a bad omen, for two days later Charles II was dead.[124]

Charles was succeeded by his Catholic brother James II. The menagerie in St James's Park probably came to an end when, on James's orders, 'several rare beasts and fowl' were transferred to the Tower. There they were joined by a new Barbary lion cub: 'Pompey... lived upwards of 70 years and died in the Tower on 10 November 1758.'[125] Women as well as men were employed in the menagerie—Mary Jenkinson, described as 'A woman that lookt after the lions at the Tower', put her hand near the oldest lion, 'he caught hold of it and grip'd it so hard that it was forc'd to be cut off to prevent a gangrene, but she died of it in a little time'.[126]

The post of 'Keeper of His Majesty's Lions' was a sinecure, held since Elizabethan times by successive members of the same family, the Gills, the last of whom, William Gill, lived in the Lion House built under Christopher Wren's supervision in the 1670s. Gill was succeeded in 1686 by Thomas Dymock, who soon had the misfortune to lose his black slave. His advertisement a year later in the *London Gazette* is a grim reminder of the times:

> On the 30th of December last, Run away from Mr Thomas Dymock at the Lyon Office in the Tower, a black Boy, with about 10 [pounds] in Silver, and one Guinea; he is aged about 16, wore three colored Coats, two grey, his uppermost Cinamon colour, lined with black, black Shagg Facings on the Sleeves, great Stockings, a Silver Collar about his Neck, Engraven, Thomas Dymock at the Lyon Office. Whoever shall apprehend him, and bring him to the Lyon Office in the Tower, shall have two Guinea's Reward, and Charges born. He speaks but bad English, and hath holes in both his ears.

Dymock found it necessary to petition James II to re-establish the Keeper's monopoly regarding the public display of lions and other fierce beasts. The *London Gazette* for 4 July 1687 printed the King's response:

> His Majesty has been pleased... To Prohibit and forbid all Persons whatsoever, except Thomas Dymock, the Keeper of His Majesties

> Lyons for the time being, to carry abroad, or expose to publick view for their own Gain, any Lions, Lionesses, Leopards, or any other Beasts *ferae Naturae*, without his Majesties special License, as they will answer the contrary at their perils.

The importation of birds on a commercial scale began with consignments of wild canaries from the Canary Islands, probably brought on the numerous ships that carried barrels of 'Canary wine',[127] but during the reign of Charles II captive-bred birds from Germany began to arrive via Holland; initially very expensive, they came in such large numbers that by 1678 prices had fallen and they could be kept by 'mean men' as well as 'nobles and great men'. In the *Ornithologia* Ray gave detailed instructions for keeping and breeding canaries and wrote that the birds from Germany 'in handsomeness and song excel those brought from out of the Canaries'.[128] In 1685 the *London Gazette* advertised the arrival of 700 'canary birds' from the Canary Islands, to be sold 'at the Sign of the Black Bull at Tower Dock' by Thomas Bland, and two years later Bland had 400 captive-bred canaries brought via Holland from 'High Germany'—that is, Austria. A year later Dowey Hobbs, the master of the ship *Prince William* from Friesland, complained that he had had his ship seized in London by the Tide Surveyor, even though he had for 'divers years past... brought over live birds for the late King and in the hold of his ship always brought slit boards for the birds to walk on...'. He promised that, if his ship were released, the next May he would bring over more birds for the new King.[129]

Commercial contacts between England and Holland had been close ever since the collapse of Antwerp as an international entrepôt in the third quarter of the sixteenth century, and colonies of English merchants became established in several Dutch ports disposing of English exports (mainly woollen cloth) and buying a great variety of imports.[130] During the last quarter of the seventeenth century some of the exotic animals that appeared in London were probably purchased at the famous *Blauw Jan* public house in Amsterdam, having been brought to Holland from the Far East by the Dutch East India Company.[131]

Dutch influences were to become even stronger in 1688, when the Protestant Willem III, Prince of Orange, the Stadtholder of the Dutch Republic, accepted the invitation to become King of England as William III, to rule jointly with his wife Mary Stuart, the Protestant daughter of James II.

3
William and Mary to George II, 1688–*c*.1760

3.1. 'Bird Men' and 'Wild-Beast Men'

As we have seen, there was a surprising amount of traffic in people and goods between London and the Netherlands in the second half of the seventeenth century, which gained momentum when William of Orange and his English wife, Mary, ascended the throne in 1689, bringing with them Dutch diplomats and courtiers, Dutch taste in gardens, tulips, and birds, and much else.[1] In 1690 the bird merchant Thomas Bland, whom we have already met, received three consignments of several hundred canaries from Holland. He was still in business at the turn of the century, at 'The old German Canary Bird house, at the sign of the George and Dragon at Tower Dock', while a rival, Nicholas Heath, was selling canaries 'at the Old Black Joe's the German Birdman in Crooked-lane'.

Birds were brought through Holland by itinerant sellers from Germany—in 1708 Safferin the Birdman brought over a parcel of canary birds for sale at the Old Bird House at the Eagle and Child in St Martins-le-Grand. Two years later he was back with 'a choice Parcel of Canary-Birds of various colours, as Good Song as have come this many Years, to be sold at Mr Hawkin's at the Dolphin in Creed Lane'. In February 1717 'Jacob, the old German, is just come over from High Germany with a Choice fresh parcel of sound Canary Birds of the best Song, and of all Colours, as White, Mottled, Lemon, Ash and Grey, are to be sold at the Eagle and Child in St Martins Le Grand, the Old Canary Bird House'. Some advertisements for the arrival of canaries from Germany specify 'High Germany', Bavaria, or 'Tirol'—in modern terminology, southern Germany or Austria.

Wild canaries are a dull yellowish-green with brown streaks on the back, but part-yellow birds had been bred in Germany by the late sixteenth century, and birds that were entirely yellow or entirely white by the mid-seventeenth.[2] John Ray described just a single kind in which the wings and tail were green and the breast, belly, and upper part of the head were yellow, more so in males than females,[3] but by 1710, if not before, there were several colour variants, bright yellow birds with jet black spots, mostly white but mottled with black or brownish spots, 'mealies', all yellow, all white, and all grey, as well as pale yellow 'Jonquil' birds said to have been bred in Paris. Canaries were valued for their song as well as their colour and were often advertised as having been taught to sing by the 'flageolet'.[4] One of the many manuals published on bird-keeping in the early eighteenth century even printed the score of a tune that could be taught to canaries.[5]

Eleazar Albin wrote: 'There are many people in England, as well as France, Germany &c, that get a good livelihood by breeding canary birds: besides a great number of people who breed them only for pleasure.'[6] By the mid-eighteenth century competitions were held for the best birds and prizes awarded,[7] and by the 1770s English-bred canaries were being offered for sale—those from London, it was claimed, 'far-exceeding' those of Colchester, Worcester, Ipswich, and even Germany.

Canaries were by no means the only exotic birds brought to England, although none excited as much 'rage'. Various breeds of domestic poultry were imported from Holland, including Hamburgh chickens and hookbill ducks, which originated there in the seventeenth century, as well as topknot chickens, known as 'crested Dutch chickens', and topknot 'Polish' ducks. With their expertise in breeding domestic poultry, it is probable that the Dutch were also pre-eminent in breeding birds of more exotic origin, such as bantams, mandarin ducks, guinea fowl, Muscovy ducks, peacocks, pheasants, and various types of doves and pigeons. As late as 1791 it was stated that 'the Dutch excel in every thing relative to birds; they breed more varieties, they keep them alive longer, and in short they are more attentive to them, than the English . . . '.[8] There may also have been some deliberate attempts at breeding in England, since Albin in the first volume of his great *Natural History of Birds* published in 1731 stated that swan-geese (also referred to as 'East India geese') interbred with ordinary domestic geese. Most of these birds were imported not for consumption, but as ornaments for the growing number of parks and gardens being developed in the new Dutch style; thus water fowl might be advertised as 'fit for a Person of Quality's Ponds or Canals'.

In the seventeenth and eighteenth centuries ships of the English East India Company brought birds from India, more parrots, ring-neck parakeets, mynah birds, and the variously spelled 'amedevats' (small amedavat waxbills that could be taught to sing), but the importation of exotic animals from the Far East was largely in the hands of the Dutch East India Company, founded in 1602. The company had special warehouses and stables on the quayside at Amsterdam for the reception of regular cargoes of animals from their trading post in Java, Batavia (Jakarta), and, as we have seen, a large collection of exotic animals was displayed at the famous Blauw Jan public house in Amsterdam from 1675 to 1784.[9] Animals also arrived in Holland from Brazil and Africa, brought presumably by the Dutch West India Company, which had trading stations in both areas dealing in gold, ivory, and slaves. Most of the animals that came on to London from Holland were shipped from Rotterdam, which was nearer and easier for access across the North Sea than Amsterdam—sulphur-crested cockatoos, brilliant coloured lories, and parrots were included in the cargoes. One of the most spectacular rarities to arrive was a 'Casheward bird' (cassowary) weighing 200–300 lbs from New Guinea, offered for sale by David Randall in 1704.

The British and the Dutch had trading posts dotted along the coast of 'Guiney' or Guinea—the term used in the seventeenth and eighteenth centuries for the coastal areas of Africa beyond the Senegal River as far south as Angola. Some very hardy birds obtained there might have crossed the Atlantic twice in slave ships, first on the notorious middle passage from Africa to the Americas and the West Indies, and then on the inward passage to Europe, leading to much confusion as to their origin.

Birds from the New World—wild turkeys and 'parcels' of Virginia nightingales (cardinals), various parrots, and mourning doves—also arrived in British naval and trading vessels; in 1736 Albin painted a 'Barbadoes Parrot', actually a yellow-shouldered Amazon parrot, which 'was as tame and good humoured as he was beautiful, suffering his Mistress to play with him, as with a Dog; he also talked very prettily; he was brought from Barbados by a Captain which traded to the West-Indies'.[10]

Exotic birds were big business. Albin saw a lory for sale at Bland's for 20 guineas, equivalent to nearly £2,000 today, and a macaw from Jamaica for 10 guineas,[11] so Richard Perry's offer of a reward of 2 guineas for the return of an African grey parrot was not overgenerous. The various mammals must have cost a lot more.

Thomas Ward first appears in 1714 as the author of *The Bird-Fancier's Delight*, a guide to catching and keeping wild songbirds from the countryside, but he also included canaries:

> The bird is called the Canary-Bird, because the Bird originally came from thence, but now with Industry they breed them very plentifully in Germany, and in Italy also; and they have bred some here in England, tho' as yet not anything to the Purpose as they do in other Countries: But they begin to esteem 'em more now than of late...

By 1717 Ward had acquired a pub in the City of London, the Bell and Bird Cage in Wood Street, periodically visited by Jan Safferen and Black Joe with their 'parcels' of canaries, and was also selling English birds and birdseed. Eleven years later he published a new edition of his book, updating the canary-bird entry: 'The birds brought from the Canaries, are not so much in Esteem with us as formerly, for those brought from Germany and France, far exceed them in Handsomeness and Song...'.

In 1732 Ward, described as the victualler of the Bird House Inn on Stamford Hill, was fined five shillings for exercising his trade on a Sunday.[12] He sold birds as well as beer, and in 1735 he published the third edition of his book. Meanwhile John Ward (his son?) carried on the business at the Bell and Bird Cage, continuing to sell English songbirds (and the occasional monkey), and taking deliveries of canaries from the German birdmen Anthony Bush, Thomas Pepperly, and George Turner. He was also breeding them himself, while running the pub, dealing in horses, and selling copies of his father's book. In 1745 his stock included cardinals from Virginia and 'as fine a parrot as ever talked'; in the same year his father advertised his own stock of birds in Stamford Hill, which included a large parcel of English-bred canaries, describing himself as 'the oldest Birdman now living'.

From 1700 onwards Thomas Bland's business at the old German Canary Bird House was being run by Michael Bland, who was probably his son, at the George and Dragon on Tower Dock, then at the Leopard and Tyger, then at the Tyger—presumably the pub's name changed from time to time. Bland and the other London birdmen imported birds in hundreds, but their shipments sometimes included a few mammals; small monkeys were the most numerous, usually of unspecified type. A shipment in February 1705 included a 'young Man-Tyger newly come from Moscow'; 'man-tigers' are usually interpreted as mandrills or baboons, so Morocco is a more likely source. A year later Bland had a large 'he Antelope from the East Indies' for sale—that is, a blackbuck from India, or more probably a nilghai.

Bland was particularly well placed to acquire exotic animals, as ships arriving from foreign parts had to unload their cargo on the Legal Quays—the strip of river frontage between the Tower of London and Billingsgate, and from 1717 to 1738 he acted as agent for many of sloops that plied to and from Rotterdam, and, occasionally, Calais, Bordeaux, and Dunkirk. The volume of trade was enormous: in a single year, for example, 1732, Bland was one of the three agents acting for about eighty sailings, all to Rotterdam, which explains why so much of his stock came from Holland and the Far East. A typical advertisement reads:

For ROTTERDAM,
The KING DAVID *Sloop*,
JOHN ALCOCK *Commander*,
Sails on Friday next
being the 14th Instant, now lying at St
Katherine's, to take in Goods and Passen-
gers, and may be spoke with every Morn-
ing at the Kings-Arms Coffee-house, by
the Custom-House, and on the Dutch-
Walk at Exchange Time, or John's Coffee-
house in Sweeting's-Alley, and at Michael
Bland's, at the Tyger near Tower Dock.
PETER VONK *for the Master*.

Bland's advertisement of 1714 lists a wild variety of exotic birds and domestic fowl, probably all brought from Rotterdam:

> Newly brought over from the East-Indies, a parcel of Ammerde-wates of divers fine colours, and sing very well, two Gavey Birds of several fine colours, the like never brought over before, several of the finest Mackaws of divers colours as ever was seen; two Cock-a-Tores, several fine talking Parrots and Parrokeets; one Lowrey; variety of Scarlet Nightingales from the West Indies; choice Canary Birds of all Colours; two small Memzetes Monkeys no

> bigger than a squirrel, and small Turtle Doves no bigger than a Lark; and other sorts of Turtle Doves, Hookbill Ducks, Borrow Ducks, large Muscovy Ducks, Whistling Ducks, East India Geese, large Hambrow Fowls; two West India Yews and a Ram with hair upon them like a Deer. Such Rarities no Man has to sell besides the Owner of these, which is Michael Bland, at the Leopard and Tyger at Tower Dock, near Great Tower Hill, where Gentlemen will be kindly used.

The three 'West India' sheep must have been descendants of African haired sheep taken to the West Indies by Dutch or British slave-traders in the previous century, which are thought to be ancestral to the Caribbean haired sheep of today. In the following year, as well as birds and monkeys, Bland had a large porcupine from Turkey (that is, the Ottoman Empire) and a 'Posum with a false Belly'—an opossum from the New World.

The year 1719 must have been a bumper one for Bland, for among the numerous birds, including king birds of paradise from the Spice Islands, he had a very spectacular animal for sale:

> There is just arrived from Beuenos Arres, in the South Seas, the greatest Rarity that ever was see in England, that is A large Hog in Armour alive, which Armour runs all in Joints all over his Body, his Head and Tail, and the Armour is so hard that a Musket Ball may be shot against him and can't hurt him...

This must have been an armadillo. One wonders who purchased it, or perhaps Bland was able to charge the curious to view it. This was certainly the case the following year when he showed a cassowary and a crowned crane from the East Indies. Bland's business received royal approval in 1723, when the Prince and Princess of Wales called at Tower Dock to see his 'collection of foreign birds and were highly delighted'. In 1726 he could boast 'the largest Musk Cat as ever was seen, and a large Wolf from Barbary'; next year a pair of wolves from Maryland was accompanied by a white wolf from Greenland, probably brought in a whaling ship.

Michael Bland died on 9 April 1737, clearly a man of some standing: 'On Monday Night last died, at his House on Tower-Hill, the noted Mr Michael Bland, who for many years past was a great Virtuoso in Birds and other Foreign Curiosities.'[13]

David Randall was in business from at least 1701 in a more fashionable part of town, at the old Bird House in Channel Row (now Cannon Row) between Parliament Street and the Thames. As with Bland, much of Randall's stock came via Holland. Indeed in 1702 he was in Amsterdam negotiating the purchase of an elephant from a Mr Barden. In his anxiety to deter a rival, Gregory Johnson, from acquiring said animal en route, he issued a notice announcing that he would pay Barden £1,000 on delivery, but nothing more is known of it. Perhaps the 'curious' of London had seen enough elephants for the time being, for earlier that year one had been on show at the White Horse Inn in Fleet Street, prior to being sold to the King of Denmark. The cassowary that Randall advertised in 1704 probably also came via Holland. Thereafter he stuck with less spectacular 'foreign birds' and the occasional mammal. In about 1714 his premises were taken over by Richard Perry, but Randall carried on dealing at the Porter's lodge at the Royal Mews near Charing Cross, where he was referred to as the 'Old Birdman'.

As well as numerous exotic birds offered for sale by Richard Perry—'a great Dealer in foreign Birds and Beasts'[14]—in 1714 were 'a large Peccary from Barbary' and a 'Tiger and Civet Cat from Guiney'. The first was probably an African member of the pig family, not a peccary, since peccaries are confined to the New World and the 'tiger' was probably a leopard. Civet cats were big business in Holland in the seventeenth century, having been imported from India, Java, and especially Guinea in their thousands for the extraction of their valuable scent. There is rather scanty evidence that civet cats were also imported in large numbers to England to be farmed—in the late seventeenth century the writer Daniel Defoe borrowed £850 in order to buy seventy civet cats, which he kept in Stoke Newington; the failure of the project contributed to his bankruptcy in 1692.[15]

In 1718 Perry advertised a more spectacular animal: a 'large white Dromedary that is 22 Hands high and 12 foot long, taken by Prince Eugene from the Turks at the Battel of Belgrade; she will travel 50 Miles in one Day and carry 1000 Weight'. Hundreds of camels, not to mention 5,000 prisoners, were seized from the Turks when Prince Eugene of Savoy captured the city for the Austrians in 1717; according to Lady Mary Wortley Montagu, it was camels that brought the news of the battle to Constantinople.[16] A Virginia deer arrived in 1722 and was offered for sale by Perry along with dozens of birds. It is not clear what he was doing for the next twelve years, but in 1734 he was said to have paid 500 guineas for a young elephant, presumably intending it for show rather than for sale, and a year later he gave a dead cassowary to Sir Hans Sloane. In 1740

Perry was showing his animals near the Exeter Change in the Strand; his advertisements of 1740 and 1743 show that these were large mammals rather than birds, big cats being the main attraction:

> This is to give notice to all Gentlemen, Ladies, and others, that Mr Perry's Grand Collection of Living Wild Beasts is come to the White Horse Inn, Fleet Street, consisting of a large he-lion, a he-tiger, a leopard, a panther, two hyenas, a civet cat, a jackal, or lion's provider, and several other rarities too tedious to mention. To be seen at any time of the day, without any loss of time. Note: This is the only tiger in England, that baited being only a common leopard.[17]

Perry's menagerie was in still in existence in Mile End in 1749, and again it was mammals that figured, including 'the magnanimous King of the Beasts' (a lion) from Persia, a tiger from Bengal, a leopard captured in the Forest of Lebanon, a pair of hyaenas from the coast of Guinea, a mountain cat from Virginia, a jackal, 'a noble Panther from Buenos Ayres in the Spanish West Indies [*sic*]', and 'a marmoset [*sic*] from China... with several other curiosities too tedious to mention'.[18] Given that Perry's tiger is stated to have been from Bengal and is distinguished in the text from the leopard, this may well have been one of the earliest, if not the earliest, tigers *sensu stricto* brought to England; it was probably a female, since those at the Tower of London in 1776 were listed as 'two royal Tygers from Bengal (the only Males that were ever brought into England)'. Mrs Perry, presumably his widow, carried on with the business, advertising in 1755 a pair of lions, two hyaenas, and a leopard, 'to be seen or sold', as well as 'several other Creatures too tedious to mention'.[19]

During the reigns of Queen Anne and Georges I and II many large exotic mammals were 'to be seen or sold' at London pubs and at fairs (especially Bartholomew Fair), and were occasionally in provincial towns, as Hannah Twynnoy's tombstone shows (Plate 2). They were usually acquired individually from seamen or dealers, the aim being to make a quick buck by showing or selling them before the animals, often already weakened by a voyage of several months, died of neglect, starvation, or both. Primates were a star attention, but were particularly difficult to keep, although a few survived long enough to be shown, among them several lemurs from Madagascar, two 'man-tygers' from Angola, a Turkish Ape 'much admired for his Ugliness', a pig-tailed monkey, a 'Satyr of the

Woods [baboon?] from Scythia', and a 'wonderful and surprising satyr, call'd by Latin authors, Pan' (probably a mandrill or a baboon, rather than a true chimpanzee).

A real chimpanzee arrived in London in May 1698 but soon died:

> The Monster, that was exposed at Moncrief's Coffee-House in Threadneedle-steet, near the Royal Exchange, died on Saturday in the Evening, and is dissected by the learned Dr Tyson, Lecturer for the Viscera in Surgeons-Hall and Physician for Bridewell and Bedlam who desiges [*sic*] to publish an Account of it, for the Satisfaction of the Curious; with Cutts, describing the Anatomy of it; Which, together with the view of its Skeleton, and the Skin stuffed, so as all its Lineaments and Proportion may still be seen, will tend as much or more to the information of those that come to see it, than when it was alive.
>
> The Account given of it by those that brought it over is as allows; That it came from Angola in Africk, and was sold to the Captain amongst a parcel of Slaves...[20]

The large ape seen by Pepys in 1661 *may* have been the first chimpanzee to arrive in Britain, but, if Tyson's was not the first, it was arguably the most important, for Tyson produced an extraordinarily detailed, well-illustrated account of its anatomy, interspersed with notes from earlier authorities, not least the Greek philosopher Aristotle, many of which he disagreed with. Although he referred to this chimpanzee as 'Orang-outang', this simply reflects the fluid nomenclature current in his day. Tyson showed that in many respects 'our pygmie' differed from both humans and Barbary apes (the only tailless monkeys), yet also bore resemblances to both, showing it to be intermediate in form between the two. Although it has been claimed that in doing this he prefigured Darwin's theory of evolution, this was not the case; he simply added an extra step to what he called 'this Chain of the Creation' (Chapter 7).

A second more celebrated chimpanzee, a young female, arrived in London from Angola in August 1738: 'There is lately arriv'd in the Ship Speaker, and to be seen at Randall's Coffee-house against the General Post-Office in Lombard-street, the Creature by the Angolians call's Chimpanze or Mock-Man... being perhaps the greatest Curiosity in the known World' (Figure 3.1).[21] Londoners flocked to view the chimpanzee at a shilling a time, making a fortune

Figure 3.1. The chimpanzee displayed at Randall's Coffee House, 1738 (*Chimpanzee*, Scotin, after H. Gravelot: BM, 1882, 0311.1183).

for her owner Captain Henry Flower, a West India merchant, the commander of the *Speaker*. Much celebrated for her gentle demeanour, the chimpanzee

> was extremely loving and fond of the Persons it was used to, particularly of a Boy about twelve or thirteen years of age, who was aboard the same ship in which it was brought over . . . The Chimpanzee was a very pretty Company at the Tea-Table, behav'd with Modesty and good Manners, and gave great satisfaction to the Ladies . . . would drink tea . . .

One of her visitors was Sir Hans Sloane, who professed himself 'extremely well-pleased'.

Soon after this the chimpanzee was transferred, 'at the desire of several Persons of Quality', to more genteel surroundings, the White Peruke coffee house at Charing-Cross, 'entirely dress'd after the newest Fashion *A-la-Mode a Paris* . . . There is a separate Apartment for the Ladies.' She did not survive the winter, dying on 23 February: 'the Owner's Loss by her Death is very considerable, as she was a kind of Estate for him.' She was 'opened' in front of Sloane and the King's surgeon John Ranby. Sloane pronounced that 'it was the nearest to the Human Species of any Creature'.[22] It was not the end. Her body was preserved in spirits and 'lay in state' at the White Peruke, displayed with a stuffed male chimpanzee, 'put in attitude', which Captain Flower had bought to London from Angola five years previously, preserved in a barrel of rum.

The most unusual animals displayed during this period were a pair of Brazilian tapirs, shown at Charing Cross in London 'By Her Majesty's Authority'; they are pictured on a handbill (Figure 3.2) and described as 'two Kaamas's . . . lately arriv'd from the Bear-Bishes . . . '.[23]

Three Indian rhinoceroses are recorded in Britain in the mid-eighteenth century.[24] The first, a 2-year old female, arrived on 1 October 1737 on board the *Shaftesbury* East Indiaman.[25] She was shown at the George Inn in the Haymarket in London, but on 1 June 1739, less than two years after her arrival, a rival appeared—a 2-year old male—on board the *Lyell* East Indiaman. He could be seen, from nine in the morning till eight at night, on payment of half-a-crown, in Eagle Street in London. The anatomist Dr James Douglas described the male rhinoceros at a meeting of the Royal Society on 24 June and showed several drawings made by his

By Her Majeſty's Authority.

Is to be ſeen at the ſign of the Coach and Horſes and Charing-Croſs *at* Charing-Croſs,

TWO *Kaamas*'s, Male and Female, lately arrived from the *Bear-Biſhes*; being the ſtrangeſt and admirableſt Creatures that ever was ſeen alive in *Europe*; being as tame as a Lamb; having a Trunk like an Elephant, Teeth like a Chriſtian, and Eyes like a *Renoſſerous*: Likewiſe wonderful Ears, with a white Fur round them like Saple; alſo a Neck and Main like a Horſe, and Skin as thick as a *Bouffler*, and has a Voice like a Bird; having ſtranger Feet then any Creature that ever has been ſeen, they live as well in Water as on Land: Theſe Rarities are to be ſeen at any time of the Day.

Vivat Regina.

Figure 3.2. 'Two kaamas's ... lately arriv'd from the Bear-Bishes'—England's first Brazilian tapirs (reign of Queen Anne) (anonymous handbill: BM, 1914,0520.691).

assistant Dr James Parsons, one of which was engraved and published as a poster in October 1739:

> An Exact Figure of the RHINOCEROS that is now to be seen in LONDON. Inscribed to Humffreys COLE Esq. Chief of The Honble East India Company's Factory at PATNA in the Empire of the Great MOGUL for the favour he has done the Curious in Sending it over to England.[26]

Parsons spent many hours observing the male rhinoceros first in Eagle Street and later at a booth at the London-Spaw (a pleasure ground in Clerkenwell) and reported that it was fed on rice, sugar, hay, and 'greens of different kinds'. The rhino's keeper, who came with it from Bengal, 'would make him thus emit his penis, when he pleased, while he lay on the ground, by rubbing his back and sides with straw'.[27] Both rhinos were on show in London in November 1740, the male at the Bell in the Haymarket, when it was promoted as 'being of the Male Kind... much more curious than the Female that has been lately advertize'd'.[28] By December 1741 the female was on display at the Unicorn in Oxford Road, before being taken north on tour. In January 1742 she was shown in Derby:

> This extraordinary animal is five years old; she is five foot and a half high, 18 foot two inches round the body, and 14 foot two inches from the Nose to the Rump, and is not yet come to her full growth: She has travell'd a Thousand Leagues by Land, thro' the Great MOGUL's Dominions to Bengal, from whence it was brought over on Board the Shaftesbury, Captain Bookey. She is upwards of Forty Hundred Weight, has a large Horn on her Nose, three Hoofs on each Foot, and a Hide thick with Scales, Pistol proof...

London may have had its fill of rhinoceroses, for by March 1742 the 'Great male Rhinoceros' was at the George and Angel in Stamford in Lincolnshire; he travelled on to Lincoln and Gainsborough in April, and then disappeared from the records. Meanwhile the female was still going strong. She was trundled through the Midlands and the north of England, and by 1747, when 'the noble Female RHINOCEROS... taken by the famous Kouli Kan from the Great Mogul', was in Edinburgh, her weight had reached 70 hundredweight; she was shown with a 'buffalo', a mandrill, a wolf, and a performing ape. Two years later she was in Nottingham with 'a surprising CROCODILE Alive, taken on the Banks of the River Nile'. By August 1751 both the female rhinoceros (now weighing in at 80 hundredweight) and the crocodile were back in London at Bartholomew Fair,[29] and at various pubs in London and Southwark throughout 1752. Two years later the much travelled female rhino 'was taken ill upon the Road from Mansfield to Nottingham, and died . . . to the great Loss of the Proprietor, Mr Pinchbeck'.

A third rhinoceros arrived in 1756—no less an animal than the redoubtable elderly female known on the Continent as 'Clara'. She was displayed in a tent at the Horse and Groom in Lambeth Marsh, where, having been seen by 'the Royal Family, and Nobility and Gentry . . . with great Satisfaction', she was on view from 8 a.m. till 6 p.m., initially at '2*s* the first Place, 1*s*. the second, and 6*d*. the third'.[30] She had arrived at Rotterdam in 1741 as a 3-year-old, and for the next seventeen years was shown all over Europe. In the course of her well-documented travels she was seen and described by several naturalists, including the Comte Georges de Buffon in Paris. Dozens of images of her were reproduced on posters, and her portrait was painted by Jean-Baptiste Oudry in Paris, Pietro Longhi in Venice, and Jan Wandelaar in Leiden. She is said to have died in London on 14 April 1758.[31]

At least eleven live Indian elephants were brought to Britain during the reigns of Queen Anne, and Georges I and II, although two died before they could be unloaded. Two of the elephants that died after having been put on show were 'anatomized'; the first in Dundee in 1706 by Patrick Blair:

> After this animal had travell'd most part of Europe, she came at last to this Kingdom; where, after some stay at Edinburgh, they conducted her to the North, and in their return came along the Sea-Coast; where being but few Places on the Road for making Advantage by long and continued Marches they hastened hither; and when they were come within a Mile of this Place, the poor Beast, much fatigu'd and wearied, fell down. They us'd many Endeavours to get her on foot again, but they all prov'd ineffectual. At last they digg'd a deep Ditch, to whose side she might lean, till she were sufficiently rested; but that prov'd her Ruin; for shortly afterwards there fell great Rains, which fill'd the Ditch with Water: So that after lying in a puddle a whole Day, she died next Morning, being Saturday April the 27th 1706 . . .[32]

The stuffed carcase and mounted skeleton were displayed in the 'Hall of Rarities' in Dundee. Blair's paper on the dissection was addressed to Sloane in his capacity as President of the Royal

Society; Sloane had been encouraging a full anatomical report in order to have a basis for comparison with the mammoth bones that were beginning to turn up in Europe and elsewhere, some of which he had described in the *Philosophical Transactions*.[33] The notion that these were the bones of an extinct form of elephant was gaining currency and heralded some notions of the theory of evolution.

A young female elephant arrived on the *Marlborough* in 1720 from Bencoolen (the East India Company's station in Sumatra); she could be viewed 'upon the paved Stones near Hosier-Lane End in West-Smithfield', but 'for want of a suitable and proportionate method of food, and from the ignorance of the keepers, who expos'd it to cold and moisture...', she died. Her corpse was transported to Sloane's garden in Chelsea to be dissected by William Stukeley (a physician, but now better known as one of the world's first archaeologists).[34] Ten years later the *Marlborough* returned with a second elephant, but when it reached Blackwall the whole ship exploded, 'by the snuff of a candle falling among some Saltpetre, of which there is some hundred bags on board', and the poor elephant—valued at £5,000 and intended for the Duke of Richmond—was incinerated.

A polar bear fared better than most—a ship, the *Walker*, arrived in London from New England in 1734 carrying 'a monstrous white bear', captured 'upon an Island of Ice 100 leagues from shore' in the Davis Strait by Captain Henry Atkins. It was seen by Thomas Boreman in London five years later reduced to obesity and docility after its diet of raw flesh had been replaced by bread and milk.[35]

In 1758 Horace Walpole wrote laughingly: 'I should tell you that one of the fashionable sights of the winter has been a dromedary [one hump] and a camel [two humps], the proprietor of which has entertained the town with a droll variety of advertisements.'[36] Not well versed in camelid taxonomy, the authors of the advertisements distinguished between the two as 'a wonderful camel and a surprising camel'; one wonders which was which.

People with bloodthirsty tastes and a propensity for gambling could watch animals being baited. The history of the baiting of brown bears and bulls is well known and continued throughout this period,[37] but other more exotic animals were occasionally used. In 1698 Sloane watched a 'larder tyger' being baited by three bear-dogs,[38] and at least three instances were advertised in London during the reign of George II. In the first, the contest was held in 'Marybone-fields' on 24 July 1721—'twelve English dogs', the property of an English gentleman, were to be set, one at a time, on a 'wild and savage panther' imported by a foreigner, the betting odds being twelve to one. The next two matches were held in Boughton's Amphitheatre (normally used for boxing matches). In one, on 29 January 1747,

> the celebrated white sea-bear, which has been seen and admired by the curious in most parts of England, will be baited...from his uncommon size, excessive weight, and more than savage fierceness, but he will afford extraordinary entertainment; and behave himself in such a manner as to fill those who are lovers of diversion of this kind, with delight and astonishment.

A few months later 'a large he tyger... being the first that ever was baited in England', suffered the same fate; despite the large size claimed, he was probably a leopard rather than a Bengal tiger.

3.2. Sir Hans Sloane and his Contemporaries

The new rationality, the questioning of received wisdom, and the emerging spirit of scientific enquiry that typified the 'Enlightenment' was accompanied in Britain by an enormous increase in wealth, generated in the main from the profits of the trading companies, especially from the East Indies (usually meaning India rather than the Far East), the West Indies (sugar and slaves), and the new British colonies in North America. The rich became richer, and both they and the increasingly wealthy middle classes were hit by a new collecting mania, establishing in their houses cabinets of curiosity, which varied from miscellaneous assemblages in corner cupboards to highly organized collections of coins, shells, fossils, minerals, gems, stuffed animals, books, manuscripts, drawings, paintings, sculpture, or ethnographic curiosities, the more exotic and expensive the better. It was a mania fuelled by a desire to show off new wealth, pride of possession, and curiosity about the expanding world. Not least among these collectible items were living animals—exotic birds and deer to ornament the parks and gardens of the wealthy or to be kept as household pets.

Pre-eminent among these collectors was, of course, Sir Hans Sloane, who amassed over 70,000 objects—the foundation collection of the British Museum—which included several thousand preserved animals and parts of animals meticulously catalogued. His large herbarium was enriched by friends who shared his interest in botany, especially Dr Peter Collinson, Lord Robert Petre, Admiral Sir Charles Wager and his wife Martha, and Charles Dubois, as well as Mark Catesby, whose plant-hunting expeditions in North America he sponsored.[39] Sloane also had a collection of living animals, mostly exotic, some of which were described and illustrated by George Edwards, the genial 'Bedell' of the Royal College of Physicians, whom Sloane employed as a draughtsman. Edwards devoted much of his life to drawing exotic birds and a few mammals, reptiles and insects, preserved or living, which he saw in the houses of Londoners or embellishing the gardens of their country estates in what are now the suburbs. His engravings of his own drawings were published in seven magnificent volumes—*A Natural History of Birds* (1743–51) and *Gleanings of Natural History* (1758–64).[40] Many of his original drawings are in an album owned by Sloane, which is in the British Museum.[41] By marrying the entries in the manuscript catalogues of Sloane's vertebrate collection[42] with Edwards's descriptions, it is possible to build up a picture of his menagerie, as distinct from, and in addition to, the preserved vertebrates in his museum collection, documented by Juliet Clutton-Brock in 1994.[43]

Sloane's interest in living animals is obvious from the second volume of his 'Natural History of Jamaica', published in 1725, long after his return from the island, where he had been employed as a private physician to the second Duke of Albemarle from 1687 to 1689. Nearly four years later he set up in private practice as a physician with many aristocratic and royal patients. He attended Queen Anne in her last illness, was created a baronet by George I, and appointed Physician-in-Ordinary to George II. He became President of the College of Physicians in 1719, and in 1727 succeeded Isaac Newton as President of the Royal Society.

Ever curious, Sloane occasionally attended animal combats. In 1698 he wrote to the naturalist John Ray:

> This day a larder tyger was baited by 3 bear-dogs; the second was a Match for him, and sometimes he had the better, sometimes the Dog; but the Battle was at last drawn, and neither car'd for engaging any farther. The third Dog likewise sometimes the better, and sometimes the worst of it; and it also came to a drawn battle. But the wisest Dog of all was a fourth, that neither by fair means, nor foul, could be brought to go within reach of the tyger, who was chain'd in the

> middle of a large cock-pit ... I am apt to think 'tis very rare that such a battle happens, or such a fine tyger is seen here.[44]

Ray replied wondering how the tiger had been obtained, and querying its identity: 'surely it is no true Asiatic, but American Tyger.'[45] Ray's doubts were fully justified, as the epithets 'tiger' (or 'tyger'), 'leopard', and 'panther' were used interchangeably, not only for leopards, but also for most of the other large and medium-sized cats.[46]

Sloane lived at nos 3 and 4 Bloomsbury Place in London, and in 1712 acquired Chelsea Manor, to where he retired in 1741. Even though he sold off much of the garden at Chelsea, what remained was large enough to provide a pond for wading birds and space for his aviaries, referred to as famous by Edwards in 1758.

Sloane's animals were well cared for; several of his birds lived for many years, among them a king vulture (a spectacularly ugly bird from North America), a 'Scotch Fishing Eagle from the Islands of Orkney that lived ... upon raw flesh and used often to dive in a Fountain in my Garden to clean itself ... ', an eagle owl that laid an egg, and a 'smaller sort of Bustard from Moca in Arabia. It lived in my garden many years and ate flesh & other foods as it had done in Mitchum in Mr Dubois' garden, who gave it me & had it brought over by one of the Coffee Ships.' Charles Dubois was treasurer to the East India Company, whose ships brought coffee to England from Mocha on the Red Sea coast of Yemen, and, like Sloane, he was one of Catesby's backers.

Sloane's golden pheasant from China, which he kept alive from about 1732 to 1747, pre-dates the bird drawn by Eleazar Albin on 21 July 1735 at a house in Windsor Park, which is usually reckoned to have been the first ever seen in Britain. He succeeded in cross-breeding, or so he thought, a golden pheasant cock with a 'hen wch is of a dark brown colour & wth tail waved black wch brought to me from China by Mr Bell', not having realized that they are actually the same species.[47]

Canada geese, first introduced in the seventeenth century, were already being bred in England in the early eighteenth. Sloane had at least one, as well as a gambo goose, which laid an egg. He kept a sarus crane—'a red headed crane from Bengall, given me by Mr Dubois, this crane lived in my garden severall years, & died by swallowing a brass linked sleeve button ... '. One of his crowned cranes died after many months in his possession. Being a physician,

Sloane was naturally concerned to discover the cause of its death. The post-mortem indicated that 'it dyed of a swelled Liver and Jaundice, with much water in severall parts of its body, especially the pericardium'. Another crane lived with him for over twenty years—it 'was given me by Mr Harrison who had it from the East Indies'.[48]

Sloane was one of the subscribers to Albin's *Natural History of Birds*, published in London in three volumes from 1731 to 1738, and he employed Albin, as well as Edwards, to draw animals for him. Albin probably also worked for Lord Petre, as it was on Petre's orders that he purchased two Nicobar pigeons that arrived from India in 1737 and presented them to Sloane, along with a large quantity of 'Rice in the husk', imported to sustain them. Albin also mentioned a mynah bird from India 'that spoke very prettily', which Sloane presented to Queen Caroline.[49]

Like many of Sloane's friends, the young Lord Petre was involved with the large-scale importation of North American trees for his estate at Thorndon Hall in Essex, where he also grew tropical plants, including pineapples, guava, ginger, limes, and bananas. He kept ornamental ducks and pheasants imported from abroad, which he hoped to naturalize on his ponds and woods. Significantly, the man he employed to take charge of the birds was Dutch. In 1736 he wrote to Sloane, 'having accidentally met with two uncommon fowls which are I believe of ye Widgeon kind, and come from ye East-Indies. I have taken ye liberty of ordering them to be left at yr house. If you find them deserving of a place amongst ye other curious birds it will be a singular pleasure to me . . . '.[50]

To his description of the African grey ('The Ash-Coloured and Red Parrot'), which was given to Sloane by Sir Charles Wager and drawn by Edwards in 1736, Edwards added a chilling reminder of the times: 'I am well assured that what we have are brought from Africa, generally by way of the West-Indies, by our Guinea traders, that supply our Sugar Islands with Negroes.'

Sloane's large egg collection contained 'parrots eggs laid in September 1724, after the parrot had been nine years in England & never bred, given me by Mr Harris'. Among his caged birds were a 'Green-wing'd Dove' (an emerald dove from India), a 'Padda or Rice-bird' from China, and an oddly named 'Spotted Greenland Dove', rather surprisingly kept alive as it was actually a guillemot brought from Greenland in 1738 by Captain Craycott, along with

Figure 3.3. Sir Hans Sloane's 'Greenland Buck' (caribou) in 1738; the antlers appear to have been added as an afterthought (from George Edwards 1743: i, no. 51).

a pair of caribou. Edwards drew the male caribou (Figure 3.3) before the pair was sent on to the Duke of Richmond at Goodwood; soon afterwards he recorded: 'I hear they are since dead, without any increase.' Their small size suggests that they were a sub-species of tundra caribou, either the barren-ground caribou, or even the extinct Arctic caribou of north-east Greenland.

Sloane had several mammals that lived as household pets; his descriptions suggest an affectionate as well as a scientific regard for them. They included a one-eyed wolverine from Hudson's Bay and a marmot with overgrown incisors: 'By being fed with soft Meats and Disuse to knaw [*sic*], its Teeth grew so long and crooked, that it could not take its Food, so to preserve its Life, they were obliged to break them out. This Drawing was taken, as it lay by the Fire reposing itself.'[51] This was probably the marmot presented to Sloane by Peter Collinson, who described it in a note written in his personal copy of volume ii of Catesby's *Natural History of Carolina* (1743):

> This animal was brought mee alive from Maryland. I gave it to Sr Hans Sloane [it] lived with him many years and became a domestic animal. Run up and down stairs like a catt or a dog but loved the kitchen best for the sake of the cook's favours. Lived on Bread, roots & greens, call'd the Ground Hogg or Monac or seven sleepers from Virginia 1733. Is neither rabit, nor rat & has some properties of each. P.C.[52]

Other live mammals owned by Sloane were a porcupine from Hudson's Bay, a beaver, and a small, blind fox Arctic fox from Greenland, whose cataracts he extracted: 'He lived many years wth me in my garden was brown in summer & turned white in winter . . . '.

As well as birds and mammals Sloane also owned several live reptiles:

> A large grayish green lizard from Malaga. It would eat flies & drink water. It would likewise drink milk but vomited it curdled & died towards the latter end of Sept. in London.[53]
>
> A large lizard said to come from Mexico. It was mostly grey with severall transverse fasene[?] Of a brown colur with white spots on them. It had on its tail many rings of [illegible] protuberances. It was brought by the Elizabeth man of war from Vera Cruz, fed by the way on cockroaches, came alive to England where it lived till killed by the cold in January, on milk a little warm'd. It appeared dead & stiff severall times but on being warm'd before the fire would [illegible] & walk about.[54]
>
> A sea tortoise which I kept alive for severall months in fresh water upon flounders or whiting putting to the water a 40th part of Salt.[55]
>
> A hawks bill turtle or tortoise from the coast of Guinea where it was taken & brought to me by Mr Harris I kept it in a tub of fresh water and made salt by the addition of 40th. part of bay salt. It fed on whitings which it wd eat with its bill, it would come up to breath frequently & [?] itselfe in the water according to its pleasure, it was killed wt the cold weather.[56]

Edwards called on Sloane in Chelsea once a week during the fourteen years of Sloane's retirement 'in order to divert him, for an hour or two, with the common news of the town, and with any thing particular that should happen amongst his acquaintance at the Royal Society'. Sloane died on 11 January 1753; Edwards retired in 1760 and died at the age of 80 in 1773.[57]

The second Duke of Richard, unlike most of the aristocrats of the time who kept deer to pose decoratively under the spreading oak trees in their parks and exotic water birds to swim on their lakes and ponds, also kept carnivores, large and small, dangerous animals, some confined in catacombs cut into the sloping garden behind his surprisingly modest house at Goodwood, between the South Downs and the Sussex coast.

The earliest reference to exotic animals at Goodwood, 'a coat for a monkey', is dated March 1726, and soon after this the Duchess wrote to her husband hoping 'that the great baboon is got safely into much's cage'. Later the same year a large iron cage was built for a 'tiger'; the cost of £93 included the transport of the cage to and from the waterside, presumably at nearby Chichester, and the hire of a barge to take the cage to the ship on which it had travelled. By 1728 the animals at Goodwood included '5 wolves, 2 tygerrs, 1 lyon, 2 lepers, 1 sived cat, a tiger-cat, 3 foxes, a Jack all, 2 Greenland Dogs, 3 vulturs, 2 eagles, 1 kite, 2 owls', which were all meat-eaters; while '3 bears and 1 white bear, 1 large monkey, a woman tygerr, 3 Raccoons, 3 small monkeys, armadilla, 1 pecaverre and 7 caseawarris' had to survive on bread. There was also 'one wild Boare, 2 Hoghs afatning, one sow and one sow with piggs'. A second list of about this date includes two 'mush-catts', a 'manligo', and five bears.[58]

Some taxonomic clarification is required. 'Tygerrs' could be any species of large cat, 'lepers' were probably leopards, a 'sived cat' a civet, a 'tyger catt' one of the many species of smaller wild cat, and a 'Mush Catt' perhaps a Muscovy cat—that is, a lynx. A 'Jack all' would have been a jackal, and the 'Greenland Dogs' must have been huskies. The 'woman tygerr' was a female mandrill, drill, or baboon; a 'pecaverre', a peccary, and seven 'caseawarris', cassowaries. The 'manligo' might have been a male mandrill or a baboon, but the identity of 'much' is a mystery.

On 11 November 1728 the Duke's sister wrote to him: 'All your relaytions this side the water are very well, & your Lion allso & I hear Lord Baltemore has Brought over a Bare for you I think a white one, but I wont be sure.'[59] The Duke's former tutor and lifelong friend Tom Hill wrote in October 1730 that 'your Bear for instance during the cold rainy weather we have had, has been in the utmost delight. The villain cares not if we were all starved to death, provided he can enjoy his ice and snow and cutting eastern winds.'[60] On 16 September 1732 the Duke wrote to Martin Folkes (later President of the Royal Society in succession to Sloane): 'I have had great augmentations to the Menagerie of late', but Indian pheasants were 'mighty scarce, and difficult to be gott'.[61] Some years later Folkes praised the Duke as one who 'cultivates and loves all sorts of natural knowledge...'.[62]

To the despair of those who managed the Duke's finances, no expense was spared in the acquisition of animals for Goodwood; these were often brought by boat from London. In 1728 or 1729 the lioness was transported from Bishopsgate Street in London, a 'great boat' brought 'ye baboon from Deptford', a fox, a mountain cat and a bear came from Southwark, a 'tiger' was collected from Tower Hill, and in 1730 two bears were brought all the way from Cadiz. But the most spectacular animal failed to arrive; Tom Hill broke the news to the Duke: 'Your Grace has heard without doubt, and wept, the misfortune of the poor Elephant that was burnt with the Vessel he came in.'[63] The young Indian elephant, valued at 500 guineas, had arrived at Blackwall Dock on the Thames, and was incinerated when the ship's cargo of saltpetre exploded; 'the poor creature roar'd out dreadfully when the Flames reached him'. Perhaps the Duke was mollified for its loss when a boat was despatched a year later to Deptford to collect another outstanding animal, a 'White Bear', probably the second polar bear in his collection.

The animals were kept in the High Wood that slopes upwards from the north side of Goodwood House, and is now the private garden of the Earl of March; some were housed in cages above ground, others confined to tunnels beneath. According to the Goodwood librarian writing in 1822, the area comprised about 40 acres surrounded by a strong flint wall, laid out in a most romantic and picturesque manner: 'Here in the very heat of summer the lover of retirement may enjoy cool sequestered shade uninterrupted by the noise and bustle of the busy world.' The Dell—'a most romantic and sequestered spot'—contained little grottos of shell work, where the family's favourite animals were buried, their names and attributes inscribed on marble tablets. The marble gravestones have all disappeared, but the grottos, though much damaged, are still visible within 'The Dell',[64] and some of the tunnels can still be walked through.

As with many great houses, Goodwood and its gardens were open to the public on days when the family were not at home. The animals were a great attraction, and so many people came to see them that Henry Foster, Goodwood's steward, complained to the Duke: 'we are much troubled with Rude Company to ye animals. Sunday last we had 4–5 hundred good and bad.'[65]

The Duke was one of the many subscribers to the two volumes of Catesby's *Natural History of Carolina* published in 1731 and 1743. An appendix dated 1747, added to volume ii, includes a plate of one of the Duke's favourite animals—his 'Java Hare' (Plate 3). Described as very tame and inoffensive, it was neither a hare nor from Java: it was the first agouti from South America to be recorded in England.[66]

The Duke was well aware of the scientific value of his animals, even when they were dead. Many were stuffed or dried, to be kept in his own museum at Richmond House in Whitehall, but others were sent to two of the leading medical practitioners of the day, Sloane and the surgeon John Ranby. After the marriage of his daughter Sarah to the Duke's young uncle-in-law Charles Cadogan in 1717, Sloane had become the family's physician, veterinarian, and friend.[67] Porters were employed on various occasions to transport the carcasses of the larger animals to London—the lion sent to Ranby in 1730 was followed by a lynx and an eagle, and Sloane received a dead pig, a civet cat, and a camel.[68]

Sloane also facilitated the supply of living animals to the Duke, who complained in 1735:

> I wish indeed it had been the Sloath that had I been sent me, for that is the most curious animal I know, butt this is nothing butt a common black bear, which I do not know what to do with, for I have five of them already. So pray when you write to him, I beg you would tell him not to send me any Bears, Eagles Leopards or Tygers, for I am overstock'd with them already.[69]

The Duke's 'sloath' was probably an 'ursine sloth', several of which were to arrive in England later in the century and excited the interest of the naturalists Thomas Pennant and George Shaw and the artists Thomas Bewick and Charles Catton, but their identity as bears, albeit sloth bears, not sloths, was not generally accepted until well into the nineteenth century.[70]

The only exotic ruminants reliably recorded at Goodwood are the short-lived caribou presented by Sloane, but in 1750, the year of the Duke's death, some 'buffalos' (probably actually zebu) were sent from Goodwood to the Duke of Cumberland at Windsor.[71]

A stone Lioness still reposes on a plinth at the top of the High Wood, a permanent memorial to the second Duke of Richmond's affection for the animals in his menagerie.

Sloane's friend Admiral Sir Charles Wager acquired a fortune from his most famous exploit, 'Wager's Action', when he defeated the Spanish treasure fleet off Cartagena on the Caribbean coast of South America in 1708, capturing a ship loaded with silver from the mines at Potosi, formerly 'Upper Peru', now Boliva. He was awarded the larger share of its treasure, valued at over £60,000. In 1720 he acquired Hollybush House at Parsons Green, then a village near London, where, apart from some naval engagements between 1726 and 1731, he settled for the rest of his life. The botantist Dr John Fothergill wrote that Wager, 'strongly excited' by Collinson, was a most generous contributor to Sloane's vast treasure of natural curiosities, 'omitting nothing, in the course of many voyages, that could add to its magnificence, and encouraging the commanders under him, who were stationed in different parts of the globe, to procure whatever was valuable in every branch of natural history'.[72] No doubt the commanders also brought back the roots and seeds of exotic plants, which Wager nurtured in his garden and greenhouse. He was no mean botanist himself—when sent to supply reinforcements to the garrison under siege in Gibraltar in 1727 he acquired roots and seeds of several previously unknown plants, which he 'communicated to several curious persons'. In 1731 Wager wrote that that there was a time to sit still and a time to be active, and that, as he was past sixty, it was time for him to be in his garden at Parsons Green. It was in his garden in 1737 that a magnificent evergreen *Magnolia grandiflora* from North America flowered for the first, or one of the first, times in Britain. A maple with scarlet flowers, introduced in 1728 from Virginia, was named in his honour: 'Sir Charles Wager's maple'.[73]

Lady Wager stocked their garden with exotic birds—sarus cranes from India, crowned and demoiselle cranes from Africa, and whistling ducks from the West Indies. Edwards wrote that this 'curious Lady[,] being a great Admirer of Birds, had by Presents and Purchase, procured a greater living Collection of rare Foreign Birds, than any other Person in London, and I owe a good part of my Collection of Drawings to her Goodness, in communicating to me the Knowledge of every thing New that came to her Hands'. Lady Wager's interests extended to mammals—in his letter accompanying the gift of the marmot to Sloane, Collinson added a postscript: 'I hope you have received from Lady Wager a She Possum with three young ones.' Collinson later wrote to a friend in Virginia thanking him for his account of a possum and adding 'had I known so much before I doubt not but That Lady Wager had sent might have been alive now Had Wee known its Food, but had yours come alive it would have been a much greater rarity having the young att the Teats'.[74]

Sir Matthew Decker, a wealthy Dutch merchant and director of the English East India Company, owned a truly Dutch garden beside the Thames at Fitzwilliam House on Richmond Green, west of London. He brought Henry Telende over from Holland to create an exceptionally fine garden:

> The longest, largest, and highest Hedge of Holly I ever saw, is in this garden, with several other Hedges of Ever-Greens, Visto's cut through Woods, Grotto's with Fountains, a fine Canal running up from the River. His Duckery, which is an oval Pond brick'd round, and his pretty Summer-House by it to drink a Bottle, his Stove-Houses, which are always kept in an equal heat for his Citrons, and other Indian Plants, with Gardeners brought from foreign Countries to manage them, are very curious and entertaining. The house is also very large a-la-modern, and neatly furnished after the Dutch way.[75]

Telende succeeded in growing pineapples in Decker's hotbeds, one of which—reputedly the first ever grown in England—was served to George I at a splendid banquet thrown in his honour by Decker in 1721—an event commemorated in his memorial tablet outside Richmond Church.

England's first mandarin ducks graced the oval pond in Decker's duckery.[76] Mandarin ducks were notoriously hard to breed, but it was said many years later that only the Dutch had succeeded, as they 'excel in every thing relative to birds; they breed more varieties, they keep them alive longer, and in short they are more attentive to them, than the English'.[77] It was not until 1834 that they first bred in Britain (at the London Zoo),[78] but with many escapes from country estates they are now feral in many parts of the country. Some of the earliest goldfish in England swam in Decker's ponds. George Edwards wrote: 'They were not generally known in England till the Year 1728, when a large Number of them were brought over in the Houghton Indiaman, Captain Philip Worth, Commander, and presented by him and Manning Lethieullier, Esq; to Sir Matthew Decker: Since which Time they have been propagated in Ponds by several curious Gentlemen, in the Neighbourhood of London.'[79]

James Brydges, Duke of Chandos, one of the richest and, perhaps, most corrupt men of the age, built one of the largest and grandest gardens in the country during the reigns of Queen Anne and George I, based apparently on the most magnificent Baroque garden of all time, Louis XIV's Versailles. He started in a small way in 1705 at Sion Hill, Isleworth, south-west of London, where, having been appointed to the highly lucrative post of 'Paymaster General to the Forces Abroad', he intended to create a 'country palace'. He planted peach and orange trees, imported flower and lettuce seeds from Antwerp and Holland, and kept a small collection of animals, including Barbary hens from Portugal, a monkey, a nightingale, a canary, and a parakeet, which he exchanged for a 'crackadore' (a pintail duck or a corncrake).[80] He wrote: 'the garden I have made at Sion Hill hath made retirement so pleasing to me that I should gladly withdraw from business when I cease being paymaster of ye foreign forces.'[81]

Brydges's letters show that he had a genuine interest in animals, for as early as 1698 he mentioned in a letter that he had seen a leopard and a 'tyger' in London; a year later he was fascinated by a porcupine in Long Acre; and in February 1702 he recorded that he had seen an elephant in London,[82] which must have been the elephant shown at the White Horse Inn in Fleet Street, mentioned in Section 3.1.

After the death of his first wife in 1712, Brydges married his first cousin, the 'excellent' Cassandra Willughby, daughter of the ornithologist Sir Francis Willughby.[83] By end of the War of the Spanish Succession in 1713 Brydges was able to retire with a fortune estimated at £600,000 (roughly £72,000,000 in 2014 money); unsatisfied with the progress of the works at Sion Hill, he transferred his workmen to Canons Park a few miles north of London, and set about transforming the house into one of the largest and most palatial buildings in England with an estate and garden to match. His ambition was to assemble an Italianate princely court, where he and his guests could enjoy splendid paintings, be entertained by musicians as they dined, browse in his extensive library, and stroll amid the beauty of the enormous gardens, with their rare plants and exotic animals. There was a magnificent chapel, with a full choir, and in 1717, the year in which he became Duke of Chandos, Brydges employed Handel as *Capelmeister*. Handel composed twenty anthems during his two years at Canons, as well as his first English oratorio, *Esther*; his opera *Acis and Galatea* received its first performance in the gardens at Canons.[84]

Although no images of the 83-acre gardens exist, it is known that they were landscaped in a grand Baroque manner, with straight radiating avenues (each more than 1,000 yards long), with walks, statues, fountains, a prodigious canal, a 7½ acre 'great bason', a two-storey summer house, a mound for viewing the gardens, parterres, fish ponds, fountains, fruit and vegetable gardens, and a physic garden. The hot houses contained coffee trees from Barbados and pineapples, and an aviary was soon added.[85]

Tortoises arrived from Minorca, white storks from Rotterdam, quail, pheasants, and partridges from Holland, wild geese from Barbados, whistling ducks and flamingos from Antigua, as well as choughs from Cornwall and 'barrow ducks' (probably goldeneye ducks) from Wales and Somerset.[86] Later they were joined by ostriches, blue macaws, more geese, Muscovy ducks, Virginia fowls, and songbirds, 'a Gold Coast red bird of peculiar prettiness', curassows and parakeets from Barbados, and an eagle. A crowned crane was killed by a Gambian gander, but Chandos had three more, 'so tame they come and eat out of my hand. I wish I could get more, or any other sort especially so large and beautiful'.[87]

Mammals too were sought. A Virginia buck and doe (probably white-tailed deer) arrived from Mr Willis in Virginia, Chiswell of Virginia sent mockingbirds, and a Mexican muskrat arrived with supplies of rice, kidney beans, and pineapples from Captain Massey of Charleston. An animal listed as a deer (but more probably an antelope or a gazelle, as it came from Wydah on the Gulf of Guinea), was frightened to death by people crowded round its cage. The Duke's only large cat was a 'tiger' from Ghana, which mauled a servant.[88]

The seriousness of the Duke's interest in plants and animals is illustrated by the fact that in 1722 he was one of the sponsors, along with Sloane, Wager, the Duke of Richmond, Mrs Kennon (midwife to the Princess of Wales and the owner of a vast collection of 'shells, fossils, ores, minerals, and natural curiosities'[89]), and others, of Catesby's visit to North America to collect plants. His letters show that he took an active role in the acquisition of plants and animals from many parts of the world, especially North America, the West Indies, and Africa. Having acquired a major shareholding in the ailing Royal Africa Company, he attempted unsuccessfully to turn its fortunes around by encouraging the search for mineral resources and exotic plants in Africa, instead of facilitating the transatlantic slave trade.[90] His position in the company enabled him to acquire plants and animals; in 1723 he wrote to Mr Potter, a company agent: 'If you meet with any Rarities of Birds, Fowl for Ponds, Deer of a small size or such curios (but no Lyons or Tygers or Beasts of Prey), I shall be obliged to you if you'll send 'em over.'[91] Albin was allowed to study the birds in the famous aviary, one of which, 'The Gamboa groasbeak ... from Gamboa on the coast of Guinea', is described and illustrated in his *Natural History of Birds* (1738) (Plate 4).

Even if one of the precious animals died en route, it was still valued and might end up stuffed or at least preserved in Chandos's collection. In 1728 he wrote to Charles Chiswell, a furnace manager with a large plantation in Virginia:

> It would be a very good way if you are so kind as to send me any more [singing birds] to make the Person who brings them over promise in his Receipt to deliver their skins stuffed in Case they happen to dye by the way; for I doubt very often they say they are dead to cover their disposing of them another way ...[92]

'Princely Chandos' died at Canons on 9 August 1744. The house was demolished in 1747; its original colonnade now fronts the National Gallery in Trafalgar Square.

Admiral George Churchill kept an aviary at the Ranger's Lodge in the Little Park at Windsor from 1708 to 1710, which was said to be the most beautiful one in England. He commissioned the Hungarian painter Jacob Bogdani, who had arrived in London from Holland in the same year as William and Mary, to record his collection. Three of Bogdani's paintings, purchased by Queen Anne after Churchill's death, show exotic and native birds and the occasional mammal grouped in imaginary Baroque landscapes. The species depicted in one of the paintings include blue-and-gold macaws, scarlet macaws, curassows, guans, and a troupial, all from South America, as well as touracos from Africa; another painting adds water birds to the list—a white stork, an Egyptian goose, a whistling duck, a shelduck, and a widgeon, as well as a red-legged partridge; also depicted are domestic hookbill and top-knot ducks from Holland, Muscovy ducks, Chinese geese, Canada geese, peacocks, a flamingo, and a few mammals—an Indian gazelle, an Indian spotted deer, and a badly drawn quadruped, perhaps a hog-deer. The third painting includes a capuchin monkey, two great curassows, and a guan, all from South America, incongruously grouped with a Cornish chough and a guinea fowl with chicks.

Christine Jackson has listed the birds in yet more paintings by Bogdani, which he may have seen in Churchill's aviaries: Amazon parrots (yellow-billed, yellow-crowned, yellow-naped, and blue-fronted), conures (sun and jandaya), all from South America; a chachalaca from Central America; a northern cardinal from North or Central America; lories (purple-naped, and chattering), a salmon-crested cockatoo from the East Indies, a ring-necked parakeet from Africa or India, and red-faced lovebirds from Africa, as well as a guinea pig.[93] It seems there were no raptors in Churchill's aviaries. The high proportion of birds from the East Indies and South America suggests that he probably obtained many of his birds from Holland, either directly, or from the dealers who brought consignments of exotic and domestic birds to London from Rotterdam.

Edwards drew many parrots and other exotic birds kept as pets by his London acquaintances; mammals were less common, but several Londoners, including Mrs Kennon, the Duke of Kingston, a Mr Hyde and John Cook, owned marmosets, and Mrs Cook even had some success in breeding them. Mr Critington, Clerk to the Society of Surgeons, had a 'Black Maucauco'—that is, a ruffed lemur, Edwards himself owned a ring-tailed lemur, and Mrs Kennon had a 'Mongooz' from Madagascar (probably a Sanford's brown lemur) that 'fed on fruits, herbs, and almost any thing, even living fishes; and that it had a great mind to catch the birds in her cages'.[94] An Indian mongoose had the run of Mr Bradbury's house, perhaps keeping it free of mice; the most entertaining mammal was a flying squirrel from North America belonging to Sloane's friend James Theobald, a wealthy merchant and antiquary. 'It was drowsy and inactive all the Day, but when Evening came on it was very lively, leaping (or flying very nimbly) from Place to Place in its cage.'[95]

3.3. The Royal Menageries: Hampton Court, Richmond, and Kew

Soon after their coronation in Westminster Abbey in April 1688, the new monarchs, the Dutch Stadtholder William III of Orange and his English wife and cousin, Mary Stuart, set about remodelling the royal gardens at Hampton Court and Kensington Palace, employing the King's favourite, Willem Bentinck, later created the first Earl of Portland, as Superintendent of the Royal Gardens. Bentinck had been responsible for the management of the gardens at their summer palace at Het Loo in the woods near Apeldoor (still the residence of the Queen of Holland), and he introduced large numbers of exotic plants, not only to Het Loo, but also to his own magnificent garden at Sorghvliet near The Hague.

William had shown an interest in exotic animals as early as 1675, when he requested that he should be honoured every year by the Dutch East India Company with animals, plants, and other curiosities from distant countries. At the time of his marriage to Mary in 1677 he had a menagerie at the palace of Honselersdyk, with 'rare Indian birds', a spotted Indian cat and an elk, as well as exotic birds, a lion and a 'tiger', gifts from Mary's uncle Charles II.[96] One of Melchior d'Hondecoeter's most famous paintings, *The Floating Feather*, depicts a pelican at William's hunting lodge at Soestdijk, along with a cassowary, two flamingos, a black-crowned crane, a sarus crane, Muscovy ducks, an Egyptian goose, a red-breasted goose, and various ducks.[97] The royal couple had a more famous menagerie at Het Loo—another painting by d'Hondecoeter shows the park with an Asiatic elephant, three white zebu, a red zebu speckled with white, a hartebeest, a white ram from Romania, a mouflon, two four-horned sheep from Iceland, and some gazelles, antelopes, and ducks.[98] Having admired Bentinck's secret *volière* or bird garden at Sorghvliet, enclosed with dense green hedges, where exotic birds sang in cages hung from the surrounding trellis work, Mary asked him to design a similar aviary for her at Het Loo.[99] Two aviaries were built, wherein, according to William's English physician Walter Harris, were kept 'curious Foreign, or Singing Birds'.[100]

As Queen, Mary spent much of her time at Hampton Court where the gardens were undergoing reconstruction; she employed a Dutch gardener to care for Bentinck's new Privy Garden (recently restored), which was disparagingly referred to by Stephen Switzer in 1718 as 'stuffed too thick with box, a fashion brought over out of *Holland* by the *Dutch* gardeners, who us'd it to a fault, especially in England where we abound in so good grass and gravel'.[101] A vast collection of exotic plants was brought to Hampton Court from Honselersdyk and Het Loo, and men were sent from London to Virginia and to the Canary Isles to collect plants for her. Naturally the collection included a large number of symbolically important orange trees; the tender plants were housed in three 'stoves' or glasshouses, heated by furnaces, designed by a Dutch carpenter. Mary also kept birds at Hampton Court, for Bentinck's accounts as 'Keeper of the Privy Purse' in 1692 include his expenses for 'the needs of the *Vesanderie*'—that is, a *faisanderie* or pheasantry.[102]

Mary created a second 'Dutch' garden, at Kensington Palace, with embroidered parterres, box hedges, a mount, a bowling green, a banqueting house, a wilderness garden, and a menagerie with 'curious wild fowl', tortoises, 'tygers', and, surprisingly, snails. Exotic birds sang in cages in many of the palace rooms, and she is also said to have owned some of the earliest goldfish in England.[103]

After Mary's death from smallpox in 1694, 'the beloved Hampton Court lay for some time unregarded: But that Sorrow being dispell'd . . . ',[104] William spent much time improving the gardens in Baroque style. He was still interested in acquiring animals, not only for London, but also for Het Loo, which he visited annually—in May 1698 the Sheriffs of Maryland were ordered to 'take charge of several birds and wild creatures, intended for the King, till they can be shipped to London' and two months later were informed that 'the King desired 100 mocking-birds for his volery at Loo, and any other birds and beasts that could be sent'.[105] In 1700 he ordered the construction of two new small gardens along the Thames foreshore on either side of a new banqueting house. The garden to the east was the King's aviary—also referred to as the menagerie, *volière*, or pheasantry—an oblong enclosure with an apsidal end, with borders, grass plots, gravel paths, and a fountain fed by piped water. Exotic birds were kept in a double row of oak-framed hen houses, each with eighteen nest boxes, cared for by 'the Keeper of the Aviary at Hampton Court', who was paid £50 a year.[106]

Neither Queen Anne nor George I appear to have any interest in exotic animals, but in 1718 the Prince and Princess of Wales (the future George II and Queen Caroline) began to use Richmond Lodge in the 'old' deer park at Richmond as their summer residence and in due course a menagerie was built there. Having acquired much extra land, Caroline employed the royal gardener Charles Bridgeman to create a large landscaped park, bordering the Thames from near Kew Green to Richmond Green, her aim being, in her own words, 'helping Nature, not losing it in art'. She was one of the subscribers to Catesby's *Natural History*, and the first volume is dedicated to her. John Rocque's *Plan of Richmond Gardens* (1734) shows the Lodge, confusingly referred to as the Palace, with parterres, shaded walks, a canal, a mount, a hermitage, a wilderness with winding paths, and an amphitheatre, all in typical mid-Georgian style. The plan includes a 'Managery', a short distance west of the Palace, which housed 'deer and wild beasts', as well as a pheasant house and a pheasant ground adjacent to Love Lane along the eastern border of the gardens. In 1740, three years after Queen Caroline's death, there were still some 'tigers' in the menagerie.[107]

Prince Frederick, her eldest son, arrived in England in 1728, not having seen his parents for fourteen years. He was soon made Prince of Wales, and in 1732 he purchased Carlton House in London, where he employed 'our Trusty and Wellbeloved William Kent' to redesign the garden in the new relaxed style that was becoming fashionable. Work on the 9-acre garden began in 1735 with the planting of many thousands of trees and flowering plants, the provision of garden sculptures, and the construction of various buildings. Intriguingly the household accounts for 1736 include a sum for 'John Glade's disbursements for buying Birds & Seeds for his Royal Higns. Aviary . . . '.[108]

Frederick also employed Kent to rebuild the house on the estate he had acquired as a rural retreat at Kew. The grounds ran along the southern border of his parents' Richmond Lodge estate, separated only by Love Lane, which is ironic, since there was no love lost between him and his parents. His mother famously called him 'the greatest ass, the greatest liar, the greatest canaille and the greatest beast in all the world'. Kent rebuilt the house as a Palladian mansion, which became known as the White House. It was completed in 1736 ready for Frederick's 17-year-old bride Augusta of Saxe-Gotha. In the 1740s Frederick acquired extra land and started 'great works in the garden', planting a multitude of 'curious and foreign trees' and creating the Great Lawn, a lake with a large island, a mound, and various follies in the garden, including a temple of Confucius and a Chinese arch.[109]

During the 1740s Frederick was sent a pair of zebras from the Cape of Good Hope. The stallion died en route, but the mare lived at Kew for several years. George Edwards drew her in 1751—the engraving, which he published in his *Gleanings of Natural History* (see Plate 5), shows her with dark brown stripes on a brownish background that fade out towards her rump—the colouration of a quagga, a much rarer animal, extinct since 1883; however, the results of recent genetic analyses of surviving quagga material have demoted it from a full species to a mere subspecies of the plains zebra.[110]

Frederick died in 1751, predeceasing his father who died in 1760. The activities of his widow, Princess Augusta (now the Dowager Princess of Wales) at Kew were to lead to the establishment of Kew Gardens, which is described in Section 4.9.

3.4. The Tower Menagerie, 1688–1757

Thomas Dymock, the Keeper of the Lions in the Tower of London, whom we met in Section 2.2, continued to be so unpopular with his servants that one even attempted to murder him—in 1692 he made a formal deposition to the Governor of the Tower at the City of London Sessions claiming that his 'blackamor' servant Edward Francis had attempted to poison him. Francis admitted that he had indeed on several occasions added rat poison to drinks consumed by Dymock's wife (who had since died), then to the new Mrs Dymock, and also to Dymock himself.[111]

Then there was the problem of unauthorized exhibitions of lions. In 1697 Dymock felt obliged to issue an advertisement reiterating the royal prohibition of 1687, and adding some notes on some new attractions in the Lion Office—a hyaena from Aleppo, presented to William IV by the Earl of Oxford, and a large 'tiger' from the East Indies.[112]

When the satirical writer Ned Ward visited the Tower in 1699, he came across a picture of a lion's head advertising 'the Royal Palace where the King of Beasts keeps his Court'. He said it smelt as frowzily as a dovecote or a dog kennel, and contained 'four of their stern affrighting catships'—that is, lions, as well as a 'tiger', a 'catamountain', a hyaena, three eagles, two outlandish owls, and a two-legged dog; he added that, if you came near the leopard, it 'would stare in your face, and Piss upon you, his Urine being as hot as Aqua Fortis, and stinks worse than a pole-cat's'.[113]

In his *Survey of the Cities of London and Westminster* John Strype listed 'the wild Beasts and other savage Animals' that were in the Tower in June 1704, with an extra entry dated 1708:

> SIX LIONS. First, a She Lion; which was presented to King William by Admiral Russel. Secondly, a He Lion, being about Six Years old; brought over by Captain Littleton. Sir Thomas Littleton presented

him to King William. Thirdly, Another Lion presented to Queen Anne by the Lord Grandville: It came to the Tower on Easter-Monday, Anno 1703. Fourthly, Two young Lions, sent to Queen Anne from the King of Barbary. They came in October, 1703. Fifthly, Yet another young Lion, brought over for the young Duke of Glocester.

In July, Anno 1708, The Emperor of Morocco's Ambassador came to London, and brought Five Lions, a Present from his Master to the Queen.

TWO LEOPARDS, or Tygers. One ever since K. Charles the Second's Time; but now in Decay: The other very beautiful and lovely to look upon; lying and playing, and turning upon her Back wantonly, when I saw her.

Three EAGLES, in several Apartments. One had been there Fourteen or Fifteen Year; called a Bald Eagle, with a white Head and Neck.

Two Swedish OWLS, of a great Bigness, called Hopkins. They were presented to K. Charles.

CATS of the Mountains, Two: Walking continually backwards and forwards. One of them was presented to Queen Anne; of the Colour somewhat of a Hare: Much larger than our ordinary domestick Cats; and very cruel.

A JACKALL. Much like a Fox, but bigger, and longer legged, and more grisled.

These Creatures have a rank Smell; which hath so affected the Air of the Place, (tho' there is a Garden adjoining) that it hath much injured the Health of the Man that attends them, and so stuffed up his Head, that it affects his Speech. And yet their Dens are cleansed every Day; and they have fresh Water set them Day and Night.

Here be also the Skins of Two dead Lions stuffed. One died two Days before K. Charles the Second. The other was the Queen Dowager's Lion.

On a Table hanging against the Wall of the House where the Keeper of the Lions dwelleth, is thus writ:

March the 29th, 1703. Then brought here a Lion, presented to the Queen by the Right Honourable the Lord Granville. October the 1st, 1703. Brought Two young Lions, presented to the Queen by the King of South Barbary.[114]

The two stuffed lions were probably those named Charles and Catherine from Morocco presented to Charles II in 1682. The fact that they had been preserved even in death underlines the veneration in which the Tower lions were held.

In 1710 a German visitor Zacharias Konrad von Uffenbach saw four lions in the same cage as a dog, two wolves, two very large and savage Indian cats, and two eagles, one of which the keeper assured him was 40 years old. It was probably the same bird, a bald eagle, included in Strype's list of 1703 as having been in the Tower for fifteen or sixteen years.

Dymock was succeeded as Keeper by his son-in-law William Gibson in 1713. When Gibson died only three years later, his children were involved in a dispute with John Martin, who had applied to succeed him. John Martin successfully petitioned the King for the post, claiming to be Dymock's nephew, and was appointed Keeper the following year.

Leopards and lions continued to arrive in George I's reign, some sent as gifts for his children. The gifts the King probably appreciated most were three Turkish horses brought by an envoy from Morocco, with several 'Saddles and Bridles, richly adorn'd with Gold and Diamonds', and, as an afterthought, a lion and several other rarities. It seems that members of the royal family showed little interest in their animals at the Tower, although the triplet lion cubs born in November 1725 were taken to St James's Palace to be admired by the young princesses Anne, Amelia, and Caroline, probably by Martin himself, who is said to have had more skill than any former keeper in rearing cubs.

César de Saussure visited in 1727 and noted that the menagerie was small, poor, and dirty, and contained ten lions, including four cubs born in the Tower, which he was allowed to fondle, one panther, and two 'tigers' (probably leopards), four other leopards, a monkey from Sumatra, which died of cold six weeks later, and a great quantity of curious birds. The latter are a mystery, since they are not mentioned in other contemporary records.[115]

After Martin's death in 1739,[116] William Hogarth's friend and colleague John Ellis (also spelt Ellys), Serjeant Painter to Frederick, Prince of Wales and 'Yeoman Arras Worker to the Great Wardrobe', was appointed Keeper, but he took little interest in the animals. He had been granted the position, said to be worth £300 a year plus 7*s.* 6*d.* a day, as a reward for advising Sir Robert Walpole on the purchase of paintings for Houghton Hall. Although known as 'Jack Ellis of the Tow'r', he lived chiefly in and around Covent Garden and was said to have been 'much happier in attending a pugilistic exhibition at Broughton's academy, than in the exercise of his profession'. He died in 1757.[117]

As far as one can tell, all the lions in the Tower during this period were African, most, if not all, Barbary lions from North Africa. Their numbers fluctuated from year to year, occasionally reaching at least ten. As most of them were given names, it is possible to trace some individuals through the various guidebooks to the Tower. A few lived to extreme old age, including Phillis, who gave birth to at least nine cubs: Nancy in August 1731, and in June 1740, when she was already aged about 18, she produced Pompey, Priscilla, and Nell (they all died young, the carcass of one of them being acquired by Sir Hans Sloane), followed by Dido and Helen in about 1746. Phillis, described as 'this old queen', survived until at least 1753, when she was about 40, and her 'consort' Marco, who was one of the five lions presented by the Ambassador of the Emperor of Morocco in 1708, died in about 1752, also aged at least 40.

If one were to take the Tower records at face value, one would think that, after lions, tigers were by far the most numerous of the animals in the menagerie. The give-away is that they are often listed as from Africa, or spotted, or both; true tigers are striped and do not occur in Africa. 'Tyger-cats' also appear frequently in the lists, and could be any of the numerous species of smaller cats.

One of the most informative, but crudely illustrated, guidebooks to the Tower is one written for children, *Curiosities in the Tower of London* by Thomas Boreman, published in 1741. As well as describing eight lions and several leopards, he notes various other animals. A 'raccoon chained in the yard where he has a box to run into when it is minded', a porcupine in an iron cage, a large ape (probably a Barbary ape), 'which at command does several tricks', two eagles kept in the Tower for thirty years, and two vultures, one there for twenty years, the other a mere six. A golden eagle was reputed in 1761 to have been in the Tower for at least ninety years.

The 'warwoven or king of varwous from the East Indies', presented in 1740 by the Duke of Montagu, the Master of Ordnance at the Tower, was probably a king vulture from America. In his *Natural History of Birds* (1734) Albin wrote that he had seen a warwovwen at the George Tavern at Charing Cross that was almost as big as an eagle: 'his Keepers call'd him the King of the Vavows, or King of the Vultures. He was brought by a Dutch ship from Pallampank in the East Indies.' Albin's detailed description and illustration of this spectacularly ugly bird, with its brilliantly coloured naked head and neck and with a fleshy orange caruncle over its beak, are of the king vulture, still extant, though threatened, in Central and South America, and certainly not a vulture from the East Indies.[118]

Another long-lived bird was a female ostrich, one of two sent during the 1740s to George II by the Bey of Tunis; the other had died after swallowing a nail. According to the *Historical Description* of 1787, 'she had a pretty large warm room to live in, which was often cleaned, otherwise she would soon have died; for the climate of this country seems by no means fitted to the tender nature of these birds, though by their large bones and vast bulk they appear to be very strong'. On 15 May 1747 she laid an egg, which ended up in the Leverian Museum; by 1753 she had laid seven, and by the time she died in 1768, fourteen. Some of the treasured eggs were displayed in the menagerie.[119]

Several apes and monkeys are mentioned in the Tower guidebooks, but they quickly succumbed to the cold and unsuitable food. In 1753 a 'Man-tyger' (a mandrill or baboon) was considered to be 'a mischievous beast, when a woman approaches lecherous to a surprising degree'. The monkeys, referred to collectively from 1754 as 'the School of Apes', included two mysteriously named 'Egyptian nightwalkers', one of which escaped from his cage and climbed on to the roof; all attempts to catch it failed, until at last, having thrown down many of the roof tiles, it came quietly down and retired into its cage. Two other members of the 'school' were a pair of monkeys (probably baboons) from Turkey (that is, the Ottoman Empire); after two years in the menagerie the female produced a baby; 'the greatest curiosity the Kingdom can provide', but by 1762 all that was left of the 'School of Apes' was a single capuchin.[120]

Almost nothing is known about the people who had the day-to-day care for the animals in the menagerie at the Tower, yet it was these keepers (with a small k) who built up a fund of expertise in handling and feeding them. Not surprisingly, many of the tropical animals succumbed to the cold, but the long survival of others shows that, despite their close confinement, they were reasonably well looked after. No doubt the fact that almost all of them were meat-eaters simplified the provision of a suitable diet. The keepers were particularly solicitous to the lion cubs born in the Tower, which were given warm milk and much attention, in the hope, sometimes successful, of taming them. A monkey in the Tower in

1741 was fed on bread, roots, and fruit, which could have been worse, and it seems to have survived for at least two years. Problems with the diet of captive primates lasted well into the twentieth century, as witnessed by the many skulls in museum collections distorted by rickets and osteomalacia.

While it is clear that the lions, leopards, wolves, vultures, and eagles that arrived in the Tower were almost invariably gifts to the monarch from foreign potentates, ships' captains, or aristocrats wishing to attract royal favour, the Tower menagerie also housed increasing numbers of smaller mammals and birds. Perhaps these less spectacular animals were purchased from sailors disembarking at Tower Dock, or from one of the bird merchants, such as Thomas Bland at the 'Old German Canary Bird House' also on Tower Dock just across the moat from the Lion Tower.

4

George III, *c*.1760–1811

4.1. A Dynasty of Animal Dealers: Joshua Brookes's 'Original Menagerie' and the 'Bird Shop' on the Corner of the Haymarket

Joshua Brookes was the progenitor of a dynasty of animal dealers that has lasted until the present day. He was the son of 'James Brookes, near Holborn-Bars, Poulterer',[1] mentioned by George Edwards in his *Gleanings of Natural History* as 'Mr Brooks, of Holborn, a great dealer in foreign birds and curious poultry'.[2] Joshua took over the business, perhaps as early as 1756, when he married Frances Paul at the church of St Andrew in Holborn, or more definitely by 1758, when 'Brooke's Original Menagerie' advertised an astonishing range of birds:

> A most curious Collection of Fowls and Birds from most of the known Parts of the World, viz. India Pheasants, py's and white, Pea fowls, English Pheasants of all Sorts, silky Fowls from Bombay, Rumking from Portugal, Hambird, Poland, and Shag-bag Fowls, bantam, African, or Guinea Fowls, large Darkings, Indian geese, white Muscovy Ducks, Spanish Ducks, wild Turkies, Virginia Nightingales in full Song, Canary Birds, turtle Doves, and all sorts of fancy pigeons, Lap Dogs, tame Squirrels, Dormice etc. Most Money given for any curious of foreign Birds, Fowls, etc.[3]

In 1767 Brookes issued a catalogue detailing over seventy species of animals; he had an ostrich for sale, as well as a great number of birds from Asia, Africa, and America, and he had also acquired some mammals—a pair of lions, a porcupine from the East Indies, a camel, a 'panther' from America, a mountain cat, several leopards, wolves from Saxony and Hudson's Bay, a marmot, some small Brazilian animals, and several kinds of uncommon monkeys.

Nervous visitors were assured that the rooms were clean and well aired and the creatures well secured. A brash young American plant collector William Young sent him plants and seeds from North America, which sometimes included Venus fly-traps, but also birds, wild turkeys, geese, and ducks, and even dried flies and other insects; the arrangement lasted until the outbreak of the American Revolution in 1775.[4]

Brookes was well known among the 'nobility and gentry' as well as naturalists. The Duchess of Northumberland, who had made several visits to the menagerie of William of Orange in Holland, referred to him as 'Brookes of Holborn'. On a visit to The Hague in 1771 she was told by the British Ambassador that Brookes went to Holland once a year to pick up a cargo of birds, often dealing with a Mr Echardt at his seat between The Hague and Rotterdam. Echardt was probably the Dutchman mentioned by Brookes to his friend N. Burt as being one of the few people in Europe who had succeeded in breeding mandarin ducks.[5]

In or shortly before 1771 Brookes acquired a partner, John Cross, the owner of a menagerie in a more fashionable part of town—the corner of the Haymarket and Piccadilly (now Piccadilly Circus). There had been a shop selling foreign birds, as well as a few dogs and unspecified 'wild beasts' on the same site since at least 1766, which within a year or two had been taken over by Cross.[6] In 1768 a newspaper reported that

> two men were leading a large Bear up the Haymarket in order to sell the animal at the Menagery at the upper end of said place, he got off his muzzle and seized one of them by the hand, and bit off one of his fingers; he next seized the other man by his leg and tore it in a very terrible manner. The Bear was soon after killed to prevent further mischief. Both men were sent to the Hospital.[7]

A year later Cross widened the range of his 'Exhibition of Birds' with some unusual animals for show, a hyaena, a porcupine, a crocodile, a leopard, an agouti, a golden vulture, horned owls, marmosets, and even 'that amazing curiosity, the Porcupine Man, supposed to be the only one in the world'.[8] His partnership with Brookes was short-lived because he died in late in 1776 or early in 1777, leaving most of his estate to his wife Mary and appointing her as the sole executor; his gold watch went to Brookes, the mark of a warm friendship. Mary and Brookes took out a joint fire insurance policy on the property in which her profession is given as a 'dealer in live fowls and birds'.[9] It seems that she carried on with the bird shop for about a year, after which the premises were taken over by a James Murdoch, who as far as one can tell was not a bird dealer.[10] However, a few years later the building was again being used to show and sell birds and 'wild beasts'.

Brookes must have been running out of space in Holborn for his ever-burgeoning collection of animals and Young's consignments of plants. While in partnership with Cross, he acquired a new, spacious site on erstwhile farmland on the north side of the New Road (now Euston Road) on the edge of town, just west of a famous pub, the Adam and Eve.[11] An advertisement issued in March 1776, reads:

> Foreign Birds. JOSHUA BROOKES having left his House in Holborn returns the Nobility and Gentry Thanks for the Favours he has received, and hopes the Continuance of them at his Menagerie, in the New-Road, Tottenham Court, where there is to be seen and sold, a great collection of very curious Birds from the East Indies, Africa and Southern Parts of America . . .

A handbill headed by a picture of his house, lists an extraordinary range of birds offered for sale (Figure 4.1).

Brookes built a fine castellated house for himself and his large family, as well as several other houses nearby. Behind and to the side of the main house was a series of small inter-communicating gardens each with its own small building; beyond the boundary wall, fields stretched far way towards the village of Hampstead.[12] In these rural surroundings away from the smoke of London, Brookes no longer needed to rely on imports of poultry from Holland and had enough space to grow plants and to breed ducks and other birds, while retaining easy access to and from town. He even succeeded in breeding mandarin ducks, which he sold to the Duke of Northumberland at Syon House.[13]

Brookes appears to have made most of his money by selling game birds, dogs, and deer to stock the estates of the aristocracy; for example, he sent animals and birds to the Earl of Egremont at Petworth in Sussex on various occasions over a period of twenty-five years,[14] and he supplied deer to the Prince of Wales. Another customer, Lord Montfort, who had had to sell his estate at Horseheath in Cambridgeshire when he got into financial difficulties, sold his entire collection of water fowl to Brookes. He must have visited the menagerie on several occasions with his son, because many

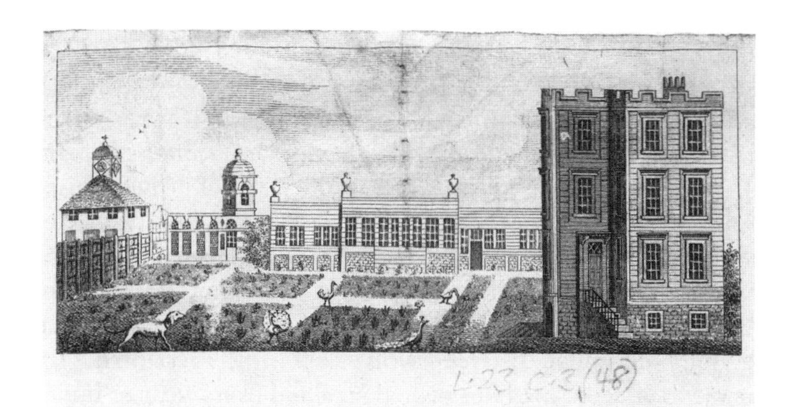

JOSHUA BROOKES,
ZOOLOGIST,

At his MENAGERY, in the New-Road, Tottenham-Court,
(Removed from Gray's-Inn-Gate, Holbourn.)
Buys and Sells and Exchanges all Sorts of FOREIGN BIRDS, QUADRUPEDS, &c. viz.

BUSTARDS.

CRANES.
Crown Bird of Africa
Cyrus from Afia
Numidian, or Demofel
Whooping American.

CASSAWAYS.
CURACOAS
CUSHUA
COCKTTOOS
Small ditto.

DAWS, (Cornifh)

DUCKS.
Aylefbury
Carolina
Dutch Topping
Hook bills
Manderil
Mule
Mufcovy
Roan
Schild
Spanifh
Turkifh
Wild
Whiftling

DOVES.
Black Ring
Cinnamon from Carolina
Cream from Barbary
Cuba Bald Pate
Cuba Blue Pate
Ground from Cuba
Nicomber
Oriental Chafer-wing'd
Paffage from Carolina
Pea from ditto
Red-Rings from Senegal
Spotted Cape Triangular
White-Wings
White from Barbary
Widow from Senegal

EAGLES.
Carolina
Guinea
Norway
Ruffia

FLEMINGOS.
Major and Minor

FOWLS.
Algerines
Bantums
Chitagalls
Hampdens
Humburgh
Large Darking white
Poland
Rumpkins
Guinea fowls
Pyde
Purple
White and Black
Shag-Bags
Spanifh
Silk Sumatra

FEN FOWLS.
Baldcoots
Dunbirds
Eafterlings
Golden Eyes
Garganaes
Shovel Bills
Teal
Widgeons
Guinea
Purple
Pyrle
White
Pyde
Common

GEESE.
Aftracan
Brent
Canada
Chinefe
Cafpian
Cape
Greenland

HERONS.
Blue
Red
White

IBIS, or Secretary.

LOWRYQUETS
LUQUORS
LOWRYS.

MINOS

MACCAOS.
Blue
Deep Red
Green
Scarlet

Small Red from the Main.

OWLS.
Great Horned, from Aunfpack
Large ditto, from Hudfon's Bay
Small ditto, from Aunfpach
Small ditto, from America

OSTRICH.
OLIVES.

ORTOLANS.
American
European

PHEASANTS.
Chinefe, Gold
Ditto, Silver
Ditto, Ring Necks
Englifh, Py'd
Ditto. White

PEFOWLE.
Japhanezes
Pyde ditto
White ditto
Wild ditto

PEROS.
European
Oriental

PARTRIDGES.
American, Barbary, French, Portugal } Red Leg'd

PARROTS.
American
Eaft India
Guinea
Portugal
Spanifh Main.

PARROQUETS.
Large Ring Necks, Ditto, with Red Wings, Small ditto, Spotted } India.
Small ditto, with Red Heads } Africa.

All Green, Do. Ring-neck'd, Ditto Grey Heads, with Yellow Breafts } Africa
All Yellow with Orange Breafts, and All Green, with Orange Heads, Large Green, with Red Heads } America

QUAILS.
American
European
Oriental

SPOONBILLS.
SWANS.
STORKS.

SONG BIRDS.
Averduvats
Blue breafted Finches
Brazil Finches
Canaries Junk
Ditto Turn-Crowns
Ditto Mealy
Ditto Fancy
Ditto Brazil
Cardinals
Carolina Robins
Dominican Widow Birds
Ditto Pintail
Ditto Common
Grenadiers
Java Sparrows
Indian ditto
Large Blue Birds
Manakins
Mocking Bird
Mule Bird
Nuns
Purple Birds
Ring Ouzell
Small Blue Birds
Schombargos
Tropial of Cayan
Virginia Nightingale
Wax Bills.

Note, Proper Bafkets, Coops, &c. provid l for their Conveyance to any Part of the World,
*** Birds, Poultry, curiou Quadrupees of all Sorts exchanged.

Figure 4.1. Joshua Brookes's new premises in the New Road (*c.*1776); note the variety of birds for sale (*Joshua Brookes Zoologist*, anonymous handbill: BL, L.23. c.3 (48.)).

years later the son was able to vouch for the good character of Brookes's son John, claiming to have known him since boyhood. The Duchess of Portland was probably also a customer, as after her death Brookes advertised for sale her collection of poultry, pigeons, and doves, as well as her Dutch belted cow and calf. The anatomist John Hunter was another useful contact, as a purchaser not only of live animals, but also of their corpses for dissection. It was probably at Hunter's instigation that Brookes became involved in cross-breeding experiments between wolves and dogs (Section 4.7).

As we have seen, Brookes visited Holland every year to buy birds and other animals. They were shipped, usually from Rotterdam, as there was little direct commerce between London and Amsterdam; nevertheless he must surely have visited the Blauw Jan coffee house and menagerie in Amsterdam to see and perhaps buy some of the more exotic animals displayed there, brought by the Dutch East and West India Companies from South East Asia and the Dutch colonies in the West Indies, South America, and West Africa—in 1775 he advertised the arrival of various birds, including flamingos from the 'Dutch settlements, in the Southern part of America'. Five years later Brookes sent his 17-year-old son Paul to South America to acquire birds, presumably buying in ports rather than venturing into the interior to collect them. The mission was crowned with success—soon after Paul's return in June 1784 father and son 'waited on the Queen at Buckingham-house with a collection of rare birds from the Brazils and the southern parts of America, at the sight of which her Majesty and the Royal Family were pleased, and gave orders for several varieties to be left'.[15]

In 1789, or a few years earlier, Brookes's menagerie was extended to include No. 1 Portland Row, a few minutes' walk further west along the New Road. It was run by his son John, who started with much the same range of birds as his father, but then began to specialize in game, especially deer. He sold deer to the Prince of Wales in 1790—fourteen and a half brace of red deer at 50 guineas a brace, together with a further four brace which he had obtained in Paris—he transported them to parks at Bagshot, Brighton, and Kempshot, which the Prince was anxious to restock.

Joshua Brookes's most ambitious project was a commission from the Duke of Norfolk to acquire reindeer. In the summer of 1799 he, or perhaps Paul or John, sailed from Hull to Lapland, where he managed to acquire six male and six female reindeer, which he shipped back to Hull, from where they were transported to the

Duke's estate, Greystokes Castle in Cumberland. Brookes took the precaution of bringing back several sacks of Arctic moss with him, to be mixed 'by degrees' with English moss and grass in order to accustom the deer to their new diet. However, like several attempts before and after, this experiment in the acclimatization of reindeer must have failed, for nothing more is known of them.

One of the commissions Joshua Brookes received was from one of his most important clients, Sir Joseph Banks, the great panjandrum of the scientific establishment and President of the Royal Society. It involved the collection of some emus from Portsmouth harbour.[16] John Hunter (not the anatomist, but the retiring Governor of New South Wales) left Botany Bay in October 1800 on HMS *Buffalo*, captained by his nephew William Kent. The *Buffalo* sailed into Portsmouth harbour in May 1801 carrying two black swans and three emus. The swans were soon unloaded and given to Earl St Vincent, who presented them as *rarae aves* and *nigro simillimae cygno* to Queen Charlotte, who sent them on to Windsor. One soon died and the other flew away across the Thames and was promptly shot by a gamekeeper. Kent was heartbroken. Not only had the birds been 'brought safe over such a vast track of ocean', but they had been reared by hand, and had become so tame that their owner in New South Wales had been reluctant to part with them, until Kent had persuaded him with 50 guineas.

The emus were intended as a gift from Captain Kent for Banks, who hoped that they might be induced to breed. They remained on board the *Buffalo* in Portsmouth harbour, or nearby at the Spithead. Crisis point was reached in November when Kent received an order to sail the *Buffalo* back to Australia with the convict ships. He wrote a polite reminder to Banks saying that the birds were in perfect health even though they been on board ship for thirteen months. Banks turned to Joshua Brookes, who sent his assistant and son-in-law Philip Castang[17] down to Portsmouth to collect them. On arrival, Castang ran into trouble. He wrote immediately to Banks to say that the customs officials would not release the birds without the payment of duty. Banks was already aware of the problem, because a customs official had written to him to say that a Special Order had been received from the Honourable Commissioners for the birds to be delivered *only* on payment of duty, the amount charged to be *ad valorem*—that is, based on their monetary value. In an attempt to have the order rescinded, Banks wrote angrily to the Office of Trade in London pointing out that the birds were to be embellishment to the country, that the Marquis of Exeter for whom they were intended had no financial interest in them, and that, if they were taken from him, they would end their lives in a show cart owned by 'persons who exhibit wild beasts'. What he did not tell them was that a person who might do just that, Gilbert Pidcock of the menagerie at the Exeter Change in London (Section 4.2), had valued them at £3,000.

Meanwhile Castang was waiting, making several fruitless trips out to the *Buffalo* to visit the birds, as well as acquiring and modifying a cart for their transport to London. After several days it transpired that a warrant had been issued for the birds to be off-loaded in the *Thames* and the confusion was sorted. The complexities of loading such large, ungainly, and precious birds from the *Buffalo* onto a small boat, then on to the quay, and then on to the waiting cart, can be imagined; they were permitted to leave only after Castang had bribed the porter at the Dockyard gate with 1*s*. 6*d*. worth of grog. At last Kent was able to assure Banks that the birds that he had cared for over so many months had been safely landed and were on their way to town. Brookes must have been heartily relieved when they arrived without mishap at the menagerie in the New Road. The emus were then taken north to join their compatriot kangaroos in a warm stable on the Marquis of Exeter's estate, Burleigh Hall in Lincolnshire, not far from Banks's own home at Revesby.[18] How long they survived is unknown. Brookes's invoice is housed in the State Library of New South Wales:

Sir Joseph Banks Bart
Joshua Brookes
1801
Nov 25 & Dec 1st (Paid coach to Portsmouth £1 6*s*.
"Supper on the road (coachman and guard) 3*s*. 6*d*.
26th Boat hire two different times 3*s*. 6*d*.; boat to ye ship 1*s*. 6*d*.;
27th Five days board at Portsmouth 5*s*.
28th & 29th Four nights lodging 4*s*.
Letter 7*d*.
Dec 1st Coach home and coachman £1 2*s*.
Mr Castang's time & attendance, five days and seven nights £2 12 6*d*.
Fees for warrants for the delivery at ye Custom House12*s*. 8*d*.
Tide waiting fees 5*s*.
Boat hire 2*s*.

Landing at the Crane 2*s.* 6*d.*
Cart to take home ye box 2*s.*
Grog to ye porter at the Dockyard gate 1*s.* 6*d.*
Four new mats and nailing in the cart 6*s.* 9*d.*
Seed for the birds some on the road and at Tottenham Court £3 11*s.*
£8 14 11*d.*

Joshua Brookes died late in 1802 or early in 1803. It is clear from his will that he died a wealthy man.[19] Having made generous provision for his three daughters and for four of his seven sons by his marriage to Frances, he left the bulk of his estate to his second wife, Elizabeth Marsden of St Mary-le-bone, who was the mother of his younger children. The animals in the menagerie were not mentioned in the will. John gave up the Portland Row menagerie after the death of his father, but, as we shall see, he had in the meanwhile taken over the menagerie in Piccadilly, where he acquired the epithet 'Wild Beast Brookes'; Paul became an inveterate traveller, and the eldest son, Joshua, became a well-known anatomist (Section 4.7).

The disposal of the Original Menagerie in August 1803 generated a great deal of acrimony. Although many of the animals were removed by Castang, who set up a rival menagerie in the pleasure gardens behind the Adam and Eve, a Mr Whitehead acquired the land and carried on with the business in the New Road. He announced:

> It having been falsely and invidiously insinuated and given out that this Menagerie has been shut up, and the business carried on elsewhere, in answer to so flagrant and pitiful an attempt to impose upon the public, E. Whitehead begs leave to repeat, that he is the only successor to, and was particularly recommended by, the late Joshua Brookes . . .[20]

By November 1804 Whitehead was bankrupt; nevertheless the business survived, and advertisements show that he carried on selling dogs as well as 'Pheasants of all kinds, Foreign birds, Water Fowl, Red, Fallow and Fancy Deer'. However, six years later the Original Menagerie was taken over by Paul Brookes (Section 5.1).

William Hone remembered the Adam and Eve as having spacious gardens at the side and a courtyard at the rear, where tables for tea-drinking were set out under large trees, 'when it could boast of a monkey, a heron, some wild fowl, some parrots, with a small pond for gold-fish'. It is the inn depicted in Hogarth's famous painting *March of the Guards to Finchley* (1750).[21] Castang's little menagerie went from strength to strength, selling a wide variety of game birds, ducks, poultry, foreign birds, and from time to time a few buffaloes (probably zebu), Alderney cows, and wild boar. In 1804 he received two black swans and 'two large Kangaroos, from Botany Bay, male and female', which were sent on to Kew and for which he invoiced Banks, stating that he also had some geese from New South Wales.[22] Two reindeer arrived in 1807, and, like his father-in-law, Castang supplied ducks to the Earl of Egremont. Despite being declared bankrupt in 1813, he carried on importing water fowl from Holland, but he specialized in breeding pheasants. John Lawrence, in his *Practical Treatise on Breeding, Rearing, and Fattening All Kinds of Domestic Poultry, Pheasants, Pigeons, and Rabbits* (1819), referred to the 'intelligent and experienced' Mr Castang, and included in the book a chapter on pheasant rearing written by Castang himself.

To return to the 'Exhibition of Birds' on the corner of the Haymarket and Piccadilly: the building itself changed hands shortly after the death of John Cross in 1775 or 1776, but by 1781 it was referred to as the 'noted Bird shop', and two years after that the premises were being used by George Bailey to show off a baboon, 'The Eastern Wonder or Child of the Sun'. He had previously displayed the baboon at the Empress of Russia's Head, near Sadler's Wells, along with a miscellaneous collection of preserved birds, beasts, insects, shells, and clothing from the South Seas. The term 'Child of the Sun' was subsequently used for many different individual monkeys imported into Britain, most often hamadryas baboons, or other pale-coloured monkeys. Bailey's monkey may have been one of the two shown in the picture of two monkeys painted by George Stubbs for John Hunter; one was a drill and the other a smaller pale animal with pink ischial callosities, which Hunter noted as 'Bailey's Monkey of which I have a painting'; it has recently been identified as a guinea baboon—its preserved parts are listed in old catalogues of John Hunter's anatomical collection in the Royal College of Surgeons in London as from the 'Child of the Sun'.[23]

Bailey's 'Child of the Sun' appears to have attracted much attention, since he took it on tour as far away as Glasgow before bringing it back to London late in 1784. A commentator (perhaps Bailey himself) wrote that the 'Siddonian rage', the 'Balloon rage', and the 'party rage' were being supplanted by

> the savage rage to behold the wondrous Child of the Sun, now exhibiting the second house in Piccadilly from the Haymarket, where all denominations of people are now flocking, from the peer to the peasant, and seem universally interested in the grand question—Whether this astonishing production is man or monster? The naturalists affirm the former, while the connoisseurs contend for the latter; but the general opinion appears to be, that it is that very important link of nature's chain, which unites the animal to the human species.[24]

Bailey's monkey may be the 'Child of the Sun' referred to in a letter from Philip Thicknesse to Lord Monboddo written in 1789:

> He understood everything said to him by his keeper, and had more sense than half the brutes erect we meet in the streets. He possessed himself of his lady's lap dog and no art or force could make him deliver it up. It lived sometime in the cage with him and when his caresses and holding the dog too close, it died, he would never eat after.[25]

Monboddo himself was much ridiculed for believing that apes, or orang-outangs as he termed them, were human, lacking only the power of speech.[26]

Within a year Bailey had taken over the lease of half of the building on the corner of Piccadilly and the Haymarket.[27] Here he continued to act primarily as a showman, exhibiting such items as the 'Grand Cassuwary', 'two wonderful Great Siboya Serpents', 'two live Rattle Snakes', a 'learned Bull-finch or Scientific Bird', a 'Living Crocodile from the River Nile', as well as 'nonpareil Parroquets from Botany Bay, more beautiful than ever was publicly exhibited in England', a talking ring-neck parakeet, widow birds, red birds, blue birds, parrots, three Dutch pugs, and an Italian greyhound. Bailey's crocodile is commemorated in a copper coin showing a crocodile on the bank of a river; the words around the circumference read: 'A crocodile to be seen alive at G. Bayly's Museum', with a rattlesnake on the reverse,[28] but presumably the animal did not last long, as its corpse was acquired by John Hunter. Hunter also received one of the half-bred puppies produced by a female jackal from Bombay, given to Bailey by a Captain Mears, which had mated with a spaniel soon after its arrival.[29]

After George Bailey's death late in 1798 or early in 1799, John Brookes took over the business, expanding it into the whole block on the corner of Piccadilly and the Haymarket, and there he showed

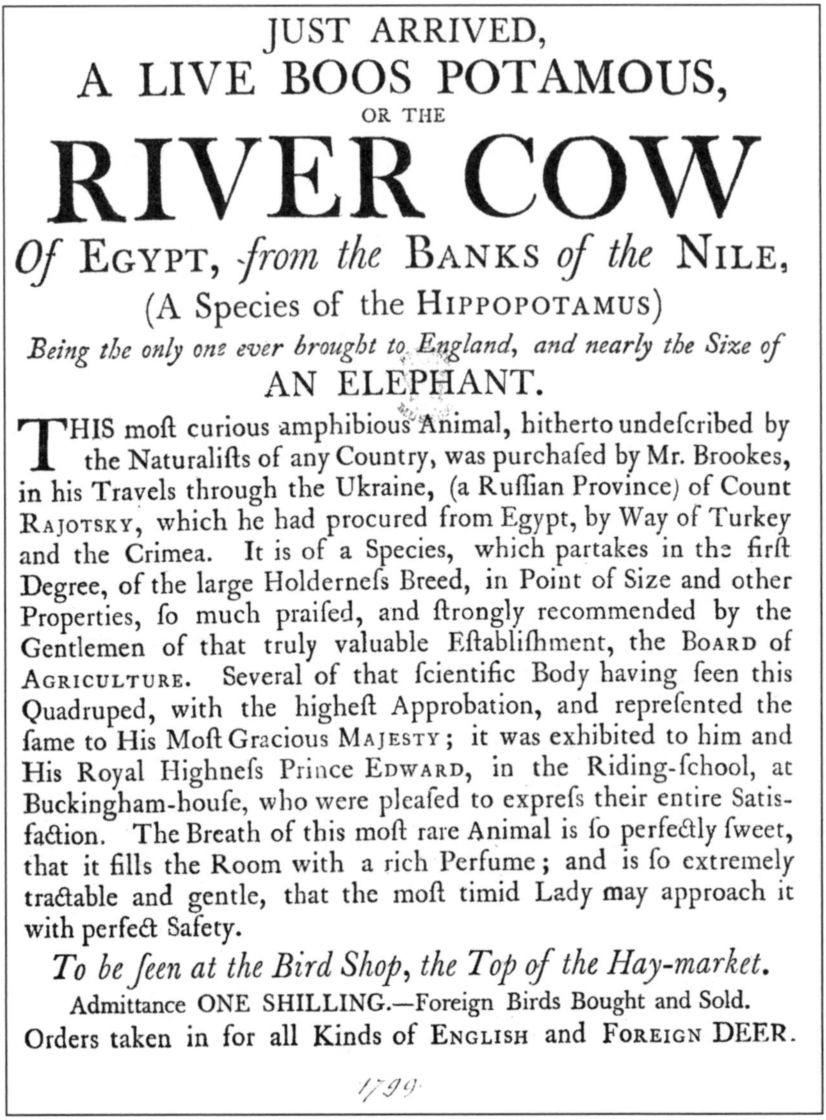

JUST ARRIVED,
A LIVE BOOS POTAMOUS,
OR THE
RIVER COW
Of EGYPT, *from the* BANKS *of the* NILE,
(A Species of the HIPPOPOTAMUS)
Being the only one ever brought to England, and nearly the Size of
AN ELEPHANT.

THIS moſt curious amphibious Animal, hitherto undeſcribed by the Naturaliſts of any Country, was purchaſed by Mr. Brookes, in his Travels through the Ukraine, (a Ruſſian Province) of Count RAJOTSKY, which he had procured from Egypt, by Way of Turkey and the Crimea. It is of a Species, which partakes in the firſt Degree, of the large Holderneſs Breed, in Point of Size and other Properties, ſo much praiſed, and ſtrongly recommended by the Gentlemen of that truly valuable Eſtabliſhment, the BOARD of AGRICULTURE. Several of that ſcientific Body having ſeen this Quadruped, with the higheſt Approbation, and repreſented the ſame to His Moſt Gracious MAJESTY; it was exhibited to him and His Royal Highneſs Prince EDWARD, in the Riding-ſchool, at Buckingham-houſe, who were pleaſed to expreſs their entire Satisfaction. The Breath of this moſt rare Animal is ſo perfectly ſweet, that it fills the Room with a rich Perfume; and is ſo extremely tractable and gentle, that the moſt timid Lady may approach it with perfect Safety.

To be ſeen at the Bird Shop, the Top of the Hay-market.
Admittance ONE SHILLING.—Foreign Birds Bought and Sold.
Orders taken in for all Kinds of ENGLISH and FOREIGN DEER.

1799

Figure 4.2. 'Just Arrived, Boos-Potamous or the River Cow of Egypt' (1799). Was this England's first live hippopotamus or a spoof? (anonymous advertisement, 1799: from Daniel Lysons, *Collectanea*, BL, C.103.k.11, vol. 2).

a most spectacular animal, purportedly the first live hippopotamus ever seen in Britain (Figure 4.2).

From what is known about Joshua Brookes's sons, it is probable that the hippo was imported by Paul, who made several journeys to Russia, rather than John, who was also managing the menagerie in Portland Row at the time. For the animal to have reached London from the Nile via the Crimea and the Ukraine sounds a touch unlikely; perhaps Count Ragousky was a travelling showman, only too pleased to part with his unwieldy charge. Despite Paul's claim to wish to improve English cattle by crossing them with a hippopotamus (!), the hippo seems to have attracted little attention from the press. Perhaps it died before the subscription could be raised or more probably the advertisement was a hoax intended to embarrass John.

In 1800 a Mr Brooks, almost certainly John, was called in to value some birds that a landlord had taken into his possession from a bird-seller tenant in lieu of rent. He arrived in the dead of night and, having valued and paid for the birds far less than their worth, took them away to be shown in his menagerie. In 1805 Brookes was fined for receiving a stolen dog, and, when he appealed, the fine was raised by £15, purportedly for costs. Two years later he took his servant Mazarine Bell to court at the Old Bailey, claiming that Bell had stolen a parasol, a hundred feathers, five pieces of bed furniture, two candlesticks, and a birdcage. He stated that he lived in Piccadilly and kept a menagerie, but it transpired that at the time of the theft he had been in Surrey on trial for stealing a dog, so the case was dismissed.[30]

In 1810 Brookes's reputation was in tatters, according to an account written ten years later:

> A white camel was imported, with an elephant, into this country. The white camel being a novelty, the proprietor (then living in Piccadilly) turned his attention to make it still more novel, caused it to be artificially spotted and produced it to the public as a camelopard just arrived. It was taken and exhibited at Windsor, and the deception there detected by our scientific naturalists.

The rate books show that in 1812 the houses at the corner of the Haymarket and Piccadilly stood empty.[31] The menagerie that had existed there in various forms since at least 1766 was finished. John Brookes may have joined forces with Edward Cross at the Exeter Change, where in 1814 (Section 5.3), he displayed an orangutan, named Britannia, a crown bird, a lion, a Satyr, a tiger, and some pelicans. But in 1816 he was on trial for murder.

While Brookes had been away on a trip to Scotland, his common-law wife Sarah Tookey, 'a woman of abandoned character', had been enticed away by a former lover. On Brookes's return, he went in search of her and was confronted by the lover, Mr Thompson, wielding a pistol. In the ensuing fracas Brookes wrested the gun from Thompson and shot him accidentally in the head; a few hours later Thompson was dead, and Brookes was remanded in custody. When he was tried at Kingston Assizes on 5 April 1816, one of the character witnesses 'who testified the general humanity and good disposition of Mr Brookes' was Lord Montfort, who said he had known the defendant since childhood and believed him incapable of wilfully doing injury to any animal.[32] Montfort must have been the son of the previous Lord Montfort, Joshua Brooke's customer, and had presumably visited the Original Menagerie with his father when he was a boy. Brookes was convicted of manslaughter and jailed for six months. He may have died in prison, as nothing more is heard of him.[33]

4.2. The Menagerie at the Exeter Change 1: Thomas Clark and Gilbert Pidcock

The story of the famous menagerie at the Exeter Change in the Strand in London begins, not with the Change itself, but with a travelling showman, Gilbert Pidcock from Derbyshire. He started in a small way showing his 'Grand Cassowar' [*sic*] at the 'Golden Horse' in Oxford Street in London in February 1778, and then toured the bird in a 'grand caravan' around England. Spectators could learn more by purchasing Pidcock's booklet, *The History and Anatomical Description of a Cassowar, from the Isle of Java in the East-Indies: The Greatest Rarity now in Europe*. In Warwick in 1779 the cassowary rewarded Pidcock by laying an egg, but when they reached Oxford

> a number of Collegians, who doubted the truth of this assertion, went in a body to see it; they had prepared themselves with a small drill, with which they made two holes, one at each end of the shell; this done, a person in company was desired to blow through the shell, and it produced a fine yolk: the Collegians, having ocular demonstration, went away satisfied. This caused a great number of the inhabitants of Oxford, to come and see the bird and the egg mentioned.

The cassowary laid a second egg at Hammersmith and a third at Abingdon, which Pidcock presented to Queen Charlotte at Windsor; she ordered it to be preserved and sent to her 'Cottage' at Richmond. In 1780 Pidcock reached Edinburgh in time for the races in July, showed off his cassowary to the Duke and Duchess of Buccleugh in the grounds of the Palace of Dalkeith, and then, having visited Glasgow, Linlithgow, and Falkirk, he turned south. But when they reached Durham, the cassowary died. Pidcock stuffed the carcass himself—it would eventually be displayed in the Great Room at the Exeter Change.[34]

Pidcock may have spent the next few years in Derbyshire, where in 1782 he purchased and furnished a commodious room at Buxton-Bath, with 'Finger Organs, Harpsichords, and Spinnets'. Ladies and gentlemen were encouraged to practise on them, or to have lessons, or to buy them. The room also contained 'a curious Museum of preserved Beasts and live Birds', admittance 1*s*. Buxton was almost certainly his home town.[35]

By 1786 Pidcock was again on tour, showing a collection of exotic animals in Chelmsford in Essex. The animals included a 'wonderful little Fairy' from Madagascar, 32 years old and 17 inches high, who was perhaps a dwarf, a porcupine—half bird, half beast—from South Africa, a female 'oranoutang' with its young purportedly from the Gold Coast, a 'mammoset' from Madras, a jackal from Bengal, a macaw, and a curious cockatoo. The most notable animal was a Bengal tiger, which had arrived the previous October as a present to the Prince of Wales, but which Pidcock had managed to purchase. Given the loose zoological terminology of the time, if the localities are correct the 'oranoutang' may have been a chimpanzee or a gorilla, but was more probably a monkey, and the 'mammoset' may simply have been a small monkey.[36] In the following year Pidcock acquired many more animals, including a very fine, young Barbary lion, reported to have been suckled by a goat: 'It is so exceeding tame, that it seems to be unhappy but when it has company. The keeper who came over with it has frequently gone into the den and played with it; so much the nature of even furious animals may be changed by custom and gentle usage.'

In February 1788 the engraver Thomas Bewick wrote excitedly from his home in Newcastle to his brother John in London:

> I am glad to find that a large collection of animals is now on its way to this Town. They are expected here at the latter end of this month.

> They consist of various kinds of the Ape tribe, Porcupine, Tiger-cat and Tiger, Greenland Bear, and one of the finest Lions (very lately brought over) that ever made its appearance on this Island so I expect to have the opportunity of doing such of them as I want, from the Life.[37]

The menagerie can be identified as Pidcock's because 'one of the finest Lions (very lately brought over)' was the tame Barbary lion that he had acquired the previous autumn. Bewick's wood engraving of the lion is included in his famous work, written with Ralph Beilby, *A General History of Quadrupeds*, first published in 1790: 'The representation we have given, was drawn from a remarkably fine one, exhibited at Newcastle in the year 1788. It was then young, exceedingly healthful, active, and in full condition.'[38]

Bewick produced another woodcut of the lion. His *Weekly Engraving Book* for 28 April–3 May 1788 reads: 'Crosse—the Royal Lion on Wood 3*s.* 6*d.*';[39] the block was used to illustrate Pidcock's advertisement for the 'Royal Numidian Lion', which is dated 'Newcastle, 3 May 1788', the day that it was printed in the *Newcastle Courant* (Figure 4.3). This not only identifies a new, if minor, wood engraving of Bewick's, it shows that someone named Cross was employed by Pidcock in 1788—probably Edward Cross, then aged only 14 or 15, the most important figure in the history of menageries in England in the late eighteenth and early nineteen centuries, whom we have encountered briefly in Section 4.1 and whom we shall meet again in many subsequent chapters.

After visiting Alnwick, the menagerie travelled on to the famous Stagshawbank Fair, just south of Hadrian's Wall, and then to Edinburgh, where, in December 1788, Pidcock announced that he was 'now on his road to London'. He was on the road to the menagerie at the Exeter Change, founded, not by Pidcock himself as often stated, but by Thomas Clark the previous spring, when Pidcock was in the north.

Clark's menagerie was housed in the Great Room over the Exeter Change on the north side of the Strand; built in about 1676, by the eighteenth century the Change housed dozens of small shops and businesses.[40] Clark had started at the Change in about 1765, selling cutlery, but by 1770 he was so successful that he was able to take a long lease on the whole building, as well as on the Lyceum a few blocks further east, while carrying on with own hardware business and letting out the shops.

An advertisement issued on Saturday, 26 April 1778, announced that the menagerie at the Exeter Change would open the following

and is well ſituate for ſporting, being in the midſt of all ſorts of Moor Game.—The tenant will ſhew the premiſſes; and for further particulars, enquire at Meſſrs Davidſons Office, in Newcaſtle.

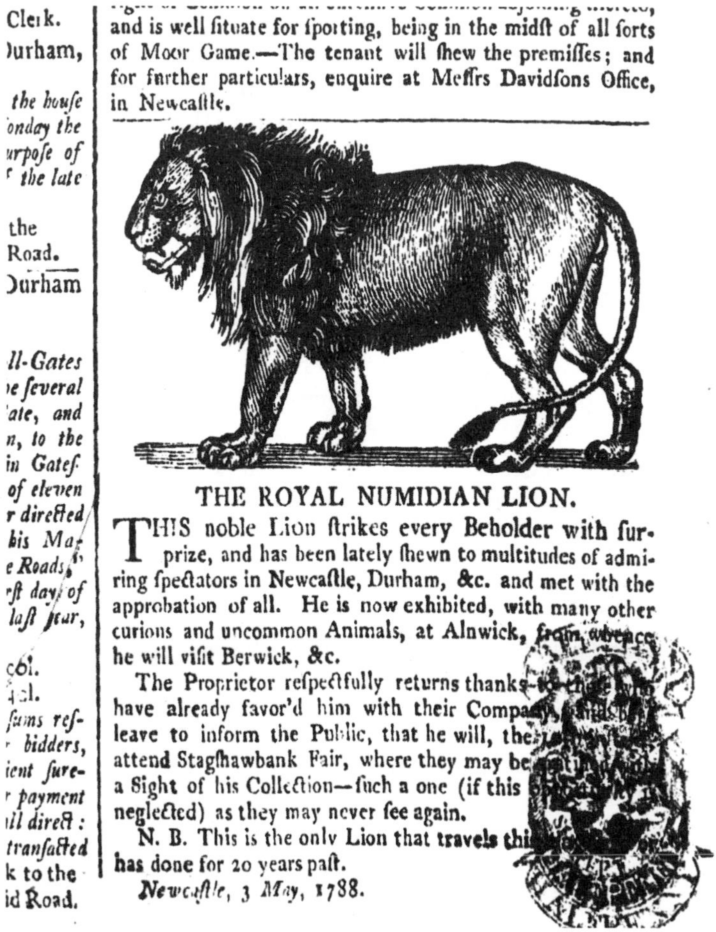

THE ROYAL NUMIDIAN LION.

THIS noble Lion ſtrikes every Beholder with ſurprize, and has been lately ſhewn to multitudes of admiring ſpectators in Newcaſtle, Durham, &c. and met with the approbation of all. He is now exhibited, with many other curious and uncommon Animals, at Alnwick, from whence he will viſit Berwick, &c.

The Proprietor reſpectfully returns thanks to those who have already favor'd him with their Company, and begs leave to inform the Public, that he will, then, attend Stagſhawbank Fair, where they may be gratified with a Sight of his Collection—ſuch a one (if this opportunity is neglected) as they may never ſee again.

N. B. This is the only Lion that travels this way, or has done for 20 years paſt.

Newcaſtle, 3 *May*, 1788.

Figure 4.3. Pidcock's advertisement in the *Newcastle Courant* (3 May 1788) with Thomas Bewick's 'Royal Lion on wood'.

Monday with 'A Grand Collection of living Beasts and Birds, selected from Asia, Africa, and America, and is allowed to be the finest assemblage offered to the inspection of the curious this twenty years . . . the creatures are well secured in iron dens—Ladies and children may see them with the greatest safety'.[41] It was open from eight in the morning till eight at night, admission 1*s*., children under 12, 6*d*. A month later the advertisements listed the animals on view, a noble lion, a lioness, jackal, antelope, leopard, panther, a pair of tiger cats, Eckneumon [*sic*], a horned owl, a golden vulture, and 'a pleasing variety of other birds and beasts' including an 'Arabian

Nightwalker' (a monkey) and a 'Gobinite'. Clark was keen to enlarge his collection, for his advertisements contain the postscript: 'The most money given for curious birds and beasts.'

In July Clark announced that the animals in the Great Room had been joined by an 'Arabian savage' or 'Child of the Sun', most probably a hamadryas baboon from Arabia, claiming that it was only the second animal of the kind ever seen in England, the first having 'died of grief'. This was probably a jibe at George Bailey's 'Child of the Sun' shown in the rival menagerie on the corner of the Haymarket and Piccadilly (Section 4.1). The Arabian savage lived up to its name a year later by tearing its keeper's arm so badly that the arm had to be amputated and the man died.[42] By December it had been joined by a 'Brazil tyger', a pair of curly-tailed leopards, a blue-faced mandrill, a porcupine 'with a long ring-tail', an ape, a silver fox, a lynx, a white-headed eagle, a young bear, and eight varieties of monkeys and squirrels, as well as a 'capital talking Cockatoo', followed in January 1789 by a pair of mountain cats from the River La Plate (probably mountain lions) and a pair of racoons.

Pidcock came to an arrangement with Clark soon after his return to London early in 1789, which probably involved the purchase or hire of some of Clark's animals to take on tour around the country. Clark's advertisement of February 1789 is informative:

MOUNTAIN CATS

Just arrived at the menagerie over Exeter 'Change a fine pair of the above cats, from the River La Plate, and reckoned a capital substitute for the Lioness.

These great Strangers, from the Land of Silver, Travellers like, are first paying their respects to the Capital which *they will soon leave, to make the Tour of England.*

Pidcock set off across the south of England with Clark's mountain cats, now referred to as 'lion-tygers', as well as his own noble lion, a lioness, his Bengal tiger, two 'Ethiopian Savages', a porcupine, and several other mammals and birds. He boasted that his macaw had conversed with 'their Majesties and the Princess Royal' at Weymouth, and that at Norwich in October the 'lion-tygers' had produced a beautiful cub. They are next recorded at Bath and finally in December in Derby in the Midlands, 30 miles from Buxton. Here Pidcock appears to have had second thoughts about running a menagerie, as he advertised the entire collection for sale, but by the end of February 1790, when the collection remained unsold, he cut his losses and set off again on his travels, being rewarded by the birth of a second 'mountain lion-tyger' cub in Coventry in June. In August Pidcock's Grand Menagerie was at the fair on Bullmarsh Heath in Reading, and, in October, augmented by several new purchases, including a young spotted 'sea lion leopard' (probably a common seal), at St Faith's Fair in Norwich.

While Pidcock was away on tour, Clark acquired a pair of bettongs, or rat-kangaroos, from New South Wales, mentioned and figured in the *The Voyage of Governor Phillip to Botany Bay* (1789) as 'Kanguroo rat. Two of the species are now to be seen alive at the curious exhibition of animals over Exeter Exchange. One of these, being a female, has brought forth young.'[43] They were not only the first live marsupials (apart from American possums), but also the first live Australian mammals ever seen in Britain. Another advertisement mentions another first—a 'fallangar opossum' from Botany Bay, presumably a flying phalanger.[44]

Clark scored a second great coup in 1790 when he acquired a young Indian rhinoceros, the first seen in England for fifty years, brought from Lucknow as a gift for Henry Dundas. Dundas gave it to the person who had been in charge of it on the voyage, who sold it to Clark for £700. The rhino was displayed in the Lyceum (Figure 4.4), along with three 'stupendous ostriches' from Barbary and the vast 'Royal Lincolnshire Ox'. The rhinoceros was extraordinarily docile, 'equal to that of a tolerably tractable pig'—it would drink three or four bottles of sweet wine in a few hours and would bleat like a calf when it saw anyone carrying its favourite food. The image on the handbill of 1790 shows that it was a very young animal in which the horn has not yet developed, but in its portrait painted at the Lyceum by George Stubbs the horn is fully developed, suggesting a later date for the painting. The portrait was commissioned or at least acquired by John Hunter Section 4.7) and is still on display in the Hunterian Museum in the Royal College of Surgeons in London.[45]

Pidcock probably spent the winter of 1790–1 in London, when his relationship with Clark seems to have been consolidated. In January 1791 he used the Lyceum to exhibit Owen Farrell, the famous Irish dwarf, a bay colt with three legs, and a heifer with two heads: 'it is the received opinion of John Hunter, Esq. Professor of Anatomy that she has Two Hearts. Admission one shilling. NB All sorts of Foreign Beasts and Birds, bought sold or exchanged by

The RHINOCEROS,

O R

Real UNICORN,

Juſt arrived at the

LYCEUM,

NEAR

EXETER - CHANGE

In the STRAND,

FROM the Empire of the GREAT Mogul, he was preſented to an Engliſh Nobleman by an EASTERN RAJAH, as a Rarity ſeldom to be met with, and His Lordſhip has complimented the curious of his native Country by preſenting him to a Gentleman who has carefully brought him home for their Inſpection

HE is about two Years old in perfect Health

THIS wonderful Beaſt with his Impenetrable COAT OF MAIL and other ſingularities is ſo fully deſcribed and admired by Naturaliſts in general, that we preſume it is ſufficient to inform thoſe who Contemplate and Admire the boundleſs Productions of the Creation, that this Herculean Quadrupede is to be ſeen as above.

Admittance One Shilling each Perſon.

1790

Figure 4.4. Thomas Clark's Indian rhinoceros on show at the Lyceum in 1790 (anonymous handbill 1799: from Daniel Lysons, *Collectanea*, BL, C.103.k.11, vol. 2)

G. Pidcock.' The heifer had been exhibited as a yearling in March 1781 'at the noted bird shop, corner of the Haymarket and Piccadilly', and for the next few years had been toured around the country before being purchased by Pidcock for the considerable sum of £150; she was a good investment.[46] Also on display in the Lyceum were Clark's rhinoceros, his newly purchased zebra—said to have been bred in the Queen of Portugal's menagerie—and 'three stupendous' ostriches.[47]

Over 400 animals could be seen in the Great Room, including the 'handsomest lion in Europe' from Algiers (that is, Pidcock's 'Numidian Lion'), a condor, a silver-headed eagle, two 'royal or crown birds', an imperial vulture brought from Vienna, a cassowary, a pelican from the Cape of Good Hope, and an 'eagle of the sun' which had been given to the Prince of Wales by General O'Hara at Gibraltar. A rattlesnake was exhibited in a box with a well-secured lid. A secretary bird was sometimes let out of his pen and 'walked about the great room in a pleasing attitude . . . like a dancing master'.[48] Other animals included a pair of curly-tailed leopards, a hyaena, a golden vulture, various macaws and cockatoos, as well as a paca (a large South American rodent), which died and was dissected by John Hunter. Copies of an illustrated book by N. Burt describing the animals could be purchased from 'Mrs Russell at the exhibition of Birds over Exeter Change' and from Pidcock at the Lyceum.[49]

Clark was not above exhibiting people; those displayed at various times between 1789 and 1791 were depicted and described by Burt along with Clark's animals—Patrick O'Brien, 'commonly called the Irish Giant',[50] and Peter Davies (another Irish Dwarf), at the Lyceum. Clark's 15-year-old indentured servant John Bobey, who had previously been exhibited as the 'Spotted Negro' in the Haymarket, was made to pose near the 'Arabian Savage' in the Great Room.

In the summer of 1791 Pidcock took Clark's 'two stupendous ostriches . . . a Bengal tyger, a young Lioness, a real laughing Hyena, a ravenous Wolf, an African ram, and Twenty other Animals and Birds; also the Royal Heifer, with two heads . . . and the double-jointed Irish Dwarf', on tour.[51] They travelled through the south of England to the great fair known as the Mart in Portsmouth, followed by the Post Down Fair above Cosham a few miles away, and reached the Lansdowne Fair on the hills above the city of Bath on 10 August. By Christmas they were in Derby, and Pidcock may have spent the holiday with family members in Buxton 30 miles away. The following spring the same animals, described as 'two of the grandest assemblages of living rarities in all Europe, from the Lyceum and Exeter Change', were on the road again, with the two-headed heifer, the three-legged colt, and the long-suffering Peter Davies. The tour was not without incident. When Pidcock's caravans[52] were travelling between Gainsborough and Brigg in Lincolnshire, a tremendous storm of thunder and lightning frightened the horses that were drawing the caravan containing the ostriches, so that they bolted, overturning the caravan and killing both birds. In another accident Davies had the misfortune to have his shoulder dislocated when his carriage overturned.[53]

By 22 November 1792, when Pidcock was back in London, he found that Clark had acquired a new and very special animal, England's first kangaroo: 'To be seen alive, at the Lyceum . . . the following wonderful productions of the Creation, amongst which is the Kangaroo from Botany Bay, the only one in the Kingdom . . . '. The strange, lively, and attractive animal had first been shown in December 1791 at the 'Trunk-maker's, No. 31 Haymarket', across the road from Bailey's menagerie on the corner of the Haymarket and Piccadilly, where it had attracted huge crowds of admirers, among them the naturalist Thomas Pennant, who described it in detail. Odd newspaper reports and advertisements suggest that only a few months later kangaroos had been imported in numbers large enough for breeding at least to be considered.[54]

By the end of the year Clark had decided to sell off all the animals that belonged to him. The auction was held on the premises on the 24 January 1793 by Messrs Humble and Henderson. Pidcock, as he proudly announced a few days later, purchased the principal part of the collection, including the rhinoceros, and continued to display the animals in the Great Room and the Lyceum. Bobey assisted at the sale by bringing forward each of the lots as it was auctioned. He was deeply offended when Clark offered him up for sale, and protested 'I can't stand that, I will *not* be sold like the monkeys', but Clark persisted and attempted to sell the remaining years of Bobey's indenture to Pidcock for 50 guineas. Having learnt that an apprentice could not be sold without his own consent, Bobey was having none of it, and instead he continued in Clark's employ as a normal servant. He also exhibited himself at Bartholomew Fair in 1795, 1796, 1798, and 1799, but he eventually he married an English woman; they set up a travelling menagerie of their own and accumulated a decent fortune.[55] Clark himself carried on with his cutlery business at the Change and died in 1816 in his eightieth year, having 'remembered all his friends and servants in a handsome manner'.[56]

The last advertisement mentioning the Bengal tiger at the Exeter Change is dated 10–12 April 1793, and it died soon after. Its carcass was acquired by George Stubbs:

> intelligence was brought to him, at ten o'clock in the evening, that a dead tiger lay at Mr Pidcock's in the Strand, and that it was to be obtained at a small expense if he thought proper to apply for it... his coat was hurried on, and he flew towards the well-known place and presently entered the den where the dead animal lay extended: this was a precious moment; three guineas were given to the attendant, and the body was instantly conveyed to the painter's habitation, where in the place set apart for his muscular pursuits, Mr. S. spent the rest of the night, in carbonading the once tremendous tyrant of the Indian jungle.

Stubbs dissected the tiger and made a large number of drawings of its musculature and skeleton, many of which were engraved and included in his magnificent work *A Comparative Anatomical Exposition of the Structure of the Human Body with that of a Tyger and Common Fowl*, which was issued in parts, but was incomplete at his death in 1806.[57]

On 23 April 1793 Pidcock took out an insurance policy 'on a Rhinoceros & Carriage for the same travelling around the country for Exhibition, not exceeding Two Hundred Pounds'.[58] Although the rhinoceros was clearly unwell, it was shown at several fairs around London. Pidcock then received a Royal summons to take it to Windsor, where it was viewed on the 3 June by King George, Queen Charlotte, and the princesses, 'who expressed themselves highly gratified'. On 11 July Pidcock announced that the rhino was in Portsmouth and would remain there for the duration of the Mart, but it was not to be: 'DIED Yesterday died at Portsmouth, the Rhinoceros or real Unicorn; it was computed to weigh near thirty hundred weight; it was the property of Mr Pidcock, of Exeter Change.' Pidcock transported the massive, stinking carcass to the Post Down Fair on the hill slopes above Cosham, but it could not remain there for long:

> on the carriage arriving at the latter place [Cosham], the stench arising from the body was so offensive that the Mayor was under

> the necessity of ordering it to be immediately buried. This was accordingly done, on South Sea Common. But it was privately dug up about a fortnight afterwards, for the purpose of preserving its skin, and some of the most valuable of the bones. The persons present declared, that the stench was so powerful that it was plainly perceivable at the distance of more than half a mile; and it was with the greatest difficulty they could proceed in their operation.[59]

Pidcock was not finished with the rhinoceros. Its corpse was brought back to London, and its skin displayed at Bartholomew Fair in September 1793.[60] Later he arranged for it to be stuffed—probably by Thomas Hall, 'the first Artist in Europe for preserving birds and Beasts', whose offices were in the City Road—to be displayed in the Great Room over Exeter Change.

Pidcock had good reason to remain in Cosham—the impending arrival of three new animals from Australia. Having retired from running the penal colony in New South Wales, Governor Arthur Phillip returned to England in the *Atlantic* and arrived at Falmouth in Cornwall in May 1793, bringing with him a kangaroo and two 'non-descript' animals. After several months the animals were transported to the Post Down Fair, where Pidcock acquired them. One of the nondescript animals may have died, but the other was a rare animal indeed, a dingo, described by Pidcock as a 'wolf-dog from Botany Bay'. The kangaroo, a female, was installed with Pidcock's male at the Exeter Change, presumably in the hope that they might procreate. Nothing more is heard of the dingo.

The loss of the rhinoceros was soon compensated for by the arrival of an elephant in the *Rose* East Indiaman on 14 September 1793. Pidcock purchased it for 1,000 guineas,[61] and immediately insured it for £950, adding £25 for each of the kangaroos.[62] The 'stupendous elephant', which seems never to have acquired a name, was Pidcock's star attraction for the next nine years and was frequently taken on tour, along with the two-headed heifer and an assortment of other animals.

Pidcock was nothing if was not a showman. In 1793 spectators at the Exeter Change could see, 'free of any additional expense', the original model of the French Beheading Machine from Paris;[63] bearing in mind that this was the height of the Terror in France, and the streets of Paris were literally running with blood, this was a singularly tasteless idea. Or was it intended as a dire warning? Other more occasional attractions included performances by the

'Celebrated Musical Children, lately arrived from Paris', and concerts held in Pidcock's new music room featuring such novelties as 'Beloudy's wonderful organ, with its musical mechanism'. An optical machine with 'moving animations' could be seen in another room for 1*s*.

The Exeter Change is usually described as a two-storey building, with many arcaded shops on the ground floor and a large room—the Great Room—occupying much of the first floor above, yet the many depictions of the outside of the building drawn during Pidcock's tenure show another large structure above the Great Room. Insurance records show that this was added by Pidcock and that it contained several rooms, including the room sometimes used for concerts. In the course of time he also acquired two rooms on the top floor of the adjacent house to use as his living quarters, which communicated directly with the new rooms over the Great Room. Animals were shown in the Great Room and at least two other rooms, admission being charged separately for each, usually at 1*s*. reduced to 6*d*. for servants and children. The main staircase had to be covered by a broad sloping platform to enable the new elephant to walk upstairs.[64]

Pidcock inserted dozens, if not hundreds, of advertisements in London newspapers, and when he was on tour, in local papers as well; those that survive and are accessible are an invaluable source of information about his animals and the goings-on at the Exeter Change and elsewhere. He frequently managed to insert puffs, thinly disguised as news items, of visits by royalty and the arrival and occasionally the death of some of the more spectacular animals. His gift for publicity extended to the well-known Pidcock tokens, which are some of the most 'collectable' items still on sale today. Each of these copper tokens has a picture of one of his most famous animals on one side and another on the reverse, often in different combinations (Plate 6). They were probably given as small change—mementoes of people's visits.

Pidcock used another innovation to attract visitors into the menagerie at the Exeter Change and when on tour. One or more of the keepers were dressed like the warders at the Tower of London—otherwise known as Beefeaters—wearing scarlet coats, criss-crossed with gold, black, and scarlet braiding, white muslin ruffs, black velvet hats, scarlet breeches and stockings, and black shoes. They were employed to stand on the pavement outside the Change, or on the steps of the caravans, to entice people in. Although like all salesmen Pidcock genuflected to the 'Nobility and Gentry', he also appealed to 'Lovers of Nature' and 'Amateurs of Natural History'; he liked to add a touch of scientific authenticity by including the names of scientists such as John Hunter and Sir Joseph Banks in his publicity material, claiming that they had authenticated his identifications, occasionally he even managed to include the names of Buffon and Linnaeus.

While the elephant was away on tour in the autumn of 1795, Pidcock acquired not only a new 'real Bengal Royal Male Tyger' but also a second elephant, a female, which had arrived in London from India on board the *Contractor*. When the male returned to the Exeter Change in December, the two elephants met for the first time. Pidcock must have been concerned about the outcome, but all went well:

> As the male animal entered the room from a tour which it has been in the country, as they had never seen each other before, the male beheld the female for a few moments in surprise, and made her a very low bow; and the female, with all the sagacity and complaisance imaginable, returned him a handsome courtesy, and they remained some time in raptures with each other, and continue kind and sociable. They are both exhibited together ... at 1*s*. each person.[65]

Pidcock's original elephant became famous for many endearing tricks, in course of time acquiring the epithet 'scientific' as well as 'stupendous':

> 'At the command of his keeper [he] will take up the smallest piece of money, a watch &c and lodge it in the pocket of any lady or gentleman ... he will likewise take it out and return it to his keeper. He will take up a tankard of any kind of liquor, particularly ale, and blow it into his mouth ... [he] will take up a broom with his proboscis and sweep after his keeper doing the same.'

Another account states that every evening Pidcock would treat himself and the elephant to a glass of spirits, always serving the elephant first. One day he served himself first—the elephant was so offended that he refused his glass with disdain and never drank with Pidcock again. Both elephants were used for children 'of the nobility and gentry' to ride on.[66]

The year 1796 was marked by an embarrassment of riches or, rather, of Indian elephants. In January a young male sent as a present to Pidcock landed at Blackwall from the *Rockingham* East Indiaman; it was followed in April by a second young male on the *Royal Admiral*. Two more presents, a 'Royal Nylghaw' and a 'ravenous Hyaena' also arrived in January, so that after they had been displayed at the Exeter Change Pidcock was able to set off on tour with 'TWO LIVING ELEPHANTS, a Cow with TWO HEADS, the ROYAL NYLGHAU, and other curious Animals', leaving the other two elephants to entertain summer visitors to the Exeter Change.

In May Pidcock had to interrupt his peregrination to appear at the Court of the King's Bench, suing a Mr Morella for damages—for during two days of the previous Bartholomew Fair Morella had placed his caravan on the pitch hired by Pidcock, thus preventing him from exhibiting his wild beasts. Although the counsel for the defence contended that it was not in the power of the jury to ascertain damages, 'because it was impossible to know the number of fools who might have paid their three-pence to see the plaintiff's wild beasts', the jury found in favour of Pidcock and awarded him £10 damages.

Pidcock's travelling menagerie returned south to Reading in August en route for Bartholomew Fair and then set off again to the south-west, with *three* elephants and a 'large and full organ' 'which plays a piece of musick, called the Battle of Prague, with all its variations, and concludes with the Grand Chorus in the Messiah'. They were in Bath in January 1797 and in Lincolnshire in March. On the way from Lincoln to Louth they were stopped by the turnpike man at Wragby toll bar, who insisted on weighing the elephants and charged Pidcock £1 19*s.* 6*d.* before allowing the caravans to proceed. After touring in the north, they arrived back in London in September. By the end of the month the much-travelled heifer, along with the original stupendous elephant and a selection of other animals, were at Stourbridge Fair in Cambridge in competition with 'W. Cross, originally from the Haymarket but lately from the Tower of London' (Section 4.3).[67]

After Pidcock's acquisition of the menagerie at the Exeter Change, it seems that he confined his tours to England, occasionally venturing as far north as Yorkshire, and usually returning to London in late August for Bartholomew Fair and the London 'Season', but after the Stourbridge Fair at Cambridge in September 1797 he took some of his more impressive animals north to Scotland. The elephant's caravan was drawn by eight horses, and the other three caravans, containing a tiger, a nylghai, a vulture, a pelican, the two-headed heifer, and various other animals 'too numerous to insert', as well as the 'large and full organ of new and curious construction', were each pulled by four horses. By December they had reached Scotland, where the animals were exhibited in Glasgow for most of January 1798; they travelled on via Falkirk and Stirling and arrived in Edinburgh in February. Here Pidcock made much use of his talent for publicity, and not only with advertisements. The magistrates of the city mounted an entertainment to honour Admiral Duncan, the commander of the British fleet that had destroyed the Dutch fleet at the Battle of Camperdown the previous October. The entertainment involved a parade of the nobility through the city streets, with Lord Duncan himself in full admiral's uniform, wearing the 'Large Naval Gold Medal' awarded to him by the King, as well as the star and ribbon of the order of St Alexander Nevsky presented to him by the Tsar of Russia. The magnificent procession, viewed by cheering crowds, was headed somewhat bizarrely by two of Pidcock's keepers, each dressed 'in the habit of his Majesty's Beef Eaters'.[68]

After a month in Edinburgh, the caravans took the Great Post Road south. At Alnwick Pidcock parted with the animal that had stood him in such good stead for many years: the cow with two heads. He had previously arranged to sell her 'to a foreigner [Antonio Alpi], who intended to exhibit her in Hamburg', and had agreed that she should be delivered to him in London. Pidcock sent her off from Alnwick, seemingly in perfect health, in a wagon drawn by four horses accompanied by two attendants. When they arrived at Durham, the cow was taken ill, but, instead of consulting 'an experienced cow-doctor', they hurried on; a few hours after reaching London she died. Pidcock had to forgo the sale price of 170 guineas, 'besides the expense of her journey from Alnwick to London a distance of three hundred and ten miles'. However, all was not lost; the cow was stuffed and joined Pidcock's treasured cassowary and rhinoceros in the Great Room at the Exeter Change.[69]

Pidcock carried on towards the south, stopping at various places along the way, and arrived in London in time for the Bartholomew Fair late in August. No sooner was the fair over than Pidcock was on the road yet again, carrying a new zebra that had arrived in London in July, the stupendous male elephant, and many other animals. After visiting Reading for the Cheese Fair in September, the menagerie made its way westward to Cirencester, where, on hearing the sound of drum beats, the zebra went berserk, foaming at the mouth. It recovered only when, following eighteenth-century medical practice, nearly a quart of blood was taken from him by a farrier. The caravans are next recorded at the Hereford Fair for 'horned cattle, cheese and Welch butter', Ross-on-Wye, and Monmouth, and on Christmas Eve they were in Leeds.[70] By April 1799 the menagerie was in Perth in Scotland, and then travelled north to reach Aberdeen; it must have been with some relief that by 3 June they were back in Edinburgh, on the Earthen Mound. They did not stay long.

On his way north in March 1799 Pidcock had commissioned Bewick, whom he had first encountered when showing his 'Numidian lion' in Newcastle in 1788, to produce wooden printing blocks of the lion, the tiger, the zebra, and the elephant. In June, on the first leg of the return journey to London, he wrote to Bewick from Haddington a few miles east of Edinburgh, mentioning the 'cuts', and adding that he intended to be at Stagshawbank Fair on 4 July, before arriving in Newcastle two days later.[71] For much of the next two months Pidcock remained in Newcastle, where Solomon Hodgson printed several hundred impressions of the blocks for him. In August Pidcock hurried back to London for Bartholomew Fair. In settling Bewick's invoice on 21 September 1799, he added a postscript: 'Mr Bewick I have purchased a fine Rhinocerous there for I think I shall want a Cut for that but I will give you further Intelligence. Hope you and the Family are well.' This suggests that a warm friendship had developed between the two men and signals a most important new acquisition, a new rhinoceros.[72]

It seems that on his journey south Pidcock had stopped off at York, where Thomas Garner wrote, or at least ghosted, the text of *A Brief Description of the Principal Foreign Animals and Birds, now Exhibiting at the Grand Menagerie*. It was printed in London in 1800 by Thomas Burton in an octavo edition and is a mine of information about the goings-on in Pidcock's menagerie.[73] It was quickly followed by a second, quarto, edition, also printed by Burton,[74] with elegant full-page engravings of the elephant, lion, tiger, and zebra—perhaps some of those printed by Hodgson for Pidcock in Newcastle—but the pages were only tipped in and may be missing from some copies. The cover is embellished with Bewick's tiger. Both versions were on sale at the Exeter Change and from the tour: 'Price 2*s*. 6*d*. with cuts, without cuts 1*s*'.

Burton also printed a magnificent poster (Figure 4.5) using Bewick's blocks of the elephant, lion and zebra with added text, some of it describing Pidcock's 'most singular quadruped', his newly acquired rhinoceros, as 'so gentle, that any person may approach him with the greatest safety'.[75]

The rhinoceros, which had arrived in September 1799, was displayed for a short while in the Exeter Change before being taken with some other animals on a tour around Kent, which culminated in Blackheath in November, where they were shown to the Princess of Wales—who 'was pleased to express her approbation of them and at the same time made the Proprietor a handsome present'.

One reason for Pidcock's anxiety to sell off some of his animals was that the Exeter Change was becoming desperately overcrowded. Among the new attractions were a lioness brought from the Cape by Lady Anne Dashwood—her head having been 'somewhat turned by the flatteries of Africa and a large Garrison', intended as a gift for the Duke of Devonshire—and two noble lions, Hector and Victory; the latter, born on 1 August 1798, the day of Nelson's victory over the French fleet at the mouth of the Nile, was named in honour of that 'glorious circumstance'. The tigress had produced three cubs, and in 1799 six new kangaroos had arrived from Botany Bay, including a 'joey' that had been born on board. One of the six, smaller and more elegant than the others, a 'musk kangaroo', was later described as a 'brush kangaroo', in modern nomenclature a whiptail wallaby. It 'liked to jump about the Great Room near the dens of two male lions, until the proprietor wisely removed him and the other five kangaroos to the same apartment as the elephant'. The remaining animals, numbering over 200 species, were crammed into another apartment; in the third was 'an Optical Exhibition, far excelling any thing of the kind hitherto invented. Admittance 1*s.* each, or the three Exhibitions for 2*s.* 6*d.*'. Space was also needed for a camel on which children were encouraged to ride.[76]

Antonio Alpi,[77] an agent of Francis II, the last Holy Roman Emperor and owner of the famous menagerie at Schönbrunn near Vienna, was in London purchasing animals for Schönbrunn and for his own collection. Alpi, whose offer for the two-headed cow had been negated by its death, now agreed to pay Pidcock £1,000 for his precious rhinoceros. But it was not to be. After only two months in a stable-yard in Drury Lane, it died, probably, but not certainly, after Pidcock had been paid. The carcass was dissected by the

Figure 4.5. Poster printed by Thomas Burton in 1799 using Bewick's blocks of the elephant, lion, and zebra; with a description of Pidcock's newly acquired rhinoceros (Victoria & Albert Museum S.516-1996). Note the handwritten incorrect date.

surgeon Leigh Thomas, who published an account of its anatomy and its appearance when alive in the *Philosophical Transactions of the Royal Society* in 1801.[78] That was not the last of Pidcock's problems. Alpi had also agreed to buy a large cat, reported to be a tiger, but more probably a leopard, but, while in the Tower of London awaiting shipment, it escaped. Persuaded by the offer of a reward of 10 guineas, the leopard's keeper, named Mason, was induced to risk his life in recapturing it; in the ensuing fracas the enraged animal sprang upon Mason, threw him down and mauled him; a sergeant hired by Mason fired, killing the leopard and narrowly missing Mason. So Alpi lost yet another animal. Mason eventually recovered, and, though scarred for life, he carried on working at the Exeter Change.[79]

Having been stymied in his attempts to acquire the rhinoceros, the two-headed cow, and the leopard, Alpi purchased the female elephant that Pidcock had acquired in 1795 and one, or both, of the male elephants which had arrived in 1796, as well as one of the zebras and some of the other animals that had been on tour that autumn. He took them across the Channel and travelled through Germany, Switzerland, and northern Italy, finally reaching his own menagerie in Turin in 1802. On the way he sold two of the elephants, a male and a female, to the Emperor for a very large sum. The pair's enduring popularity attracted a large number of paying visitors to the zoo at Schönbrunn; when the male died in 1811, his stomach was found to contain a large number of copper coins that had been thrown at him by his adoring public. His stuffed carcass was placed in the Zoological Museum in Vienna and his skeleton in the Imperial and Royal Veterinary School there. The female died in 1845 aged 53 years.[80]

Pidcock also sold a leopard, a wolf, a bear, two monkeys, and some other animals to an impecunious showman, Abraham Saunders, who refused to pay for them, because when he inspected the wagon carrying the animals he found that the leopard and two of the monkeys were dead. In July 1800, when 'Pidcock's Grand Assemblage of Curious Foreign Animals and Birds' was at Portsmouth Free Mart, Pidcock himself was in London suing Saunders in the Court of Common Pleas. The court ruled in Pidcock's favour, *because there had been no warranty that he was selling the animals alive*!

Pidcock had another reason to be in London. The newly reconstituted menagerie of the Jardin des Plantes was in need of animals;

in July 1800, despite the state of war between Britain and France that had had existed since 1793, the French Minister of the Interior sent the Superintendant of the menagerie, Citizen Delaunay, to London with a budget of 17,500 francs. Delaunay returned to Paris in November with Pidcock's tiger, his newly pregnant tigress, a male and a female lynx, a mandrill, a leopard, a panther, a hyaena, and some birds.[81]

After having had a large quantity of cash stolen from his booth at Bartholomew Fair in September,[82] Pidcock decided to sell off 'Several Close-bodied Travelling narrow and broad Wheel WAGGONS, two Close-bodied CARTS, ten well-seasoned DRAUGHT HORSES',[83] harness, and other items. It is obvious that he must have had stabling and accommodation for his various vehicles away from the Exeter Change, so it is no surprise to learn that the sale was to be held at his 'Repository' in the Borough Hay-Market. The sale suggests that Pidcock really had at last decided to cease long-distance touring and was intending to concentrate his energies on the menagerie at home. He may have felt exhausted after the strain of two major excursions to Scotland, and the growing success of Stephen Polito's travelling menagerie, which had been on the road since at least 1797, may have been perceived as a threat (Section 4.3).

Sales or no sales, Pidcock continued to acquire exotic animals to display at the Exeter Change, and the menagerie continued to receive favourable notice in the newspapers. A description of 1800, probably written by himself, declared: 'Pidcock, the Terrestrial and Aerial Fascinator, is truly called the modern Noah, and Exeter Exchange his Wonderful Ark . . . '. Aristocratic visitors included the Prince of Orange, Lady Hamilton, Nelson's mother, and the Duke of Kent, as well as 'several parties of the nobility who 'expressed the highest approbation of that extensive and most wonderful Collection'. Whether any of the nobility attended Pidcock's 'public experiments of animal magnetism' in 1801 is not, however, recorded. He also showed off some of his animals to members of the royal family at Weymouth, Kew, and Windsor, and it was perhaps on the strength of those visits that he was able in April 1802 to name the collection at the Exeter Change 'The Royal Menagerie'.[84]

Pidcock wrote that 'scarce a ship comes home but what brings Animals or Birds for that truly grand and unique collection . . . '. All the way through the records of his menagerie there are references to elephants and some of the big cats brought over in particular ships, especially those arriving from the 'East Indies'—usually meaning

India rather than South East Asia. But hundreds of other smaller animals were brought over on spec by sailors involved in trading, especially slave-trading, and by sailors, passengers, and troops returning from India and North America, as well as by servants of the various trading companies, all in the hope of making a quick sale when they finally reached port. Many animals arrived in London, or further downstream in the Thames estuary, especially at Blackwall, and Pidcock must have employed scouts to await their arrival. There were other points of entry on the south coast, of which the most important was Portsmouth, and no doubt Pidcock scouted round the docks there when showing his animals at the Mart each July—in 1804 the local paper wrote that Mr Pidcock had arrived 'to view the Beasts and Natural Curiosities, which were brought home in the Calcutta from New South Wales'. Occasionally he managed to acquire animals taken 'in a French Prize'—that is from boats captured by the Navy, like the lion and lioness mentioned earlier in this chapter and some of those intended for Napoleon's brother Jérôme in 1805. Like several other animal showmen, he acquired animals that were said to have been obtained 'as curiosities by private gentlemen, or servants of the [East India] Company' from Seringapatam—most were cheetahs, but in November 1803 Pidcock acquired a miniature zebu bull which had been 'honoured by drawing the Princes, sons of Tipoo Sultaun at Seringapatam'.

The Napoleonic War ceased briefly with the signing of the Treaty of Amiens in March 1802, and a few months later another keeper from the menagerie at the Jardin des Plantes took advantage of the peace to approach Pidcock in the hope of acquiring a pair of kangaroos in exchange for a lion. Nothing came of this idea. Pidcock had recently acquired two new lions, and anyway kangaroos were a far greater novelty. He was, however, persuaded to sell his most famous animal, the 'stupendous', 'sagacious', 'scientific' male elephant, acquired from on board the *Rose* in 1793; the price agreed was £750, but shortly before the sale was due to take place the elephant became ill and died. According to its obituary, 'there being a suspicion of poison, the huge animal was dissected by several professional gentlemen; but there was no evidence of such a fact'. The only elephant now remaining at the Exeter Change was the female, who now weighed 'upwards of three tons'.[85]

At the end of the year Pidcock acquired an animal that the keeper from Paris would have given his eyes for—a black swan from

Botany Bay, sent to him 'by a gentleman of the first science, and being the first time that such a prodigy has been exhibited in England'. The claim was correct: the only black swans to have been brought to England before 1802 were the pair presented to Queen Charlotte the previous year, but one of those had been shot and the other had soon died. The 'gentleman of the first science' must have been Sir Joseph Banks, who had been involved with the importation of Queen Charlotte's swans and indeed many of the other animals brought alive or dead from Australia (Sections 4.1 and 4.9).[86] Another bird acquired in extraordinary circumstances arrived in March 1803:

> A prodigy!—A Stupendous Condor, from the Cordilleras, with most uncommon spread of wing, extending about ten feet; driven, as supposed by an hurricane across the vast Atlantic Ocean; and was discovered near Seville in Spain, almost exhausted, feeding on the carcass of a dead horse, from which he was taken after a desperate resistance, and is now, for the first time, to be viewed at Pidcock's Royal Menagerie...

In August 1804 Pidcock was in Portsmouth, where he acquired his first emus or 'ostriches from Botany Bay', two more black swans, and yet more kangaroos. The emus were not the first to have been brought to Britain; as we have already seen, a pair had arrived at Portsmouth in 1801, when Sir Joseph Banks had obtained a valuation from Pidcock of £3,000 for the pair, though whether Pidcock had actually managed to clap eyes on them is uncertain. Pidcock's emus settled in so well that some months later one laid an egg, 'which is of a beautiful grass-green colour, mixed with white, weighs two pounds...'.[87] The kangaroos continued to produce young, not only at the Exeter Change, but elsewhere as well, so that Thomas Smith, writing in about 1805, could remark that 'these animals may now be considered as in some degree naturalized in England'. But, more than that, one of the largest kangaroos had 'remarkable pugilistic skills': Smith wrote:

> I saw this noble quadruped wrestle with the keeper for the space of ten or fifteen minutes, during which time he evinced the utmost intrepidity and sagacity; turning in every direction to face his opponent, carefully watching an opportunity to close with him, and occasionally grasping him with his fore paws, while the right hind leg was employed in kicking him upon the thigh and hip, with equal force and rapidity.

Figure 4.6. Pidcock's boxing kangaroo, perhaps the first depiction of a national symbol of Australia (from Thomas Smith 1806: i, opp. 285).

The description is illustrated with a depiction of the keeper boxing with the kangaroo (Figure 4.6). This may very well be the first picture of what has since become a national symbol of Australia—the boxing kangaroo.[88]

In an advertisement in 1806 Pidcock reiterated his statement that he had for 'some time ceased travelling', omitting to add that his last remaining elephant had died: 'A Death, the magnitude of which will be admitted by all, took place yesterday; it was no less than that of the huge Elephant of Exeter 'Change. Like many other *beasts*, it was of *great weight in*, but of *no benefit to, the country*'. The carcass was boiled down by a 'soap boiler' on Millbank in order to obtain the skeleton, which Pidcock mounted and then placed on display.[89]

In or shortly before 1806 Pidcock sold a great many animals to his former employee Thomas Miles, who took them on tour around the country:

> FOREIGN BEASTS AND BIRDS. The largest collection in the known world, *from Exeter 'Change*, Strand, London, in five commodious caravans . . . At Weymouth they were attended by his grace the Duke

> of Cumberland . . . His grace could scarce express his gratification at being so highly delighted with the great Bengal Tyger devouring a whole bullock's head, horns and all . . .

In later advertisements Miles added that Pidcock's animals had cost him several thousand pounds. Miles had fathered an illegitimate child in 1795 and shortly afterwards had begun to show kangaroos around the country;[90] indeed, Pidcock may given him the kangaroos and sent him on his way in order to avoid a scandal. Miles continued to tour his 'Grand Menagerie of Living Curiosities' around Britain until at least 1812 in competition with Stephen Polito (Section 5.2).

In 1807 Pidcock acquired the first wombat seen in Britain. It had been brought from an island in the Bass Strait by Robert Brown, the botanist who had accompanied Matthew Flinders on his 'Voyage of Discovery' around Australia in the *Investigator*. On Brown's return in 1805, the wombat was given to the anatomist Everard Home, John Hunter's brother-in-law and first president of the Royal College of Surgeons, who kept it in his house for two years. It was said that many famous anatomists of the day made its acquaintance there and found that

> it was not wanting in intelligence, and appeared attached to those to whom it was accustomed, and who were kind to it. When it saw them, it would put up its fore paws on the knee, and when taken up would sleep in the lap. It allowed children to pull and carry it about, and when it bit them did not appear to do it in anger or with violence.

After its death the wombat was dissected by Home, and its skeleton donated to the Museum of the Royal College of Surgeons.[91]

In August 1807 Pidcock acquired another very unusual animal, a Brazilian tapir, described as a kind of hippopotamus; it could be seen in a separate compartment on payment of 1*s*., but did not live long. A few weeks later a lion and lioness who lived in the same den and were said to be brother and sister, arrived on board the *Serapsis*; they had been 'taken out of a French vessel' and had been intended as a present for Napoleon. The following December Pidcock's 10-year-old lion named Victory died, 'despite the attendance of seven physicians'; a few days later it was announced that 'his skin is stuffing and in a few days will be ready for inspection', adding that Pidcock had recently refused £800 for the animal. Another lion was acquired by a group of Charles Bell's surgical students, including the American physician William Gibson, who lugged it one night from the Exeter Change to the dissecting room in Bell's house. Next morning, when Bell saw the lion, he exclaimed: 'My dear Doctor, I must have a sketch of that fine fellow's head before you cut him up.' When Gibson visited his old teacher in Edinburgh in 1839, he saw Bell's first oil painting, the life-size head of the dead lion.[92] Bell also acquired the body of a lioness from Pidcock's, which he gave to the artist Benjamin Robert Haydon, who wrote excitedly in his journal in 1810: 'I dissected her and made myself completely master of this magnificent quadruped. It was while mediating on her beautiful construction, and its relation in bony structure to that of man, that those principles of form since established by me arose in my mind.'[93]

Pidcock died unmarried in January 1810 at the age of 67 and was buried on 1 February at St James's Church, Pentonville. His will is most revealing.[94] He left £5 each to William Williams and Jonathan B[?]unyard, who were probably his employees at the Exeter Change, and £10 to Sarah Farnley, who was living with him at the Change, though in what capacity is not known. His executors were charged with selling his collection of animals for the 'most money', which was to be invested and the income used to pay for the education and maintenance of three children—Sarah's son William in London, and Charlotte and Mary Ann Wilkinson, who were living with Pidcock's sisters in Buxton. The girls were to receive £1,000 each when they reached 21. They were probably two of the four 'bastard daughters' to whom, according to Wirksworth Parish records, Gilbert Pidcock 'bequeathed his property'.[95] Identification of Mary Ann as one of his daughters is confirmed by a news item in the *Hull Packet* on 31 March 1818: 'Mary Ann, daughter of the late Mr Gilbert Pidcock of Exeter Change married John Storer Moore of Buxton.'

Pidcock's menagerie and his household goods were put up for auction by Messrs King and Lochee. The first day's sale on Monday, 19 March 1810, consisted mainly of household items, paintings, transparencies, and tools from the joiner's shop. The second and third days of the sale were 'attended by a number of gentlemen, who devote much their time to the study of natural rarities. Lord Stanley, and several persons of consequence, were among the amateurs.' On the second day, devoted to the dispersal of skeletons and stuffed animals, the skeleton of the elephant that had died in 1806 was knocked down for £57 15*s*. to 'a gentleman of the faculty', probably the surgeon Joshua Brookes; the stuffed rhinoceros, deprived of its horn, went for only £5, the horn itself for £1; Pidcock's iconic two-headed cow raised only £1; a stuffed monkey—'denominated the satyr'—was bought by a lady for £2 4*s*., and a preserved baboon bearing a watchman's lantern and rattle sold for £1 12*s*.; its 'effigy appeared as capable of performing his duty as many of our nocturnal guardians'. The whole collection, which included many stuffed birds and five organs sold in 205 lots, raised only about £140.

The sale of the live animals on the third day was more successful. Two lions were sent to the Tower; another lion and a lioness were bought by Thomas Miles for 260 guineas. A Bengal tiger went for 85 guineas to Stephen Polito, who purchased almost all the remaining animals, and also spent £9 5*s*. on Pidcock's 66-foot-long whale skeleton. An elephant that Pidcock had acquired only a few days before his death was disposed of by private contract to Miles for 1,500 guineas. The *Sporting Magazine* remarked: 'Our artists have occasionally derived great information by copying the living subjects in their dens; and on the whole, its dissolution may be considered a loss to the metropolis. The purchasers in general are showmen, who intend to exhibit their savage companions at country fairs.'[96]

The *Sporting Magazine* was wrong; there was to be no loss to the metropolis. The menagerie at the Exeter Change was to remain one of the sights of London for many years to come, run next by Stephen Polito and then by Edward Cross.

4.3. Rivals on the Road: William Cross and Stephen Polito

For much of the period when Gilbert Pidcock was on the road or at the Exeter Change, there were two other travelling menageries in Britain that can be considered to have rivalled his, those of William Cross and Stephen Polito.

William Cross (not to be confused with the William Cross who ran a menagerie in Liverpool) was a rather shadowy figure reputed to have been for many years 'the most formidable oppositionist to the late Mr. Pidcock'. One of the few advertisements that bears his name states that Wm Cross was 'originally from the Haymarket but lately from the Tower of London' and that he had 'the most valuable collection of animals that ever travelled this country'. He showed his animals at Bartholomew Fair almost every year from 1790 to 1802 and toured the country with an elephant (which arrived on the *Earl of Oxford* in 1794) and a great many other animals from 1794 to 1799, advertised as 'the largest collection that ever travelled this country'. In the note of his daughter Mary Ann's marriage to Edward Cross in 1810 he is referred to as 'the late Mr Wm Cross, of Bow, Middlesex'.[97]

Stephen Polito's career is easier to follow; born in 1763/4 at Moltrasio on Lake Como, he was in England from at least 1787 exhibiting three people, whom he named the 'Monstrous Craws' (on account of their large goitres) at various places in London and the south of England until their last appearance at Bartholomew Fair in September 1790.[98] A few days later he married Sarah White at St Andrew Church in Holborn.

Polito exhibited a few 'wild beasts' at Bartholomew Fair almost every year from 1792 until 1809, often in competition with Pidcock, but by January 1797 he was on the road with a substantial number of exotic animals travelling in two 'safe and commodious caravans'.[99] Three years later his 'Grand Collection of living Birds and Beasts' was transported in three caravans; it now included a lion, a Bengal tiger, a leopard, a pair of pumas, a wolf, and an 'ursine sloth . . . a most surprising animal lately brought from the interior parts of Bengal, yet undescribed by naturalists'—that is, a sloth bear. He exhibited the same animals in Hereford in April 1801, and by 1802 he needed 'four commodious caravans' to carry his burgeoning collection from Bartholomew Fair north to Leeds[100] and then on to Edinburgh, where his book, *Description and Natural History of S. Polito's Collection of Living Beasts and Birds*, which is clearly modelled on Pidcock's *Brief Description*, was published in 1803. Polito's *Description* mentions several of his sources—his stork came from Elsinore and his male pelican 'particularly attached to Mrs Polito' from the Emperial [*sic*] Aviary at Vienna (the Schönbrunn menagerie) in exchange for two beautiful fancy birds. Several of his animals came from the Tower, including a lion 'whelped at the Tower when the late George Payne Esq was shepherd to his Majesty's Menagerie there' and a ferocious wolf. Ever anxious to dispel any rumours of disagreeable odours, Polito described how the cages were cleaned—fresh straw was put into them every evening and removed the following morning with specially designed scrapers, after which the floor was washed with warm water using long-handled mops and then well dried.[101]

By the time it had reached Lancashire in March 1806, the collection had grown so much that it was accommodated in six caravans; it now included a lioness from the Tower menagerie and a lion, named Nelson, which had arrived in Liverpool from Senegal,

> whose, limbs [are] superior in size and strength to any horse or ox; he is of the finest and largest kind in the known world, and absolutely the only male lion that travels in his Majesty's dominions. To the great astonishment of every beholder, he is accompanied by a beautiful pointer bitch which he has been suckled by ever since he was taken from his own dam, and he now not only shews great affection for her, but suffers her to rule quite over him.

Polito also had a pair of noble 'panthers' (melanistic pumas?) from South America, a pair of leopards, 'the finest in this kingdom', and 'the most industrious of all animals', a beaver, several kangaroos, a new 'ursine sloth', and 'upwards of fifty other quadrupeds'. Among his rare and curious birds were a cassowary, an emu 'or ostrich of New South Wales', a pair of the only pelicans alive in Great Britain, a 'king of the vultures' and 'a pair of those singular rarities of the New World, black swans *rara avis in terris, nigroque simillima cygno*', as well as 'a variety of other birds of the most splendid plumage in the known world'. In Oxford in 1807, like Pidcock in 1800, he described himself as the 'Modern Noah', and later that year, as Pidcock had done before, he acquired a 'Royal Tyger', the second tiger to have been brought to England as a present to the Prince of Wales. In August at Peckham Fair on the outskirts of London, there was a disaster. A boy 'got under a caravan of wild beasts belonging to Mr Polito . . . when a panther seized his arm, and lacerated it in such a shocking manner, that it is feared amputation will ensue'.

In December 1807 Polito's menagerie reached Newcastle, where it remained for Christmas and the New Year. Like Pidcock in 1788 and 1799 (Section 4.2), Polito contacted Thomas Bewick with a view to obtaining woodblocks to enhance his publicity material. Edward Cross, who had met the engraver during Pidcock's visits, was now employed as Polito's 'Keeper of the Animals'. Bewick was pleased to renew their friendship and, having tipped Cross 5*s*., brought his three daughters to see the animals, once in the morning and again in the evening at feeding time.[102]

Sarah Polito seems to have had an active role in managing the menagerie—Bewick cut a woodblock of a lion for her, for which he charged £1 5*s*.; while Polito himself commissioned a larger cut of a lion and a smaller one for use in advertisements. However, these had not been completed when the menagerie moved on towards the south.[103] One of Polito's ostriches died on the way; Bewick passed the news to his friend, the local ornithologist and 'bird stuffer' Richard Wingate, who was keen to acquire the dead bird and agreed a price with Cross of three guineas.[104]

Soon after Polito had reached York, the large cut of the lion arrived, and he wrote to Bewick saying how happy he was with it. The only known poster in which the block was used reads: 'Nelson, a noble young lion from Senegal in the menagerie of S. Polito—What is very remarkable, it was suckled and brought up by a small pointer bitch. Drawn from life, and engraved by Mr T. Bewick of Newcastle-upon-Tyne, January 1808.' In fact it did not depict Nelson; it is a direct copy, in reverse, of the woodcut that Bewick had produced for Pidcock in 1799, embellished with a palm tree.[105] The small woodcut intended for 'some country newspapers' arrived in due course and was used in several of Polito's subsequent advertisements; however, it is a version of the image of Pidcock's lion drawn by Bewick in 1788 and used in his *Quadrupeds* (1790), albeit reduced in size and reversed.[106]

In his letter of 20 March Polito passed on Cross's particular respects to Bewick and his family: 'he is happy to say he really experienced more civility and friendship from you than from any other person he ever met on his travels—the weather coming on so severe as it has been for some days past rather effects him but he is in good spirits that the fine weather will perfectly restore him to his health.'

In York Polito's advertisement mentioned that the lioness from the Tower of London had been born 'on the memorable day that Lord Howe obtained his victory over the French force'—that is, the 'Battle of the Glorious First of June' at Ushant. The same lioness and her sister are featured in various handbills relating to the Tower Menagerie, as 'Miss Howe and Miss Fanny Howe, whelped lst June 1794, and named after the gallant conqueror of the French on that day' (Section 4.10). According to Polito's advertisement, she was 'the only survivor of the original breed of lions that has been in the Tower near a hundred years'.

Having manœuvred some of his caravans over the Pennine Hills, Polito took a selection of his animals across the Irish Sea to Dublin.

Presumably it was left to Cross to take the remainder of the menagerie south to London, where he acquired *four* new tigers, brought to England in the *Marquis Wellesley*. By the time of the Stourbridge Fair in Cambridge in September, the travellers had been reunited; Polito announced that 'a few articles which have just been added to this superb Menagerie, have been exhibited to crowded audiences in the Irish metropolis'.

Polito's menagerie toured East Anglia in the autumn of 1808 and then went on to Oxford in December, and to Bath for the season (January and February). By the time he reached Southampton the following April, he had acquired a 'great polar, the most tremendous of all Quadrupeds near the North Pole'—that is, a polar bear. One of the kangaroos had given birth, and its joey could be seen climbing in and out of her pouch. Royal recognition followed at Windsor in June 1809, and Polito made the most of it in his advertisements:

> On Monday her Majesty, the Princesses Sophia, Elizabeth, and Mary, honoured Mr Polito's menagerie of birds and beasts, &c, with their presence, attended by Lord Paulett, Ladies Macclesfield and Thynne &c at a booth erected near the Market place in Windsor, which was prepared for her majesty's and the Princesses' reception, with carpets and chairs, &c. The Royal Party were highly pleased. Her Majesty and the Princesses, after viewing, and having explained to them, every beast and bird, &c for about an hour, walked back to the Palace with their attendants.

In October 1809 Polito was at last able to add an elephant to his collection. The ship that brought it from the east, the *Winchelsea*, had arrived in Portsmouth on 9 September, and the elephant was landed on the 20th; three weeks later it was in Bury St Edmunds in Suffolk. Polito claimed that the 'sagacious animal' was the only living elephant in Europe, omitting to mention Pidcock's 'scientific elephant' at the Exeter Change. Perhaps having decided that an elephant would be too much to transport around the north of England in the winter, in November and December Polito showed off his new acquisition, transported in its own caravan, all around Kent.

Stephen Polito had good reason to stay near London; as we have seen, Gilbert Pidcock was ill, and he died on 26 January 1810. When his possessions were put up for auction in March, Polito purchased many of the other animals and acquired the lease of the premises 'over the Exeter Change', complete with the 66-foot whale skeleton. As well as owning a large travelling collection, Stephen Polito was now the proprietor of one of London's most famous, and lucrative, public attractions, the menagerie at the Exeter Change (Section 5.2).

4.4. The Duchess of Portland, the Third Duke of Richmond, and Lord Shelburne

On 11 July 1734 the second Duke of Portland married 19-year-old Lady Margaret Cavendish Harley, the daughter of Edward Harley, second Earl of Oxford. As a present for his bride, the duke set up an aviary at the estate of Bulstrode in Buckinghamshire, which he had inherited from his father. It was not a success, as all the birds died from the cold. His father had fared better: 'Macaws, Parrots, and all sorts of foreign birds flying in one of the woods; he built a house and kept people to wait upon them . . . '.[107] The huge formal garden at Bulstrode was created in the Dutch style by the second duke's Dutch grandfather, Hans Willem Bentinck, first Earl of Portland, who many years earlier had created the famous garden furnished with several much admired aviaries at Sorgvliet in Holland (Section 3.3).

Lady Margaret had grown up in a cultivated household in London and at Wimpole Hall in Cambridgeshire, where her father had an immense library of books and manuscripts (the famous Harleian manuscripts, now in the British Library) and a large collection of pictures, medals, and curiosities.[108] Harley was passionately interested in gardening and in horses; he also had a large aviary and a menagerie at Wimpole, which included at least one wolf and two antelopes given to him by Robert Williamson, who had received them from his son in Bengal. Portraits of the wolf and of one of the antelopes (a male blackbuck) painted by John Wootton in 1720 are still on view at Wimpole.

As a child Margaret had developed an interest in natural history and had been encouraged to start her own collection. In the years following her marriage she built up an immense number of specimens, mainly shells, but also crustaceans, insects, birds' eggs, fossils, and drawings. This was no cabinet of curiosities; it was a carefully arranged systematic, scientific collection. She also became a serious botanist, creating a herbarium of dried plants, which is now at Kew. No mean zoologist, she collected snails in and around the Bulstrode estate, and shells and crustaceans on the coast at Weymouth. She was in touch with many of the leading naturalists of the time, including Thomas Pennant, who dedicated the fourth volume of his *British Zoology* to her, 'as a grateful acknowledgement of the many favors conferred by her grace', as well as Joseph Banks, who visited her with Daniel Solander on their return from Cook's first voyage round the world. In 1766 the Duchess and the philosopher Jean-Jacques Rousseau, who had fled to England after warrants had been issued for his arrest in Switzerland, went botanizing together in the Peak District.[109]

The first mention of an exotic animal at Bulstrode is dated 1740, when the Duchess's bluestocking friend Elizabeth Robinson (later Mrs Montagu) mentioned in a letter that there were pheasants in the woods and some birds in the house, including a macaw. Much of what is known about life at Bulstrode is contained in the letters written by the Duchess's rather sycophantic friend Mrs Delany, best known for her wonderful 'paper-mosaicks' of flowers, although she was also a talented draughtswoman. A few of her drawings of the animals at Bulstrode survive in a private collection, including the 'Java Hare, drawn from the Life by Mrs Delany at Bulstrode 1755'. The Java hare was first mentioned in a letter written by Mrs Delany in 1749, and was presumably identified by comparison with the description and picture of the Duke of Richmond's so-called Java hare published by Mark Catesby in 1747, which was actually an agouti, from South America (Section 3.2).[110]

The first reference to the actual 'menageries' [*sic*] dates from 1753. Mrs Delany wrote:

> I went on Saturday to see the menageries and saw such beauties of foreign birds as gave me great pleasure... and there is the most extraordinary bull I ever saw; it came from some part of the East Indies; it is as round as a ball, and looks as if it was bursting with fat: it is not so high as some dogs I have seen, the colour a pretty grey, between its shoulders rises a hump, in camel fashion much higher than its head, and looks soft and dark like sable, it is as tame as a lamb and has a very good-humoured countenance: his horns were broken off in a duel with an animal of his own kind.

And, eleven months later, 'the Duchess of Portland, D.D. and I walked out: first fed all the birds of the air and water, visited the

Indian bull and his fair lady, who it is hoped will bring him an heir.' The bull is immortalized in a charming sketch, 'The Fort St David's Bull, drawn from the Life by Mrs Delany at Bulstrode 1755'.[111]

On the death of her mother in 1755, Margaret inherited Welbeck Abbey, and much else besides, becoming one of the richest women in England, but she insisted on remaining at Bulstrode, and spent her summers there for the rest of her life. Henceforth she was known as the Dowager Duchess of Portland, or simply, *the* Duchess of Portland.

The history of the garden and park at Bulstrode is not well known, except that in due course the Dutch gardens were remodelled in the more relaxed style that was becoming fashionable in the mid-eighteenth century. This probably happened soon after 1755, the year in which Walpole described Bulstrode as 'a melancholy monument of Dutch magnificence', and certainly before 1776, when William Gilpin visited Bulstrode in his search of the picturesque. He wrote approvingly: 'The approach, which was formerly regular, winds now, in an easy line, along a valley. Behind the house runs the garden; where plants, and flowers of every kind, find their proper soil and shelter. One large portion is called the American grove; consisting of the plants of that continent. Here too the duchess has her menagerie . . . '.[112] Dowland's map of 1784 shows that, while much of the Earl's formal garden had been replaced by sweeping grassland, the wooded areas to the north still had the same pathways, ponds, canal, and other features marked on the plan in *Vitruvius Britannicus* dated 1739.[113]

One of the menageries at Bulstrode was described in 1757 as a 'dairy adorned with a Chinese front, a sort of open summer house'. Gold and silver pheasants were kept in the American grove, small birds in an aviary, and larger birds perhaps in an enclosure, while hares, sheep, guinea fowl, and peacocks roamed the lawns. A reference to a 'Duck island' confirms the presence of a lake or at least a pond, and the canal is said to have been covered with water fowl, some of which would be the Brent geese noted by William Hayes in his book *The Natural History of British Birds*. A picture of a 'cushew bird'—that is, a helmeted curassow from South or Central America—'drawn from life by a gentleman in the service of his Grace the Duke of Portland', was engraved by George Edwards and published in his *Gleanings of Natural History* in 1760. Idyllic as it all seems, there were constant problems with cold, predators, and thieves. Many birds were kept in cages indoors, including a little Jonquil parrot from Indonesia, named Caton, 'the prettiest good-humoured little creature I ever saw', and Mrs Delany's favourite, a weaver bird.[114]

The most comprehensive list of the birds at Bulstrode was written by Caroline Powys, who visited on 13 July 1769. After having been duly impressed by the paintings and curiosities in the house, she complained:

> The menagerie I had heard, was the finest in England, but in that I was disappointed, as the spot is by no means calculated to show off the many beautiful birds it contains, of which there was a great variety as a curassoa, goon, crown-bird, stork, black and red game, bustards, red-legged partridges, silver, gold, pied pheasants, one of which is reckon'd exceedingly curious, the peacock-pheasant. The aviary too, is a most beautiful collection of smaller birds—tumblers, waxbills, yellow and bloom [*sic*] paraquets, Java sparrows, Loretta blue birds, Virginia nightingales, and two widow birds, or, as Edwards calls them, 'red-breasted long-twit'd finches'. Besides all above mention'd, her Grace is exceedingly fond of gardening, is a very learned botanist, and has every English plant in a separate garden by themselves. Upon the whole, I never was more entertain'd than at Bulstrode.[115]

The most unusual bird, the peacock-pheasant from the Far East, is not closely related to either pheasants or peafowl. The goon was probably a guan from South America. For black and red game read black and red grouse, both native to Britain, although the red grouse was confined to Scotland; the ornithologist John Latham wrote: 'It does not bear confinement well, yet has been known to breed in the Menagerie of that noble and intelligent Naturalist, the Duchess Dowager of Portland, who informed me, that it was effected in part, by causing fresh pots of ling, or heath to be placed in the Menagerie almost every day.' Tumblers are a type of pigeon, Virginia nightingales cardinals from North America, Loretta bluebirds perhaps eastern bluebirds. Mrs Delany's letters add 'grues' (demoiselle cranes), which in 1783 had been at Bulstrode for over thirty years; she fed them by hand and was much amused by their jumping and dancing.[116]

Although most of the animals recorded at Bulstrode were birds, the Duchess also kept bees, goldfish, and various quadrupeds, including deer, mouflon, buffaloes (perhaps more zebu) and a porcupine, 'who bristled out all his fine quills and is a fine creature'.[117]

The Duchess was seriously interested in acclimatization, at least for decorative purposes; she turned out red-legged partridges (probably imported from France) in the hope that they would breed in the wild, but, like many other landowners who attempted the same, she was unsuccessful. She had more luck with Chinese ring-necked pheasants, which interbred with common pheasants in the woods.[118]

Having decided to investigate the possibility of importing reindeer, the Duchess consulted Peter Collinson, who contacted the Bishop of Bergen Eric Pontoppidan, the author of a book on the natural history of Norway, for advice. Pontoppidan was discouraging; his book includes the warning: 'Tho' the naturalizing them [reindeer] has often been attempted, and they have been transported abroad to the great and rich for their curiosity, and to propagate their kind in other parts. This will always be a vain attempt.'[119]

The Duchess had more success with mouflon from Corsica; another bluestocking friend, Mrs Boscawen, wrote to Mrs Delany: 'Mad[e] Mouflon is better perhaps with one son than two, tho' it should retard a little the peopling of the Welsh mountains.' General Paoli, 'the illustrious defender of the liberties of his country', who imported an unknown number of mouflon from Corsica, made several visits to Bulstrode (and to Wales) during his exile in Britain in 1769–90. One of his mouflon, a ram named Martino, ended up stuffed in the Leverian Museum in London, and a pair was recorded at Shugborough in Staffordshire in or before 1768. There are several subsequent references to mouflon at Bulstrode, suggesting some success in breeding, which would not have been difficult, since mouflon are feral sheep—that is, primitive domestic sheep that have gone wild.[120]

The Duchess died in 1785, when it seems her menagerie was disbanded, for in May the following year Joshua Brookes advertised for sale 'a curious sheet cow and calf, late the Duchess of Portland's'.[121]

On the death of his father in 1750, the young third Duke of Richmond left England on the Grand Tour. He surveyed battlefields, under the tutelage of Guy Carleton, studied biological and geological specimens in Europe's museums, and seduced women in its most illustrious drawing rooms, pleasure gardens, and brothels. He was accompanied by his tutor Abraham Trembley, the well-known Swiss biologist who had been tutor to Count Bentinck's sons at Sorgvliet. Trembley wrote of the young duke: 'he loves dogs prodigiously; he also loves the human race and the feminine race.'[122]

Although much occupied with his career as an army officer and politician, the third Duke was deeply involved with the management of his Goodwood estate, where, like his father, he planted many hundreds of exotic trees. However, despite the education in zoology that he had received in Leiden under Professor Allamand, he seems to have been interested only in animals that might be acclimatized, bred, or cross-bred with other species for hunting or consumption. He achieved some success with acclimatization—in 1764 the Duke's good friend the botanist Peter Collinson wrote to Georges Buffon about some zebus:

> A great number of these animals in the Duke of Richmond's, and also in the Duke of Portland's parks, where they every year bring forth calves, which are extremely beautiful. The fathers and mothers were brought from the East Indies. The bunch on the shoulder is twice as large in the male as in the female, whose stature exceeds that of the male. The young zebu sucks its mother like other calves; but, in our climate, the milk of the mother soon dries up, and the suckling of the young is completed by the milk of another female. The Duke of Richmond ordered one of these animals to be slain; but its flesh was not so good as that of the ox...[123]

A year later Collinson wrote to Buffon again to tell him that the Duke owned several spotted deer from India, and that 'they lived familiarly with the fallow deer, did not form separate herds, but even propagated together, and from that intermixture beautiful varieties were produced'.[124] At about the same date the Duke owned a young female caribou from Canada, which lived for a long time at Goodwood. He arranged for an artist to draw it and sent the result to Allamand, who published it as an engraving in his additions to the Dutch volumes of Buffon's *Histoire naturelle*, as 'Le Caribou'.[125]

The Duke also owned several of the earliest moose to be imported from North America; the first two were females, sent by his friend and erstwhile tutor Guy Carleton, who owed his position as Lieutenant-Governor of Quebec to the Duke's influence.[126] Although one was said to have arrived in 1766 and the other in 1768, the following newspaper report suggests that in fact they arrived together in August 1767: 'Yesterday two fine *mouse deer* arrived in the river from Canada, as a present from the Governor of

Quebec to his Grace the Duke of Richmond; they are sixteen hands high; the two horns of each animal weigh 100 lb weight when shed; they are esteemed to be the most beautiful beasts ever seen in this Kingdom.'[127] At 16 hands (about 1.6 metres high) these were certainly not mouse-deer, but presumably moose-deer—that is, moose. The note on the weight of horns must refer to the antlers of male moose, not of these females, as only the males have antlers. One of the females died a few months later, but the other lived long enough for the Duke to arrange for a drawing of it to be sent to Allamand. The engraving entitled 'l'Orignal' (the French Canadian name for the moose) appeared in the Dutch editions of Buffon's *Histoire naturelle* with a description.[128] Allamand's entry in Smellie's translation of Buffon's *Natural History* reads:

> The Duke of Richmond, who delights in collecting for public utility, every thing that can contribute to improve the arts, or augment our knowledge of nature, has a female orignal in one of his parks, which was conveyed to him by General Carleton, Governour [*sic*] of Canada, in the year 1766. It was then only one year old, and it lived nine or ten months. Some time before it died, he caused an exact drawing of it to be made, and of which I have given an engraving as a supplement to M. de Buffon's work.[129]

This moose is also known from one of Gilbert White's letters to Thomas Pennant. On 29 September 1768 White went to Goodwood to see it, but was disappointed to discover that it had died. He found its carcass in an old greenhouse, supported in a standing position by ropes under the belly and chin. Although it had been dead for only a short while, it was already putrid, but, despite the appalling stench, White examined it with characteristic care. Its height at the withers, measured in the same way as for a horse, was 16 hands, higher than most horses. He mentioned that it had had a female companion, which had died the previous spring. There had been hopes that it would mate with a red deer stag, but, as White wrote, 'their inequality of height must have always been a bar to any commerce of the amorous kind'.

Two years later Carleton presented a third moose, a male, to the Duke of Richmond, who allowed the physician and anatomist William Hunter, elder brother of John Hunter, to examine it and also 'gave leave to have a picture made of it by Mr Stubbs'. The result was the famous oil painting *The Duke of Richmond's First Male Moose*, now in the Hunterian Museum in Glasgow. It shows a young, rather plaintive-looking animal with antlers in the earliest stage of development posed against a romantic stormy sky. The moose excited Hunter's particular interest because he was eager to compare its antlers with those of the extinct Irish elk, which were sometimes dug up from Irish bogs. He wanted to check the possibility that they were of the same species—that is, to discover whether the huge Irish elk, although extinct in the British Isles, had survived in North America. The Duke's moose was immature and had disappointingly small antlers, but Carleton had also provided a pair of antlers of a full-grown male from Quebec, and under Hunter's direction Stubbs included it in his painting, lying on the ground (Plate 7).[130]

The Duke was anxious to breed from his male moose, but, as his females were both dead, in October 1770 he wrote to the second Marquis of Rockingham, who owned a female (which, according to William Hunter's notes, was the first moose to have been imported into England), suggesting that arrangements should be made to bring the two together. Although Hunter noted that this female was kept at Rockingham's seat at Wentworth in Yorkshire, in 1770 Thomas Pennant saw it at Rockingham's house at Parsons Green on the edge of London; he wrote: 'it seemed a mild animal; was uneasy and restless at our presence, and made a plaintive noise.'[131]

Carleton presented a second male moose to the Duke of Richmond in 1773, and the Duke again allowed Hunter access to it. Hunter, accompanied by Daniel Solander and presumably Stubbs, travelled down to Goodwood on 4 October 1773 carrying the painting of the first bull moose with them for comparative purposes. In Stubbs's new drawing, recognized by Ian Rolfe among the drawings in Glasgow University Library, the antlers are a little more developed than those of the first bull moose.[132] Hunter drafted a long account of the animals, which he intended to read to a meeting of the Royal Society with a view to publication in *Philosophical Transactions*, together with engravings of Stubbs's images, but he changed his mind.[133] However, he lent the painting and the drawing to Thomas Pennant, and engravings of both images by Peter Mazell appeared in Pennant's *Arctic Zoology*, published in 1784, the year after William Hunter's death. The same animal was also depicted by John Frederick Miller, but in his drawing the antlers have a few more 'points on top', so it was probably executed at least a year later.[134]

There were still moose at Goodwood in 1790, for on about 25 October John Hunter was sent for to attend the Duke's brother-in-law William Ogilvie, who had been gored nearly to death by a ferocious male. Perhaps John was rewarded with a carcass of a female deer—his collection in the Museum of the Royal College of Surgeons in London formerly contained 'the uterus of one of the Duke of Richmond's deer'.[135]

William Petty, the second Earl of Shelburne, inherited the estate of Bowood in Wiltshire in 1761, but in 1763, in temporary retirement from the Cabinet, he acquired the lease of Whitton Park on Hounslow Heath, a few miles west of London, where he kept a lion. Its cage was placed next to a gravel path and it was frequently roused to fury by the sight of passers-by. George Stubbs made numerous sketches of it 'by permission of his Lordship's Gardner' and used them as the basis for several of his famous paintings, including *Lion Attacking a Stag* (1763) and *Horse Frightened by a Lion* (1765?), both commissioned by Lord Rockingham.[136]

Shelburne had also inherited the estate of Wycombe, sometimes referred to as Loakes Manor, in Buckinghamshire; in September 1765 the Earl of Sandwich offered Shelburne a 'wild beast'—that is, another lion—for the menagerie at Wycombe in the unfulfilled hope that Shelburne would reciprocate by giving him some manuscripts from his collection. Soon after this the animals in the menagerie were transferred to Bowood, where on 10 July 1768 Lady Shelburne's diary recorded that their number had hugely increased.[137] Shelburne set about improving Bowood, choosing only the most fashionable: Capability Brown to landscape the park and Robert Adam to decorate the grander rooms in the house and to build a low wing, 300 feet long, modelled on Diocletian's Palace at Spoleto, containing a library, an orangery, and dens for wild animals.

Lady Shelburne wrote on Christmas Day 1768 that 'the intense cold killed in one night our poor ourangoutang, or man of the wood'; this was probably the 'large black monkey from Lord Shelburne, of the spider monkey kind, but [with] two thumbs', whose corpse was given to John Hunter and identified many years later by Richard Owen, the Conservator of the Museum of the Royal College of Surgeons, as a gibbon. If that is correct, this was the first live gibbon imported into Britain; its tongue and larynx still survive in the Hunterian Museum at the college.[138]

Shelburne also owned a 'tiger' (probably a leopard), maintained on a diet of horseflesh and beef, which had the honour to be stroked by Jeremy Bentham in 1781. Other animals recorded in the menagerie are a beaver, a wild boar sent to Shelburne by Count

Schaumburg-Lippe-Bückeburg in 1769, a white fox from Archangel given him by Bentham,[139] and a bear:

> On Tuesday sevennight a wild bear, that a man was conveying in a cart from Lord Shelburne's menagerie in Wiltshire, broke loose and jumped out of the carriage... the horses, as soon as they saw the bear, broke loose and galloped off, and the Driver fled to alarm the country, which was soon assembled with every kind of hostile weapon. The bear in the mean time ran for shelter to an adjacent tree, which he ascended, and sat there till he was shot by Capt. Bayntun.

'The skull of a lion—the sole relic of his peers—now wonders at the strange company it keeps among bookshelves and parchments' in the library, wrote Shelburne's great-grandson in 1912. The skull was still extant in the estate office in 1922.[140] The main house at Bowood has been demolished, but other buildings, including Adam's 'Diocletian's Palace', survive, with views across Brown's sublime romantic landscape to the lake below.

4.5. Memories of India: The Nabobs

Many of the men employed in the East India Company's trading posts, provided they survived the heat and disease of India, returned to England as rich men, ambitious to succeed in politics and often given to conspicuous displays of wealth, with luxurious living, country estates, and the like. They were despised by England's upper classes, as vulgar, corrupt upstarts—the Nabobs. Three of the best known were Robert Clive ('Clive of India'), Warren Hastings, and Elijah Impey; all three suffered impeachment on account of their ill-gotten gains and all three were finally acquitted. Both Clive and Hastings imported animals from India, and all three of their wives kept collections of animals in Calcutta, Mary Impey's menagerie being of particular importance.

Robert Clive and his wife were living in Cuddalore, 100 miles south of Madras, where he was Deputy Governor of Fort St David in 1756, when news reached him that Calcutta had fallen to the Nawab of Bengal, Siraj-ud-daula, and that the British survivors of the battle had been imprisoned in the infamous Black Hole. He led an expeditionary force north and in January 1757, after he had succeeded in retaking Calcutta, he sent for Margaret, whom he had left at Cuddalore awaiting the birth of their third child. But on 11 June, fearing a renewed attack on Calcutta, he led a force of about 3,000 men up the River Hooghly to attack the Nawab; his resultant victory at Plassey owed much to the treachery of one of the Nawab's generals, Mir Jafar, who, having made a personal payment of £234,000 to Clive, was installed as the new Nawab under Clive's control. The victory established Clive as absolute ruler of Bengal, ensuring British rule not only over Bengal, but ultimately over much of India. He returned to Calcutta in triumph with an astonishing quantity of 'rewards' and gifts from Mir Jafar, among them a 'strange Indian beast, the Shah-goest'—that is, a caracal cat.[141]

Margaret and the new baby duly arrived and were installed in a spacious colonnaded mansion within Fort William, with a country house at nearby Dum Dum.[142] With dozens of servants, her library, and her harpsichord, she settled in happily enough, though always concerned about Clive, who had been formally appointed Governor of Bengal by the Court of Directors of the East India Company, and was mostly away on various military excursions. The baby died, and Margaret consoled herself with a small menagerie of pets, a tame mynah bird, a young tiger, a bear, two porcupines, three 'new-fashioned' birds, and an owl 'as big as herself'. Her letters to Major John Carnac, Clive's colleague and a family friend, show the interest and amusement that her pets afforded her; she wrote that the peacock had 'expired by the bite of a Jackal' and that the mynah bird 'is half the day on a table in the hall where I sit learning of us women to talk nonsense'.[143]

The news of Clive's victory at Plassey reached London late in 1757, and on 1 January 1758 Clive's father heard William Pitt declaim to the House of Commons: 'We had lost our glory, honour, and reputation every where but in India. There the country had a heaven-born general.' Clive's reputation seemed assured. When he received the news, he wrote a long letter to Pitt saying in essence that the task that faced the British in India was too great for the East India Company alone and needed the nation's assistance. The letter, entrusted to Margaret's cousin John Walsh, took nearly a year to reach England. As well as the letter, Walsh brought a gift for Pitt from Clive—Mir Jaffir's caracal cat. Pitt seems not to have been overjoyed, for he quickly offered it to the King. The Keeper of the Lions at the Tower

Figure 4.7. Lord Clive's caracal cat in the Tower of London (*The Shar Goest*, from *The London Magazine*, 28 (1759), 664).

> was ordered to attend at St James's to receive a very beautiful and uncommon animal, lately arrived from the East Indies [Figure 4.7]. It was presented by Jaffier Ally Kawn, Nabob of Bengal, to General Clive, who sent it the Right Hon. William Pitt, Esq; and of which that Gentleman had the honour to obtain his Majesty's acceptance. It is called in the Indostan language a Shah Goest... It is now in the Tower attended by a domestick of the Nabob's, who was charged with the care of it to England.

Horace Walpole wrote delightedly:

> There are some big news from the East Indies. I don't know what, except that the hero CLIVE has taken Mazulipatam and the Great Mogul's grand-mother. I suppose she will be brought over and put in the Tower with the Shahgoest, the strange Indian beast that Mr Pitt gave to the King this winter.

Alas, the caracal 'could not support the cold of this climate, though near a good fire, and under the care of an Indian attendant, but died in a few weeks'. John Hunter acquired its corpse, dissected it, and wrote: 'In a cat I had from the Tower of a brownish dun'; its skeleton was in his collection at the Royal College of Surgeons, labelled 'bones of a Shargoss', but has not survived.[144]

Meanwhile, back in Bengal, Clive's health was giving way, and on 1 February 1760 he set sail with Margaret for England on board the Royal George, bringing with him a fortune of at least £300,000, a *jagir* of £27,000 a year, diamonds, and some animals, including a large snake, 'with a large stock of poultry &c for its subsistence on the voyage, but it could not bear our climate, and died the first winter'.[145]

In February 1762 the *London Evening Post* reported: 'Thursday his Grace the Duke of Marlborough went to Lord Clive's House in Berkeley Square, and took possession of the fine Tyger that his Lordship some Time ago made him a Present of; and his Grace ordered the Beast to be carried to his seat in Oxfordshire'. Little would be known today about this tiger were it not for Judy Egerton's fine research into its portrait painted by George Stubbs in three different versions; the fact that the animal was a female was not made apparent until a print by John Dixon was published ten years later, with the inscription *A TIGRESS. In the Possession of His Grace the Duke of Marlborough* (Figure 4.8). This is the earliest well-documented record of a true tiger, otherwise known as a 'Royal Bengal tiger', in Britain. Marlborough's tigress was well fed at Blenheim; a local butcher delivered 24lb of meat every two or three days, at a cost of 3*s*. a time, with the occasional head thrown in for 4*d*. In Egerton's own words:

> Though the tigress is depicted reclining and apparently replete (unsurprisingly on the meat supplied by the local Oxfordshire butcher), Stubbs leaves us in little doubt that, if menaced, the tigress would spring to attack with one lithe and supremely-coordinated bound. This inherent sense of danger in one of nature's most formidable creations links Stubbs's 'Tiger' with subjects in the natural world which contemporaries deemed to be 'sublime', since they aroused emotions of awe and fear.[146]

Awesome or not, by 4 April 1772 the 'famous Marlborough Tyger... brought over by Lord Clive... at Blenheim House near Six Years', was on show along with the 'Queen's Ass' and other animals in John Pinchbeck's menagerie in Oxford and was last heard of in March 1774 in Edinburgh (Section 4.9).

In 1762 Clive wrote to the new Governor of Bengal, Henry Vansittart: 'I must again repeat my desire of having a large elephant

Figure 4.8. The tigress presented to the Duke of Marlborough by Lord Clive (*A Tigress*, John Dixon, 1773, after George Stubbs: British Museum 1917,1208.2441).

embarked for his Majesty, if the thing be practicable, of which you must be a better judge than I, who are upon the spot; and if you can send me any curiosities, such as antelopes, hog-deer, nilgows or lynxes, I shall be much obliged to you.' A young elephant duly arrived in August 1763 on board the *Hardwicke*; to Clive's intense fury it was presented to George III and Queen Charlotte in the name of the East india Company, not his own (Section 4.9).[147]

Early in 1764, shortly after Clive had acquired Walcot House and its 6,000-acre estate in Shropshire, the arrival of news from India of several serious military setbacks experienced by the company at the hands of Indian forces prompted calls for his reappointment as Governor of Bengal and commander-in-chief of the army. Fearful of losing his *jagir*, Clive set out again to India, leaving Margaret, who was pregnant, at Westcombe, a pretty villa in Blackheath, near enough to the centre of London to receive the news from India as soon as it arrived and also near her brother Neville

Maskelyne, the Astronomer Royal in Greenwich. Margaret kept some animals at Westcombe, for her household accounts include the sum of £31 9*s*. 'for Bran for Birds and Beasts'. Her other brother Mun accompanied Clive on the long voyage back to Calcutta; their ship was delayed for two months at Rio de Janiero, where either Clive or Mun acquired a lion monkey for her—that is, a marmoset—as well as a 'Pretty Pol'.[148]

Clive arrived in Calcutta on 3 May 1765 determined to rout out what he referred to as the Augean stable of corrupt company servants and to make peace with the Indian rulers. He set off to meet the Mogul Emperor and the Nawab of Oudh, including in his huge retinue his elephant, hawks, and a pack of hounds he had had sent out from England, as well as some 'curious deer' and a tiger received as gifts along the way. The Mogul was persuaded to grant the *Diwani* of Bengal, Bihar, and Orissa to the company—that is, the company would collect the vast revenues of the three provinces in return for protection and various annuities.[149] In celebration Clive held a 'fandango' on the Maidan in Calcutta that lasted for four days; the guests were entertained by wild beast fights—tigers *versus* buffalo, an elephant *versus* a rhino, as well as elephants and camels fighting among themselves; one of the elephants ran amok, killed seven spectators, and removed the roof of a house. In contrast, Clive spent much of his time acquiring presents to send home to Margaret, including a hoopoe, a 'gold bird', a pair of blackbuck, and 'a deer no bigger than a cat' (probably a hog deer). After much wrangling in his attempts to sort out the company's affairs and a bad bout of illness, Clive left Calcutta in February 1767 on the *Britannia*, bringing with him his friend and servant 'Black Robin', a slave boy, several dogs and birds, and a monkey complete with clothes. The rest of his menagerie followed in a later ship.[150]

Among Clive's animals on board the *Britannia* was a turtle purchased on the island of Ascension, which he sent in a 'fish machine' to the 'Gentlemen of the Corporation of Shrewsbury' for a feast. The *Derby Mercury* reported that three zebras had been offered to him in Bengal (presumably a mistake for the Cape); however, the first zebra had lost one eye, the second was so wild that it was impossible to tame it, and the third 'he genteely purchased' in order to present it to an unnamed royal personage. The quadrupeds were probably kept at Walcot, and were of such interest that the young Joseph Banks arranged for his protégé Sydney Parkinson to draw them. Parkinson had been introduced to Banks early in 1767 and was soon hired by Banks to accompany him as the artist on the voyage of the *Endeavour*. The naturalist Thomas Pennant was curious and wrote to Banks asking to see Parkinson's drawings of Clive's quadrupeds, which seem to be otherwise unrecorded. The hog deer described in Pennant's *Synopsis of Quadrupeds* (1771) as in Clive's possession may have been Margaret's 'deer no bigger than a cat'.[151]

Clive also received a pair of nylghai, a present from Charles Crommelin, the retiring Governor of Bombay, who returned to England in August 1767. Clive's nylghais bred regularly for some years, probably at Walcot, where an aerial photograph shows that behind the house there is a large enclosure, surrounded by a very high wall, which was probably where the nylghai and deer were kept.[152]

In the summer of 1769 Clive purchased the estate of Claremont, near Esher in Surrey, from the widow of the Duke of Newcastle. He employed Capability Brown to remodel the magnificent park, which had been landscaped first by Charles Bridgeman during 1715–26, and then by William Kent in the 1730s. Despite Brown's activities, much of the unique early Georgian garden survives to this day, including the magnificent amphitheatre designed by Bridgeman. The Duchess kindly 'gave up' to Clive 'without any consideration of value, the White Ducks, the Wild Ducks, Muscovy Ducks, Brand Geese, Chinese Geese, and Wild turkeys in the Park'.[153]

Various 'foreign animals', including spotted and hog deer, the nylghais, Cyrus cranes, Cape geese, and guinea fowl, were transported from Walcot to Claremont, where Clive issued strict instructions that they were to be kept warm. The deer and the Cyrus cranes (sarus cranes) were from India, but the Cape geese are more problematical: they may have been Cape shelducks or Chinese geese, which by 1785 were common at the Cape of Good Hope and 'sufficiently common' in England, or the misnamed Egyptian geese, which are not actually geese, *sensu stricto*, and are native to much of Africa including the Cape, 'from whence', wrote Latham in 1785, 'numbers have been brought into England'.[154] Clive ordered that antelopes be sent to him by every company ship; as a result, curcurras (demoiselle cranes) and more sarus cranes arrived from India. The wall of the park was raised in height to enclose the animals, although the locals said it was to keep the Devil out.

One of Clive's nyghai escaped, getting as far as Cobham, and a female died, much to Clive's irritation. While staying in his recently acquired property at Oakley Park in Shropshire, he wrote to his secretary Henry Strachey in London:

> I am sorry to hear one of the female Nylgaws is kill'd by the Amorous Violence of Dr Hunter. I cannot help thinking Larkin might have consulted men well vers'd in these matters & avoided the Misfortune, it is not likely the Female Nylgaw would be amorously inclined while she was suckling her Young Ones. There are Forrest Men who perfectly understand the business, pray Enquire if the Cape Geese & Guinea hens have produced any Young Ones this year . . .[155]

William Hunter's paper describing two other nylghais, which had been presented to the Queen in about 1769 by Clive's enemy Laurance Sulivan, a director of the East India Company, published in the *Philosophical Transactions of the Royal Society* in 1771, includes observations that he had made on Clive's animals as well as the Queen's: 'Its manner of fighting is very particular, it was observed at Lord Clive's, where two males were put into a little inclosure . . . Lord Clive has been so kind to give every help that he could furnish me with, in making out their history . . . '.[156]

Having circumnavigated the world, the *Endeavour* returned to England in July 1771. Parkinson had died on the voyage, but Captain Cook and Banks were received as heroes. It was probably Hunter's presentation at the Royal Society that reawakened Banks's interest in Clive's animals. Clive, Banks, William and John Hunter, Nevil Maskelyne, and Clive's political 'henchman' John Walsh were all Fellows of the Society. When Clive was abroad in December 1773, Strachey wrote to him: 'I am desired by Mr Walsh to tell your Lordship that Mr Banks is very desirous of having the Nylgaws propagated and therefore wishes that you would either do it Yourself by ordering the Two which you have at Walcot & Claremont to be put together, or give them both to Banks who will keep them in his Park.'[157]

Among the animals in Clive's collection were two 'antelopes', known only from John Hunter's notes. They were probably gazelles, pretty delicate animals, each about the size of a goat; he wrote: 'This animal I have called Le Corine, as it resembled a print which is so called by Buffon; but from comparing that with the male antelope, two of which I had from Lord Clive, it would appear to be the female . . . '. Another of Clive's antelopes may have been the subject of Stubbs's painting, originally entitled *Antelope*, commissioned by William Hunter and now in the Hunterian Museum in Glasgow.[158] The subject was originally identified as a pygmy antelope, but, as noted by Ian Rolfe, has been reidentified as blackbuck. Rolfe suggested that its Latin name *Antilope cervicapra* echoed the belief that blackbucks lay somewhere between deer and goats and were therefore of particular interest to those who were carrying out experiments in cross-breeding.[159]

Clive's zebra, a mare, became the best known of his animals when she was subjected to a bizarre experiment in cross-breeding at Claremont in June 1773; John Hunter wrote: 'Lord Clive's zebra mare was in heat, but refused the advances of an ass. Lord Clive ordered that the ass should be painted in stripes similar to the female zebra; whereupon she accepted him.'[160] Banks took the matter seriously and sent a questionnaire to Clive's steward, Mr Parkin, from whom more details, published in 1799, were forthcoming. The zebra had been covered by an Arabian stallion and then by a jack ass; to both she 'shewed great disgust' and did not conceive. They then tried painting a jack ass with zebra stripes and lo! she produced a mule. The mule resembled its father rather than its mother, although faint stripes could be discerned. It was 'sullen, vindictive, and untractable', and, although it did permit itself to be ridden,

> a considerable time generally elapsed before the mule and the rider could agree about the direction in which they were to move; and when that point was in some degree settled, the labour, to the rider, of impelling and guiding his companion, was found so much to exceed that of walking on foot, that the services of the mule were not much in repute, or often called for.[161]

All this time Clive was dealing with the building, decorating, and landscaping of Claremont and the acquisition of suitably costly works of art from Italy, not to mention his political ambitions, his trial by Parliament on charges of corruption, and his deteriorating health. He died, probably by his own hand, on 22 November 1774 at his house in Berkeley Square.

An inventory of the animals at Claremont prepared soon after Clive's death included the zebra and her foal, two spotted deer, two antelopes, six hog deer, and a very troublesome African bull. After a while the zebra died and her foal was sent to Walcot.[162] Writing about the experiment in his *History of Quadrupeds*

(1781), Pennant mistook the gender of the zebra, but in the appendix at the end of the book, presumably added as it was about to go to press, he revised his account and also described the foal:

> The zebra which is now at Walcot I find is the product of a painted Jack Ass and a female zebra, foaled at Lord Clive's at Claremont.—The legs alone are spotted like the foreign ones.—It is a very vicious animal. They are obliged to confine it in a paddoc [*sic*]. Some time since, they suffered him to return to the park, to the great annoyance of passengers. It is a fine-made beast, and seems remarkably strong.[163]

Clive's eldest son Edward rebuilt the house at Oakley Park for Margaret, and she lived there, the centre of a large circle of friends and family, enjoying her cats, her birds, and her telescopes, until her death aged 82.[164]

Warren Hasting was another key figure in the establishment of the British Empire. After his arrival in India in 1750, he worked his way up through the East India Company hierarchy, becoming a member of the Calcutta Council ten years later. In 1764, when his complaints about fraud and corruption among the company staff were unheeded, he returned to England. However, after running up enormous debts in London, he decided to return to India to restore his finances and applied to the company for employment; with Clive's support he was appointed deputy ruler of Madras.

Two years after having been made Governor of Calcutta in 1771, Hastings was appointed the first Governor-General of India. He was considered by some to have ruled with deviousness, cruelty, and a total lack of scruple, and by others to have been the creator the Indian Empire. Autocratic, but sensitive to the cultures around him, he struggled in a web of conflicting economical, legal, and cultural issues, only to be undermined by enemies both at home and in India. A highly intelligent man with a great respect for Indian culture, he had learnt Urdu and Bengali during his first term in India and had a grasp of Farsi, the language of the Moghul court. He consulted Brahmin scholars when drafting laws, encouraged orientalist studies, founded a madrassa in Calcutta and another in Varanasi, was patron of the newly founded Asiatic Society, and set up an experimental farm at Suksagar for the cultivation of coffee, sugar cane, and other useful plants.

In 1774 Hastings sent an embassy through Bhutan to the Tashi (or Panchen) Lama in Shigatse in Tibet. He privately commissioned its leader George Bogle:

> 1. To send one or more pair of the animals called Toos which produce the shawl wool. If by a Dooley, Chair, or any other contrivance they can be secured from the Fatigues and Hazards of the way, the expense is to be no objection.
> 2. To send one or more pair of the cattle which bear what are called cowtails [that is yaks].
> 3. Walnuts . . . rhubarb . . . ginseng
> 4. [A long list of curiosities 'acceptable to Persons of Taste'] . . .

He added the postscript: 'Animals only that may be useful, *unless remarkably curious* . . .'.

The Lama sent '8 goats which produce the shawl wool', eight sheep, eight yaks, and eight dogs—but no 'remarkably curious animals'—care of Lieutenant Williams at Cooch Behar just south of Bhutan, who was the link man between Bogle on his travels and Hastings in Calcutta, but the only animal to survive the journey was a single shawl goat.[165]

Hastings set up a menagerie in the compound of Buckingham House (the Governor's official residence in Calcutta, today the site of Government House), which contained several species of deer, including a musk-deer,

> armed with teeth . . . with which he wounded every other kind of deer in the same inclosure with him; rising on his hind legs and striking downwards. He was smaller than a common goat, yet had scored deep gashes in the tough skin of a Ghouz, which is the largest species of stag known in India . . .

and several kiang (onager-like equids from Tibet). Hastings's cousin Samuel Turner wrote: 'I have heard, that four of these animals were once in Mr. Hastings's possession: three of them were vicious, stubborn, and untameable; the fourth, a female, was of a different disposition, following a camp, loose, and perfectly familiar with every body, and every thing around it.'[166]

Lady Mary Impey was the wife of Sir Elijah Impey, the Chief Justice of Bengal, who arrived in Calcutta in 1774. He had been a friend of Hastings since their schooldays, and, when Hastings married Marian, the former Baroness von Imhof, in 1777 at the church of St John in Calcutta, he was chosen to give the bride away.[167] The Impeys' home was a mansion in a park (commemorated today by Park Street), a short distance inland from Buckingham House. Lady Mary is well known because from 1777 to 1782 she employed three Indian artists from Patna, including Shaikh Zayn al-Din, to produce paintings—'197 of birds, 76 of fish, 28 reptiles, 17 of beasts' and '8 of flowers'. She kept a substantial collection of living birds, particularly aquatic birds, as well as a few live mammals—a pangolin, a great Indian fruit bat, a dwarf flying squirrel, a 'mountain rat', a 'musk weesel', a monkey, a sambar fawn, a mouse-deer, a musk-deer, and (less certainly Lady Impey's) a serow from Sumatra, as well as a shawl goat painted by Zayn al-Din in 1779.[168]

After his appointment as Governor-General, Hastings found his influence was undermined by a council of four men, appointed by Lord North the Prime Minister, whose intention was to remove him from power. In 1775 Sir Philip Francis, the most vituperative among them, connived with Raja Nand Kumar to accuse Hastings of fraud; in the process Nand Kumar acted fraudulently himself and was tried for forgery in the Supreme Court, found guilty by Impey and the other judges, and hanged. While the verdict appears to have been correct, it was claimed that the sentence amounted to nothing less than judicial murder and was to haunt both Hastings and Impey for many years. For this and several other reasons, Impey was recalled to England and embarked with Mary and their children on 3 December 1783.

As well as Mary's precious collection of paintings, they took with them a large number of living birds, including some rare and exotic pheasants with iridescent feathers from the Himalayas, which she had nurtured in her menagerie, but after two months there was a disastrous outbreak of disease amongst the 'poultry', and all the birds died. She managed to preserve the corpse of at least one—John Latham was later able to describe it and to name it in her honour, the Impeyan pheasant, now usually referred to as the Impeyan monal (*Lophophorus impejanus* Latham 1790).

Mary's collection of paintings was much admired by natural historians in England, especially Latham, whose *Supplements* to his *General Synopsis of Birds* (1787 and 1801) contain dozens of references to her paintings and name some of the birds for the first time, aided by information supplied by Mary herself.[169] In his *View of Hindoostan* Pennant thanked 'Sir Elijah Impey and his lady for liberal access to their vast and elegant collection, made with much fidelity on the spot . . . to them I was indebted for permission to have several copies made by my paintress Miss Stone . . . '. He referred to several paintings of mammals, including one of 'a black variety of the white eye-lid monkey . . . the thumb of that in Lady Impey's collection had no flat nail, the rest of the toes clawed'. Pennant also refers to five birds from the 'cabinet of Sir E. Impey', which suggests that they brought some stuffed birds back to England with them.[170]

Following the Nand Kumar affair, political machinations in London between the government and the company, and criticism of his activities in India, the pressure was mounting on Hastings to resign. After hesitating for several months, he returned to England; he disembarked from the *Barrington* at Plymouth on 13 June 1785 and hurried to London, having no inkling of the storm that awaited him. The papers reported: 'The presents brought over by the Governor-General of Bengal for the Royal family, and some other distinguished personages, are said to exceed, in rarity and richness, all that have hitherto been produced from India,' and: 'Mr Hastings remittances already safely arrived in France and Holland, amount to one million and a half sterling, independent of jewellery, the value of which is said to be immense,' and: 'The Amount of the Value of the Cargo and other Things brought over in the Barrington East Indiaman (including the baggage of Mr Hastings and other passengers) is rated at two Millions of Money; which is richer freight than ever was brought to Europe in one Ship.'

The *Barrington* sailed on up the English Channel to moor at Blackwall in July, when one would suppose his treasures, including some of his precious Indian animals, were unloaded. One of the animals, probably a present for the King, was the 'all-black tiger'—actually a melanistic leopard—which ended its days in the Tower (Section 4.10).[171]

Two years after Hastings's return, he was impeached, charged with 'high crimes and misdemeanours', and put on trial in Parliament. The trial, which became a giant survey of British rule in India, dragged on a few days at a time from 1788 to 1795, when, exhausted and in ill health, Hastings was finally acquitted. Needing a county home to use as a refuge during the trial, he had purchased 'a very pleasant little estate of 91 acres at Old Windsor called Beaumont Lodge', for the then vast sum of £12,000,[172] where he attempted to grow exotic fruit, including mangoes, and where he kept several Indian animals. When he quit India, Hastings had left his faithful friend and secretary George Thompson in charge of his interests there. Turner sent Hastings a 'Bootan bull, of the Chowry species'—the first yak ever brought to Britain, which he had obtained when travelling through Bhutan in 1783–4 on a mission to Tibet. The yak arrived in fine health, but a yak cow and some shawl-goats sent at the same time died at sea; Thompson was asked to send more: 'Pray get me some, and send one, or at most two, but not more, by every ship that will take them, and bespeak an inclosed berth in each for them; otherwise they will die of ill usage.' Later that year Hastings reported that he had one female shawl-goat and that he had received news that a male was awaiting collection in Dublin.[173]

Hastings had long wished to buy back the family home, Daylesford House in Gloucestershire, which had been sold in 1715. In 1788 he managed to persuade the owner to sell it to him; he sold Beaumont Lodge and instead rented Purley Hall, a few miles north of Reading, as a temporary home while the house at Daylesford was being rebuilt.[174] A rather primitive oil painting by an unknown artist, *Purley Hall, Berkshire, View of the House with Warren Hastings' Rare Eastern Species of Animals*, includes the yak in an enclosure with a stable, an Arab horse being led by a groom in the park, a zebu bull and cow, and a shawl-goat. The images of the animals may have been superimposed in about 1790 on a much earlier, primitive painting of the house and grounds.

Hastings's animals were soon moved from Purley Hall to Daylesford, and on 29 June 1791 Hastings and his wife moved into their new, though uncompleted, house there. He had at last come home. He commissioned Stubbs to paint a portrait of his precious 'Bhootan bull'—the yak is depicted on a bluff above a lake surrounded by mountains; on the edge of the lake there is a palace, which has been identified as the Punakha Dzong, the summer palace in Bhutan.[175] The landscape is based on a drawing, *A Scene on the Frontier of Bhutan* by Samuel Davis, the draughtsman on Turner's mission (Figure 4.9), which Turner probably gave to Hastings when, after his own return to England, he retired to an estate not far from Daylesford.[176] The purpose of Turner's mission had been to promote British–Indian trade across the Himalayas by strengthening relationships with the regency of the Panchen Lama in Tibet following the reincarnation of the Lama in the form of an 18-month-old boy. Turner's official report, *Account of an Embassy to the Court of the Teshoo Lama in Tibet*, was published in 1800

Figure 4.9. Warren Hastings's yak (*The Yak of Tartary*, William Delamotte, 1800 after George Stubbs: British Museum 1881,1210.272).

and is illustrated with engravings based on views 'taken on the spot' by Davis.[177] One of the illustrations has the caption 'An engraving of this Bull, from a picture in the possession of Mr Hastings, painted from the life by Stubbs, is annexed; the landscape was taken from a scene on the frontier of Bootan, by Mr Davis'. The book contains much information about yaks in general, this yak in particular, and, as already noted, Hastings's own animals in Calcutta. Turner wrote:

> I had the satisfaction to send two of this species to Mr Hastings after he left India, and to hear that one reached England alive. This, which was a bull, remained for some time after he landed in a torpid languid state, till his constitution had in some degree assimilated with the climate, when he recovered at once both his health and vigour. He afterwards became the father of many calves, which all died without reproducing, except one, a cow, which bore a calf by connection with an Indian bull.[178]

The last twenty-four years of Hastings's life were passed as a country gentleman at Daylesford. He sent for seeds of a custard-apple tree that grew at his erstwhile home at Alipore and tried to naturalize lychees. He also attempted to breed from his Indian animals. Having lost the yak cow *en voyage*, he had to cross the bull with local cows, or more probably zebus from India, in the hope of propagating 'Bhootan cattle' in England. Domestic yaks and zebus are interbred by farmers on the lower reaches of the Himalayas; the resulting hybrid males are sterile, but the females are fertile. Hastings's ambition was not idle; yaks and yak crosses are extremely hardy, and versatile; though nimble-footed, they can carry heavy loads and can be ridden, they produce rich milk, and their long hair can be put to many uses. Although none of the first consignment of shawl-goats survived the voyage over, Hastings did eventually acquire more and attempted, unsuccessfully, to cross them with local goats. Shawl-goats from the Himalayas are so named because the soft, silky down that grows under their long hair is the material from which fine Pashmina or Cashmere shawls are woven; clearly they are animals with much economical potential, but not one to be fulfilled in England, despite several attempts at acclimatization early in the nineteenth century.[179] Hastings died in 1818; his wife Marian survived until 1837; she was buried beside her husband in Daylesford Church.

4.6. Arcadia

During the reign of George III many formal gardens, with their clipped hedges, straight lines, and regimented trees, were transformed, often by Capability Brown, into naturalistic parkland, with great lawns sweeping down to sinuous lakes, and some nicely placed trees to guide the eye towards classical garden buildings. It seems that no park was complete without an aviary of exotic and native birds, or a menagerie containing birds or a few exotic mammals, but, with a few notable exceptions, information about the animals that they contained is scanty. Nevertheless it seems there was much exchange of birds and information among their aristocratic owners, usually women, including the Countess of Chatham, the Countess of Essex, and Lady Tyntes.[180] Mrs Delany visited the Duchess of Norfolk's menagerie at Worksop Manor in Nottinghamshire in 1756, but saw only a crowned crane and 'a most delightful cockatoo with yellow breast and toppings'.[181] The second wife of the first Duke of Northumberland is said to have had a menagerie with gold and silver pheasants, black swans, and, most unusually, gazelles, probably at Syon House in Middlesex. It must have been the Duchess, rather than her husband, who, having purchased mandarin ducks from Joshua Brookes in the New Road (Section 4.1), was one of the few owners to succeed in breeding them.[182]

In 1760 Mrs Delany visited Longleat in Wiltshire and described how the gardens had been transformed by Capability Brown: 'Lady Weymouth carried Miss Chapone and me all over the park, and showed us her menagerie: I never saw such a quantity of golden pheasants; they turn them wild into the woods in hopes of their breeding there for they are as hardy as other pheasants.'[183] More than twenty years later Mrs Boscowen was equally enthusiastic about Longleat:

> The menagerie there is delightful. Out of the green wood sally'd so many golden birds, so many that had white muslin cloaks over their mourning cloathes, these you know to be pheasants, but there were also many winged families of most respectable size and beautiful feather, especially Carolina ducks, wch I judged to be the Summer duck, so rare and beautiful.[184]

Little is known about Lady Ailesbury's 'famous aviary' at Park Place, Henley-on-Thames, where she lived from 1752 to 1793, although it is on record that her husband, Henry Seymour Conway, gave her a gerbil in 1752 and that Conway's cousin Horace Walpole sent them bantams and Chinese pigs from his home at Strawberry Hill in Twickenham as a house-warming present. Walpole himself kept bantams and goldfish and complained in 1753 that all his Turkish sheep, which he had kept for fourteen years, had been torn to pieces by foxes.[185]

The magnificent Palladian building at Horton in Northamptonshire, restored in the 1970s and known today as the Menagerie, was built by Thomas Wright of Durham in 1750s for the second Earl of Halifax. Although used to house the Earl's art collection and for banqueting, the building was the centrepiece of a semi-circular enclosure with ponds and cages for exotic animals, the whole being surrounded by a moat.[186] Horace Walpole visited in July 1763 and wrote:

> In the Menagerie which is a little wood, very prettily disposed with many basons of gold fish, are several curious birds and beasts.

> Storks; Racoons that breed there much, & I beleive [*sic*] the first that have bred here; a very large Strong Eagle, another with a white head; two hogs from the Havannah with navels on their backs; two young Tigers [perhaps leopards or smaller cats]; two uncommon Martins, doves from Guadaloupe, brown with blue heads, & a milk white streak crossing their cheeks; a kind of Ermine, sandy with many spots all over the body and tail.[187]

The Revd William Coles, who visited with Walpole, added some detail:

> In the garden or park, is a good artificial river, in which is a sloop and 2 or 3 boats, there are 2 temples, one with a portico, the other round, a triumphal arch, a Gothic bridge and a menagerie at a good distance, in which are kept a variety of wild beasts and birds: a young tiger not bigger than a cat, a bear, and a monstrous large eagle, and an eagle with a large head.[188]

In 1783 Sir Richard Hill set about augmenting the 'beauties and wonders' of Hawkstone in Shropshire, creating an extraordinary romantic garden, which included a menagerie building on the south side of the Elysian Hill, 'where Nature is aided by Art, without seeming to be her debtor'. By 1802 it contained a 'choice collection of beasts and birds, both foreign and domestic', various eagles, a macaw, several parrots, small monkeys, and rabbits. Those that died were stuffed and displayed in the menagerie. The couple employed to clear the ground and feed the poultry were celebrated in a verse over the door.[189]

In June 1763 Walpole wrote to George Montague from Hinchingbroke, the fourth Earl of Sandwich's country seat: 'There was formerly a little chapel there, with the Admiral's heart embalmed in a box . . . but I hear it is converted to an aviary.'[190] It was about this time that the Earl first met his future mistress, the young Martha Ray, on whom he is said to have showered presents of parrots and dogs. As Lord of the Admiralty, he had approved funding for the fitting-out of Captain Cook's ships, so on his second voyage Cook named the Sandwich Islands in his honour. In 1775, when the *Resolution* returned to the Thames, the Earl went to Woolwich with Martha to inspect it. On board were the mammals and birds that the expedition's naturalist, Johann Reinhold Forster, had purchased in South Africa as gifts for the Queen. When Martha saw the birds, she 'manifested an unbounded affection towards the pretty creatures, and a violent longing to be made mistress of them', and threw a tantrum when her demand that they be given to her was refused.[191] The animals were indeed taken to Kew, and Martha had to go without (Section 4.9).

The Earl was a patron of William Hayes and the owner of a folio volume dedicated to himself and to the Countess of Northumberland, *The Portraits of British and Exotic Birds, Coloured from Nature, by Mr Hayes, Late of His Majesty's Household. In the Course of the Work will be Given the Superb Collection of the Earl of Sandwich, the Menageries of the Earl Spencer, Robert Child Esq.* ..., produced and illustrated by Hayes and his daughter Ann, of which only a single copy is known.[192] Although the title page bears the date 1778, some of the plates were not produced until 1780. Inside the book was a manuscript letter from Hayes to the Earl stating that another copy of the book was intended for the Queen and a third for Martha Ray. But alas it was too late;, the Countess had died in 1776 and Martha had been murdered by a jealous lover in April 1777. Hayes also produced many embossed pictures of birds for his various patrons; at least three depict birds owned by Sandwich: a 'Grillan sparrow', an 'American Grossbeak', and a 'Green Parrot'.[193]

The collection of birds at Osterley Park, to the west of London,[194] owned by the banker Robert Child and his wife Sarah, is known from another book by Hayes, *Portraits of Rare and Curious Birds... from the Menagery at Osterley Park*, published in two volumes (1794 and 1799). The illustrations in the book are hand-coloured etchings based on paintings of over ninety species, which had been commissioned by the Childs in the 1780s; depictions of eighteen additional species are included in a collection of etchings and watercolours by Hayes in the Natural History Museum.[195] Hayes was recommended to Sarah by the Duchess if Portland, for whom he had also worked (Section 4.4).[196]

The birds were housed in enclosures and buildings known today as Aviary Farm on the north shore of the lower lake,[197] but it is clear from the Hayes's text that other birds were allowed the freedom of the park; no doubt the water birds decorated the serpentine lake, along with the family's pleasure boats. There were other aviaries and conservatories nearer the mansion, some designed by Robert Adam, who had been called in to remodel the house. One recently restored building, sometimes referred to as the 'Aviary or Little Orangery', is thought to have been designed by William Chambers and 'improved' by Sarah; it probably housed both orange trees and caged birds.[198]

The Childs employed a 'Menagerie Man', Jonathan Chipps, who was paid £31 a year, assisted by a 'Boy' at 8*s*. a week,[199] who had some success in breeding—a Numidian crane, some currasows, and ring-necked pheasants were raised there—but he had no success with red-legged partridges or mandarin ducks. As well as birds chosen for their beauty, the collection included raptors—a king vulture, a white-tailed eagle, a gyr falcon, a Virginian eared owl, and a buzzard. The collection was much admired; in 1772 Lady Agneta Yorke wrote that Osterley was the prettiest place she ever saw—'an absolute retreat'—and mentioned two Numidian cranes that would follow visitors like dogs.[200] A year later Walpole was delighted with 'a menagerie full of birds that come from a thousand islands, which Mr Banks has not yet discovered . . . ', less so with the park itself. In 1788 Caroline Powys was also unenthusiastic: 'the situation dreary and unpleasant, and the menagerie, which I had heard so much of, fell far short of my expectation; that of Lady Ailesbury's at Park Place is vastly superior in elegance; nor were there so many different birds as I had seen at others.'[201]

4.7. Two Scientific Menageries: John Hunter and Dr Joshua Brookes

Two London scientists, both surgeon-anatomists, kept small menageries: the famous John Hunter, FRS, from about 1769 to his death in 1793, and then Dr Joshua Brookes, FRS, FLS (the eldest son of Joshua Brookes the animal dealer) in the early nineteenth century.

John Hunter built up his menagerie at what he called his 'country box' at Earl's Court—then a rural hamlet surrounded by market gardens, now the site of a mansion block named Barkeston Gardens close to Earl's Court tube station—which he had acquired in or shortly before 1769. There he would go at weekends and holidays from London to escape the demands of his patients and pupils and the hospitals to which he was attached.

All around the house was a covered cloister dug six feet into the earth, where the smaller animals were kept. A passage led from the house to a long range of outbuildings, on the one side, and, on the other, the garden merged with a large field, with a mound housing some of his animals (Figure 4.10); the whole property was surrounded by a high brick wall. He kept all the usual farm

Figure 4.10. The 'den' at Earl's Court, where some of Hunter's animals were housed (anonymous water-colour, 1886: Royal College of Surgeons RCSSC/P 326).

animals, which could be slaughtered to provide food for both his households, cows were milked, and live animals studied or experimented upon. There were greenhouses, a conservatory with bee hives, and a fish pond in which he cultivated leeches for medical use and freshwater mussels from Scotland in the hope that they would produce pearls for a necklace for Lady Banks, the wife of his friend Sir Joseph Banks. Small wild animals were trapped locally and eels purchased from the fishmonger. But the pride of his collection were animals from overseas, including opossums and an agouti from America, a tame spotted hyaena, leopards, a jackal, an ostrich, a cowardly ratel named Moses, some zebu from India, several dozen golden pheasants from China, and various birds of prey. Any animal that died was dissected, and preparations of its various organs pickled in spirits to join his ever-burgeoning anatomical collection.

Hunter owned a beautiful small bull (probably a dwarf zebu from India) given to him by Queen Charlotte, with which he used to wrestle, but on one occasion it overpowered him, and, if one of the servants had not happened to come by and frightened the animal away, he might have been killed. On another occasion two leopards that were kept in an outhouse broke their chains and escaped into a yard, where they were attacked by Hunter's dogs; hearing the hullabaloo, Hunter ran into the yard and caught one of the leopards, which was escaping over a wall, and managed to get hold of the other and to secure them both.[202]

Hunter's activities epitomize eighteenth-century curiosity about the natural world, in his case curiosity about the structural and functional manifestation of life itself.[203] To this end he conducted innumerable physiological experiments on animals and plants and dissected thousands of animals, both exotic and native. Many of his dissections of individual organ systems were preserved and kept in his enormous anatomical collection, classified into separate hierarchical systems; those that have survived the depredations of time and the London Blitz may be seen in the Hunterian Museum in the Royal College of Surgeons in London. He also carried out hundreds of autopsies on people who had died in the London Hospitals and dissected bodies stolen by the 'sack-em-up men' from London graveyards. They were used for teaching anatomy to his students and to investigate causes of death.[204] It was at Earl's Court that he secretly boiled down his most notorious acquisition—the body of Charles Byrne, the Irish Giant—in a domestic boiler in a dungeon under one of the outhouses and then proudly displayed the skeleton in his museum in Leicester Square. It still occupies pride of place in the Hunterian Museum, 7 feet and 7 inches high.

Once a week a cart drawn by a large white zebu and two smaller zebu or buffaloes brought fruit and vegetables from Earl's Court to Hunter's household in Leicester Square, returning with stable dung and other rubbish for disposal, including, in the winter season, the 'off-fall' from the dissecting-rooms. Hunter's loyal assistant, William Clift, who took charge of the anatomical and pathological collection after Hunter's death, wrote a hilarious account of the riot that ensued when some boys investigating the contents of the cart uncovered the putrid half-dissected arms of a man and some internal organs.[205]

Among Hunter's multifarious interests was the possibility of grafting an organ from one animal to another, and he was not above removing a healthy tooth from one person and grafting it, or attempting to graft it, into the jaw of a wealthier person. In 1771 he procured two 'Jack Apes' and invited several gentlemen to watch him graft the horn of a calf into the forehead of one and the antler of a young deer into the other. This was said to be for the benefit his impecunious friend Gavin Pettigrew, so that he could retrieve his fortune by showing them up and down the country. The result of the operation is not recorded; presumably it failed.[206] Hunter transplanted some human teeth into cocks' combs, one of which appeared to 'take'—the resulting preparation is displayed in the Hunterian Museum. He carried out a great many other experiments on domestic animals—on bone growth in pigs, on exfoliation in donkey's feet, on the development of chicks and goslings; the most cruel involved the insertion of corrosive sublimate into a cow's vagina, not to mention surgical experiments on the nervous system in living dogs, whose yelps of pain could be heard all over the house.

Hunter had many opossums, brought or sent to him alive by friends from North America; not yet aware of the marsupials as an infra-class of the Mammalia, he wrote that 'this animal is distinct from all others, so far as I know'. He managed to get his opossums to mate and was aware that the embryos made their way from their mother's uterus to her pouch, but none survived thereafter.[207]

Like many others, Hunter was fascinated by the possibility of crossing animals of different kinds to produce offspring that might prove more useful than their parents, the obvious example being mules, the result of mating a male donkey with a mare. He crossed a small buffalo bull (probably a dwarf zebu) with an Alderney heifer, another 'nice little heifer' with a buffalo (zebu?) bull kept by the Dowager Marchioness of Rockingham at Parsons Green, and a South American nanny goat with an aggressive billy goat (shawl-goat or Pashmina) from the Himalayas.

Several landowners as well as Hunter were involved in crossing wolves and dogs, often with the assistance of the animal dealer Joshua Brookes. Lord Clanbrassil offered Joshua Brookes money to cross a male wolf with a dog bitch, which produced a litter of nine. Lord Monthermer bought one; Clanbrassil got at least one bitch, which he crossed with a male pointer; the resulting bitch he gave to Lord Pembroke. Pembroke's wolf–dog was buried in the garden at Wilton in Wiltshire under the inscription:

Here lies Lupa,
whose grand-mother was a Wolf,
whose father and grand-father were Dogs, and whose
mother was half Wolf and half Dog. She died
on the 16th of October, 1782, aged 12 years.[208]

Brookes crossed a Pomeranian[209] bitch with a wolf; one of the ten resulting puppies was acquired by Lord Gordon and taken to Gordon Castle in Scotland, where it was seen by Thomas Pennant. He wrote that it resembled a wolf, and, 'being slipped at a weak deer, it instantly caught at the animal's throat and killed it. I could not discover whether this mongrel continued its species: but another of the same kind did; and stocked the neighbourhood of Fochaber with a multitude of curs of a most wolfish aspect'.[210] Brookes gave another puppy from the same litter to Hunter. Hunter also purchased a puppy from George Bailey, the bird merchant on the corner of Piccadilly (Section 4.1), which was said to be the result of a cross between a female jackal and a spaniel, which had mated on board an East Indiaman on the long voyage from Bombay (Section 4.1). He took the half-bred jackal pup to Earl's Court and crossed her with a terrier, producing five puppies. It was probably one of these puppies 'being three parts dog' that Hunter gave to his friend and erstwhile pupil Edward Jenner. In 1785 William Gough, an animal dealer on Holborn Hill much patronized by Hunter, crossed a male greyhound with a very tame, though unwilling, wolf bitch. There were four puppies, one of which Hunter was to have had, but it and two others were killed by a leopard and the fourth was sold to a man who took it to the East Indies; however, on 24 February 1787 the wolf produced six more puppies by another dog. Hunter's account of cross-breeding of wolves, dogs and jackals, 'Observations Tending to Show that the Wolf, Jackal, and Dog are all of the Same Species', appeared in the *Philosophical Transactions of the Royal Society* in 1787.[211] He claimed that the results showed that the wolves, dogs, and jackals were closely related and that wolves were ancestral to dogs, thus prefiguring Darwin on the origin of species.

Gough was quick to capitalize on the publicity:

> To the CURIOUS, particularly the lover of Natural History. Mr Gough's famous Wolf Bitch has pupped, for the fourth time, and has a prodigious fine litter, got by the largest dog in Great Britain.—She is the famous Wolf which is mentioned in the Transactions of the British Society [*sic*], as having proved the Wolf to be the original of the Jackals and Dogs, from the Greyhound to the Mastiff, to all the varieties of Pugs and Lap Dogs; them which nothing can be more remarkable and curious, as has been strongly testified by Sir Joseph Banks and John Hunter, and all who are true lovers of Natural History. To be seen at Mr Gough's menagerie, No. 99, Holborn-hill.

N.B. All kinds of Pheasants and Fowls, for land or water, to be sold as above.

The source of these wolves is not recorded; they were extinct in England and Wales and rare, if not extinct, in Scotland, but they were still common over much of Europe and elsewhere.

Hunter's most spectacular animal was a stuffed giraffe. It had been acquired in 1779 by William Paterson when employed by Mary Eleanor Bowes, the widowed Countess of Strathmore, to collect plants from the Cape of Good Hope for her hothouses in Kensington. Just before Paterson sailed, the Countess's life changed dramatically—she was tricked into marrying a bounder, Andrew Stoney, who added her surname to his own and in due course divested her of her estates and her fortune. When Paterson returned to London with the skin and bones of the giraffe, the couple refused to reimburse him for the expenses he had incurred. Hunter saved the situation by buying the first giraffe, alive or dead, to be seen in England.[212] He had the skin stuffed—it must have been a wonderful sight, standing in the stairwell of his previous home in Jermyn Street and then in his museum in Leicester Square. The skin no longer exists, although some of the bones survive in the Hunterian Museum, along with a drawing, *The Camelopardalis in an upright position.*

On the day of Hunter's death from apoplexy in 1793, Mr Gough sent a pair of golden eagles to Earl's Court 'intended for the rock-work in front of the house'; but, while they were still in their cages, Clift had the melancholy task of countermanding the order. He wrote that Gough also took all the other living animals from Earl's Court 'at a valuation', though it seems he paid only for a vulture, some sheep, and Hunter's great dog.[213]

Dr Joshua Brookes was an anatomist, surgeon, and teacher very much in John Hunter's mode. From about 1787–1828 he amassed an enormous anatomical collection at his house in Great Marlborough Street, which included the remains of bodies of criminals executed at nearby Tyburn, or obtained from the 'sack-em-up' men.[214] He is said to have taught 5,000 pupils over a period of about forty years, sometimes with as many as a 150 in a class, ploughing back the profits he made—about £30,000—into his museum. Much loved by his pupils, but shunned by the surgical establishment, he was referred to by one of his colleagues as 'without exception the dirtiest professional person I have ever met... all and every part of him was dirt'.[215]

Brookes became an important figure in the development of comparative anatomy, for, as well as human specimens, his collection included a vast number of animal preparations. He coined the generic names *Lycaon* for the African hunting dog and *Lagostomus* for the plains viscacha (Section 5.3). Two animals that he dissected were particularly well known. One was 'the celebrated parrot universally known, formerly the property of the late Colonel O'Kelly, of which the oviductus contains a defective egg'; this was Polly, famous for her rendering of the 104th psalm. The other was Chunee the elephant that died in the Exeter Change in 1826 (Section 5.3). Brookes also owned the skeleton of one of the quaggas that had drawn Sheriff Parkins's curricle around Hyde Park.[216] It was a collection unrivalled anywhere else in the country, exceeding even the ever-burgeoning Museum of the Royal College of Surgeons, but, with the failure of his practice as a teacher of anatomy, Brookes was forced to put the collection up for auction; the first sale, in 1828, lasted eighteen days, the second, in 1830, twenty-three.[217] Many of his specimens were purchased by the Museum of the Royal College of Surgeons, where some still survive.

Joshua had also a small collection of living animals, some obtained from his brothers Paul Brookes and John Brookes and probably also from his father, which he kept in a vivarium next to his house. One night in January 1792, when the nearby Pantheon went up in flames, the heat was so violent that the paint blistered on the doors and windows of his house and the terrified animals in the vivarium tried to break away from their chains. The mob assembled and threatened to pull the house down around Brookes's ears.[218] Nevertheless he continued to keep living animals, which at various times included a coypu (presented by John Hunter's brother-in-law Everard Home), a racoon, a pair of collared peccaries, a dwarf porcupine, a paca, a pair of opossums, and a condor, all from South America and probably collected by Paul. He also had a sloth bear from India, an Arctic fox, and a lark bunting (a present from Mr Kendrick, the dealer in Piccadilly), as well as a badger, a marten, and, less certainly, a Roloway monkey 'bred in England'. The remains of all these animals, apart from the condor, ended up in his museum. According to an article in the *Field*, it was popularly believed that Brookes kept several vultures 'to consume the mortal remains of those human subjects that he employed for his anatomical demonstrations'.[219]

Figure 4.11 probably relates to the Brookes's vivarium well before 1830, as in May 1826 he gave the condor (initially described as a vulture) and a white-headed eagle to the newly founded Zoological Society of London, although he offered to keep them until

Figure 4.11. *A View of the Vivarium Constructed Principally with Large Masses of the Rock of Gibraltar in the Garden of Joshua Brookes Esq.* Note the hermit's cell with a pair of whale's jaw bones outlining its frontage; inside are a human skull and an hour-glass on a table in front of a large cross (George Scharf, 1830) (Royal College of Surgeons, RCSSC/P 320a).

the zoo's establishment was ready to receive them, as it did a year or two later (see Chapter 6). The zoo keepers named the condor 'Dr Brookes', and it survived there for nearly forty years.[220] Brookes carried out autopsies on various animals on behalf of the zoo, including George IV's ostrich (Section 5.6).[221]

4.8. The Duke of Cumberland: The 'Experiment with the Hunting Tyger'

William Augustus, Duke of Cumberland and Butcher of Culloden, the fourth son of George II, was appointed Ranger of Windsor Great Park in 1746, three months after his victory at Culloden.[222]

After an unsuccessful military campaign in Flanders in 1747 he spent the next ten years in England, mostly at Windsor, improving the park, dabbling in politics, breeding horses, and building up his menagerie. In September 1749 his secretary received a letter from Colonel Lockart in Marseilles saying that 'the She Wolf' had been dispatched and in his opinion if she were mated with 'a Staunch and Large Hound, the Duke would have a Noble Pack of Fox Hounds'; he hoped that he would soon be able to send a dozen 'Partridges, & other Curiosities, for Windsor Lodge and some very fine curious fowl from Barbary'. Plans of Windsor show a falcon house at the west end of the Duke's kitchen garden. The deer in the Great Park were joined by 'buffalos' (more probably zebu cattle from India, given the loose zoological nomenclature of the time) from the Duke of Richmond's menagerie at Goodwood, and a consignment of 'wild beasts', probably lions or leopards, reached Windsor by October 1752, and the following year a newspaper reported that 'there is in his Royal Highness's Menagery on Windsor Forest, the finest Collection of foreign Animals and Birds in Europe, there being four Ostriches, several Eagles, three Lions, and three Wolves, besides a great Number of other Beasts and Birds'.

In August 1754 Horace Walpole visited Windsor and admired the Duke's 'delicious lions and tigers' (more probably leopards).[223] Two more 'tygers' arrived the following year, a present from Lord Anson. Some other animals are depicted in Thomas Sandby's drawings of the east front of Cumberland Lodge—fallow deer and ostriches feeding among the trees as well as three eagles, each with its own hutch.[224]

In June 1757 Mrs Delany paid a visit from Bulstrode:

> The menagerie is not stored with great variety, but great quantities of Indian pheasants, the gold kind, blue and white, and the common sort. The wild and foreign beasts are all sent to the Tower.
>
> A terrible accident happened not long ago—the tiger got out of his den and tore a boy of eight or nine years of age to pieces; the mother was by and ran upon the beast, and thrust her hands and arms into its very jaws to save her child, but the poor child was destroyed! Upon which accident the Duke sent them to the Tower, as the only fit place for such fell beasts. There is a dromedary—an ugly creature; it is kept in a yard by itself; it made a hideous noise and frightened the horses. I forgot to name among the birds two eagles, a young eagle of the sun (not come to its beauty) and a horned owl, that looked as wise as a judge in his robes.[225]

The Duke of Cumberland's 'experiment' with the 'hunting tyger' is well known; the animal was actually an Indian cheetah, a gift from George Pigot. But what is much less well known is the story of its voyage to England. Pigot actually brought two cheetahs; they arrived in Plymouth Sound on 6 June 1764 on board a Spanish galleon, the *Santíssima Trindada*, a prize captured from the Spanish in the Philippines, the largest and richest ship ever brought into an English port. In August 1762 the *Santíssima* left the port of Cavite, near Manila in the Philippines, and sailed across the Pacific towards Acapulco in New Spain (modern Mexico), but in October it was caught in a typhoon, which brought down two of its masts, so its captain, not realizing that in the meantime the British had captured Manila, decided to return there, and was intercepted by two English ships, HMS *Panther* and HMS *Argo*. The *Panther* opened fire, and, although it did little damage and caused few casualties, the disheartened crew of the *Santíssima* surrendered. The ship was loaded with treasure, probably porcelain, lacquerware, spices, ivory, and silk, to the value of two million dollars. It was escorted to Madras, where a 'vast collection made by Governor Pigot of foreign curiosities, particularly wild beasts' was added to the cargo. The *Santíssima* was brought to England, under convoy with two men-of-war. It was a long, slow voyage: 'It was with great difficulty they steered her, being very foul,' and most of the animals died. When the *Santíssima* finally reached Plymouth, people crowded on board to view the ship and the surviving animals, which included the two cheetahs, a huge snake, and an Indian zebu cow. The cheetahs and the cow were loaded into a wagon and sent to Windsor, intended, according to the *London Chronicle*, 'as a present to his Royal Highness the Duke of Cumberland'. 'Mr Pigot, late Governor of Madras', had arrived earlier in greater comfort in April 1764 on board the *Plassey* East Indiaman; later that year he was made a baronet.

The Duke decided to stage an experiment in order to see how cheetahs attack their prey; at noon on 30 June 1764:

> A stag was inclosed by toils [nets] in the Duke's paddock in Windsor, and one of his Highness's tygers let loose at him; the tyger attempted to seize the stag, but was beaten off by the horns; he made a second attack at the throat, and the stag toss'd him an astonishing height; a third time the tyger attempted to seize him, but the stag threw him as before, and then followed him. The tyger faced him no more, but run

> under the toils, and pursued a herd of deer, one of which he instantly killed; but while he was devouring a part of him, two Indians followed him, threw a kind of hood over his head, and then fastened a chain about his neck, let him fill his belly and led him quietly to his den.[226]

The cheetah, the two keepers who had accompanied the animals from India, and the stag are immortalized in George Stubbs's famous painting *Portrait of a Hunting Tyger* commissioned by Pigot (Plate 8).[227] The stag, however, appears to be a curious hybrid; it has a red deer's body, but its three-tined antlers resemble those of an Indian sambar.[228] Stubbs's rocky, romantic landscape bears no resemblance to Windsor Park.

One of the cheetahs' keepers, Abdullah, who used the name John Morgan, spent the night of 30 October 1764 in a tavern, or bawdy house, in Kew, where he was robbed. The next morning he sought the help of a local constable, and the inn-keepers John and Mary Ryan were arrested. They were tried at the Old Bailey on 27 February 1765. Morgan swore on the Koran that he had been forcibly robbed of 'a pair of silver shoe-buckles, value 10*s*. a pair of cotton stockings, value 1*s*. one silk purse, value 2*d*. one linen purse, value 1*d*. one piece of silver coin, called a rupee, value 2*s*. two pieces of coin called Fernams, value 1*d*, five guineas, and fourteen shillings'. He told the court that he had been born in Bengal and that this was his third visit to England. He had come over with a 'tyger' for Sir George Pigot and was still 'attending' it at Pigot's house in Soho Square. He had gone to visit a friend in Kew and, finding it late, he had looked for a lodging for the night, and was recommended to the Ryans. John and Mary Ryan were found guilty, and Morgan's property was restored to him.[229] All the newspaper accounts of the arrival of Pigot's animals mention two 'tygers', whereas those of the famous experiment mention only one, so it seems that Pigot gave one to the Duke of Cumberland and kept the other for himself. Morgan returned to India in the spring of 1765 with Pigot's other Indian servants.

A year later a cheetah, described as 'The Hunting Tyger, that killed the Deer in Windsor Forest', could be seen or sold at the Admiralty Coffee House in Whitehall, and in December it was on sale at the City Menagerie in the City Road, near Dog-House-Bar in London. The other, a female named Miss Jenny, was transferred to the Tower.[230]

4.9. Queen Charlotte: Her Elephants and her 'Painted Ass'—the Menageries at Kew and Richmond

George III ascended the throne in 1760 on the death of his grandfather George II. A year later he married a German princess—17-year old Princess Charlotte Mecklenberg-Strelitz, for whom he purchased Buckingham House (the site of today's Buckingham Palace), then known as the Queen's House, Pimlico. In July 1762 the heavily pregnant Queen received a most unusual present, a female zebra, presented by Sir Thomas Adams, Captain of the *Terpsicore* Man of War, who brought it from the Cape of Good Hope (Figure 4.12). Thousands of people came to see the 'painted ass' grazing in a paddock near the Queen's House.[231]

A year later the royal couple received, not one, but two elephants. The first arrived in June, and the King ordered it to be sent to the Tower 'to be kept as a rarity'. The second, brought from India by Brooke Samson, Captain of the *Hardwicke* East Indiaman, arrived in September.

Figure 4.12. The Queen's zebra (*The Sebra or Wild Ass*) (George Townley Stubbs, 1771, after George Stubbs) (BM 1862,1011.557).

> Capt Samson had the honour to present the elephant brought by him from Bengal to his Majesty at the Queen's House. It was conducted from Rotherhithe that morning at two o'clock and two blacks and a seaman rode on his back. In their way, their strange appearance so surprised a young woman, who happened to meet them, that she fainted away.[232]

When an elephant kept in the Queen's stables happened one morning to be walking in St James's Park, an Englishman meeting it enquired: 'Where the elephant was going?' 'No doubt it is going... to the Parliament House to get itself naturalized.'[233]

Initially tractable, by October the elephant at the Queen's house had become unmanageable. Capt Samson was sent for and managed to secure it by jumping on its back, and 'running a tuck thro' the fleshy part of his neck, after the East Indian manner... which rendered him quite flexible' (or utterly inflexible, depending on which account is to be believed). Later that month the elephant was well enough behaved to parade through the City in the Lord Mayor's Show as an emblem of the Cutlers' Company.[234] In December the three Indian keepers who had arrived with Samson's elephant and cared for it in London were granted leave to return home.

Visitors flocked to see the Queen's elephants, and complaints were made that admission to view them was granted only to those attired in court dress, though a back-hander to an officer of the 3rd Regiment of Guards would usually secure admission. That the Queen's animals were among London's tourist attractions is nicely illustrated in Smollett's *The Expedition of Humphry Clinker*—Winefred Jenkins writes to Mary Jones:

> O Molly! what shall I say of London?... there's such a power of people, going hurry skurry! Such a racket of coxes! Such a noise, and haliballoo! So many strange sites to be seen! O gracious! my poor Welsh brain has been spinning like a top ever since I came hither! And I have seen the park, and the paleass of Saint Gimses, and the king's and the queen's magisterial pursing, and the sweet young princes, and the *hillyfents*, and *pye-bald ass*, and all the rest of the royal family.[235]

More presents arrived for the Queen in May 1764—another elephant and several 'buffaloes' (more probably zebu cattle) from India, a fine male zebra, and some 'hoggs' from the Cape of Good Hope, followed in July by yet another 'fine Male Elephant, which has been presented to her Majesty and placed among the Wild Beasts behind the Queen's Palace'.

Oliver Goldsmith saw both of the Queen's zebras and described the male as 'even more vicious than the former; the keeper who shows it takes care to inform the spectators of its ungovernable nature. Upon my attempting to approach it, it seemed quite terrified, and was preparing to kick, appearing as wild as if just caught, although taken extremely young and used with the utmost indulgence . . . '.[236] The male zebra was described as 'twelve hands high, of a milk white colour streaked with black, its body and legs are finely turned; but it has long ears, and in other respects resembles an ass'.[237] The intention was to send it to the Tower, but it may have died before that could be done. In any event, the Queen presented its corpse to her 'Extra Physician', Dr William Hunter. It was stuffed, and ended its days on display in the Hunterian Museum in Glasgow, whither it was sent with the rest of William Hunter's anatomical and other collections in 1807.[238]

In November 1771 the Queen gave the female zebra to Mr Forbes, who had succeeded his father three years before as 'Keeper of the Queen's Elephant and Zebra'. Forbes sold her to John Pinchbeck, who took her off on tour, progressing from Brentford to Birmingham and showing her at various towns along the way. The advertisements announcing her progress making much play on the 'astonishing zebra . . . generally called the Queen's Ass . . . This extraordinary animal has been seen by four Crowned Heads, sixteen Princes, several Princesses, many foreign Embassadors, and more of the Nobility and Gentry . . . '. But a year later the Queen's long-suffering, long-lived female zebra died at Long Billington, near Newark. Not one to be disadvantaged by the sad event, Pinchbeck had her stuffed, stating proudly that she could be seen at closer range than when she was alive. He exhibited her at the Blue Boar in York, along with an oriental tiger, a magnanimous lion, a 'real' Bengal tiger (Marlborough's tigress (Section 4.5)), a beautiful leopard, a voracious panther, a 'man-tyger' (probably a mandrill), and a porcupine.[239] The show went on to Durham and is last heard of in March 1774 in Edinburgh at the Black Bull at the head of the Canongate. The stuffed zebra ended up in an outhouse of the Leverian Museum.[240]

In 1773 a fifth elephant arrived. Brought from Bombay in one of the East India Company's ships, it 'was conducted to the Queen's menagery; he is so tame that he follows the young man who has the care of him like a dog'. It was closely followed by a sixth, a very young female, presented by Sir Robert Barker after his retirement in July 1773 as Commander-in-Chief of the East India Company's army.[241] However, all was not well with the royal elephants. In 1775 Barker's elephant died; the Queen presented its corpse to William Hunter. It was dissected by Hunter's assistants—his brother John and William Cruikshank; Cruikshank's notes on the dissection survive in Glasgow University Library.[242]

A year later William Hunter received yet another corpse. The *St James Chronicle* reported among its obituaries: 'Tuesday evening died at Pimlico, the oldest of the Elephants belonging to his Majesty: the King made a present of it on Wednesday to Dr Hunter, for dissection, and Yesterday morning the Doctor attended by near twenty of his pupils, began to work upon the Beast. The other is to be removed to Kensington next week.'

The fate of the carcasses of three more of the Queen's elephants is known. One of largest ended up stuffed in William Hunter's collection, occupying pride of place, along with her male zebra, in the Elephant Hall of the newly built Hunterian Museum in Glasgow in 1813. John Hunter received another, parts of which may still be among the numerous bits of elephant in the Hunterian Museum in the Royal College of Surgeons in London. The third elephant, a male calf, still slightly hairy, was given to Sir Ashton Lever and displayed in his museum in Leicester Square from 1777; its portrait drawn by Sarah Stone in 1786 is in the Natural History Museum.[243]

In 1769, or shortly before, the Queen's elephants were joined by a pair of nylghai presented by Laurence Sulivan, a director of the East India Company. The Queen graciously allowed William Hunter to borrow them to keep for a while to study in his own stables in Windmill Street. As the recently appointed Professor of Anatomy at the Royal Academy and the owner of a serious collection of paintings, Hunter was well aware of the importance of accurate visual records of animals, and so he employed Stubbs to 'take its likeness'. When he presented his findings to a meeting of the Royal Society in February 1771, propped on an easel beside him was Stubbs's portrait in oils of the male nylghai.[244] The opening paragraph of the resulting paper, published in *Philosophical Transactions of the Royal Society*, outlines the two benefits that might accrue if nylghai could be acclimatized in England: 'Among the riches which of late years, have been imported from India, may be reckoned a fine animal, the Nyl-ghau; which, it is to be hoped, will now be propagated in this country, so as to become one of the most useful, or at least one of the most ornamental beasts of the field...', and 'Good paintings give much clearer ideas than descriptions. Whoever looks at the picture, which was done under my eye, by Mr Stubbs, that excellent painter of animals, can never be at a loss to know the Nyl-ghau, wherever he happens to meet with it...'. The painting is in the Hunterian Museum in Glasgow along with many other items from William Hunter's various collections.[245]

Princess Augusta, Queen Charlotte's mother-in-law, must have shown an interest in natural history from the early years of her marriage, for in 1743 Mark Catesby dedicated the second volume of his *Natural History* to her. After the death of her husband, Frederick Prince of Wales, in 1751, with Lord Bute as her adviser (and rumoured to be her lover), Augusta set about redesigning her garden at Kew, the kernel of what was to become the present Royal Botanic Gardens; 1759—the year in which an item for 'cultivating the Physic Garden' first appeared in her household accounts—is taken as the date of its establishment. Bute's intention was to create a garden that would 'contain all the plants known on Earth'. Augusta also kept animals at Kew, for as early as 1752 her household accounts mention a pheasantry and an 'antelope ground'. William Chambers was employed to build various temples, bridges, and follies scattered romantically through the gardens, including the Chinese pagoda (still extant) and an aviary, which is best described in Chambers's own words,

> continguous to the [Physic or] exotic garden is the Flower Garden... the end facing the principal entrance is occupied by an *Aviary* of a vast depth, in which is kept a numerous collection of birds, both foreign and domestic. The Parterre is divided, by walks, into a great number of beds, in which all kinds of beautiful flowers are to be seen, during the greatest part of the year; and in its center is a bason of water stocked with gold-fish.
>
> From the Flower-Garden a short winding walk leads to *The Menagerie*. It is of an oval figure: the center is occupied by a large bason of water, surrounded by a walk; and the whole is enclosed by a range of pens, in which are kept great numbers of Chinese and Tartarian pheasants, besides many sorts of other large exotic birds. The bason is stocked with such water-fowl as are too tender to live

> on the lake; and in the middle of it stands a pavilion of an irregular octagon plan, designed by me in imitation of a Chinese open Ting, and executed in the year 1766.[246]

The identity of the Chinese and Tartarian pheasants is uncertain, they are probably silver pheasants and golden pheasants in modern terminology.

In 1764 George III and Queen Charlotte took over nearby Richmond Lodge; Capability Brown was commissioned to landscape the grounds in the new picturesque style—acres of lawns swept down towards the River Thames, trees were planted in naturalistic groves, and the former straight avenues, clipped hedges, and parterres disappeared along with most of the garden buildings. The plan was to replace the old Lodge with a new more suitable palace to be designed by Chambers. While these works were going on Princess Augusta was still living at Kew, separated from Richmond Lodge estate by only by a narrow lane, but when she died in 1772 the plan was abandoned and instead the royal couple took over her estate, living in the White House (then renamed Kew House) and the nearby Dutch House (now Kew Palace), where they stayed from the middle of May to the middle of August each year; Richmond Lodge was demolished. Thus the royal estates of Richmond Lodge and Kew were united, although they remained physically divided by Love Lane until 1802.

As the botanists to Cook's first voyage of 1768–71, Joseph Banks and his friend Daniel Solander had collected a vast number of plants from most of the places where the *Endeavour* anchored. Queen Charlotte and her elder daughters took a serious interest in botany, and in about 1773 Banks, who was on good terms with the King, became involved in the development of the Royal Gardens at Kew and Richmond, acting as unofficial director. Banks held the view that scientific knowledge should be useful knowledge and was particularly concerned with the naturalization of exotic plants and the occasional animal that might be of economic benefit.

Queen Charlotte's menagerie at Richmond is first recorded in 1771 in a clearing in the woods in the grounds of Richmond Lodge. The 'new menagerie', as it was called, perhaps to distinguish it from Princess Augusta's, was a 3-acre paddock containing a circle of pheasant pens and a small house that was soon rebuilt as a two-storey, thatched *cottage orné* known as the Queen's Cottage: 'In the S.E. quarter of these enchanting grounds, a road leads to a sequestered spot, in which is a cottage, that exhibits the most elegant simplicity. Here is a collection of curious foreign and domestic beasts, as well as many rare and exotic birds. Being a favourite retreat of her Majesty's, this cottage is kept in great order and neatness.'[247] It is still there, recently restored, hidden deep in the woods. Once a week the royal children, accompanied by their parents, visited the menagerie to play with their pets and to view any new additions. It was probably there that the older children were instructed in 'practical gardening and agriculture'.

Johann Rheinhold Forster and his son were the naturalists on Cook's second voyage round the world. In July 1772 they set sail on the *Resolution*, and returned in 1775, reaching Woolwich in August. On board were several live animals listed by Daniel Solander in a letter to Joseph Banks—three Otaheite dogs, 'the ugliest and most stupid of all the canine tribe', a springbok, a suricate, several small birds and two eagles, all from the Cape of Good Hope, 'I believe Capt Forster intends them for the Queen'. The fate of the dogs from Tahiti is not known, but the springbok and the eagles were presented to the Queen; during its two years in the menagerie at Kew the springbok gave birth to a still-born calf, which may have ended up, stuffed, in William Hunter's collection.[248] The 'eagles' were seen by the ornithologist John Latham, whose description shows that were actually secretary birds.[249]

The Prince of Wales (the future George IV) is said to have given his mother an extravagant gift of exotic animals in or before 1784;[250] more certainly, in 1784, she was given a bull 'not bigger than an ass' (probably a dwarf zebu) from India—perhaps the bull she gave to John Hunter for his menagerie at Earl's Court.[251] Ray Desmond in his history of the Botanic Gardens at Kew adds birds with colourful plumage, cattle from Algeria and India, and a 'hog like a porcupine in skin; with a navel on back' (the much repeated description of a peccary, which has a scent gland on its back just above the tail). The great historian of menageries Gustave Loisel wrote that at the end of the eighteenth century there were seven large enclosures for deer in the adjacent Richmond estate and that the animals at Kew included Chinese pheasants, Tartarian pheasants, many small exotic birds, goldfish, and kangaroos.[252]

As is very well known, several kangaroos were seen by Banks and his colleagues on the eastern coast of Australia during the *Endeavour* expedition in 1770. The expedition's artist Sidney Parkinson sketched two kangaroos, Lieutenant Gore shot a small one and an

84lb male, and Banks's greyhound brought down a female. A skin and parts of the animals were brought home—Stubbs's famous picture *The Kongouruo from New Holland* was probably painted from the stuffed skin.[253] The skull of the largest, probably a great grey kangaroo, was presented to John Hunter; it was destroyed when the Royal College of Surgeons was bombed during the Second World War, but a photograph of it survives. More skins and skulls were brought to England in the intervening years, but it was not until December 1791 that the first live kangaroo arrived. It was shown first at the 'Trunk-maker's, No. 31 Haymarket' and then at the Exeter Change (Section 4.2). Four months later, when Lieutenant Henry Ball anchored his ship HMS *Supply* in Plymouth Sound, he had three live kangaroos on board, several others having been consumed on the voyage. He presented at least two of them, a male and a female, to the King. More followed—on 1 June 1793 the *Oxford Journal* reported that Arthur Phillip, the Governor of New South Wales, 'has brought home with him two of the Natives of New Holland, a man and a boy. The Atlantick has also on board four kangaroos, lively and healthy, and some other animals peculiar to that country.' Phillip presented two of the kangaroos to Banks, who 'made a present of them to his Majesty. They are now at Kew. Their food is Indian Corn.' The royal kangaroos settled down peacefully at Kew and within a few months they had begun to reproduce; by March 1795 for the very first time in England some of the females had produced young.[254]

The royal family began to spend more and more of their time at Windsor, and in 1804 it was suggested that some of Queen Charlotte's animals at Kew should be transferred to Bulstrode; two years later William Townsend Aiton was ordered to replace the paddock with a flower garden, and by 1808 the dispersal of the Queen's collection was complete.[255]

4.10. The Tower Menagerie: Enhancement and Decline

Little is known about the two Keepers who succeeded John Ellis—John Bristow, a steward to the Duke of Newcastle, in 1757, and Henry Vaughan in 1771—apart from some information about their finances. George Payne, the Keeper in 1775–1800, seems to have been well connected, for in 1783, while in office at the Tower, he was appointed 'His Majesty's Consul General in all the dominions of the Emperor of Morocco'.[256] The only names of the under-keepers at the Tower—that is, those who actually cared for the animals—that have come down to us from this period are William Cross and his future son-in-law, the young Edward Cross, both of whom went on to run their own menageries.

When Commander Keppel returned from the Mediterranean in 1751, having 'settled the peace, and ransomed the Bey's English prisoners', he brought with him two lion cubs, Zara and Caesar, as presents to George II from the Bey of Algiers. Caesar fathered Zara's twins Pompey (3) and Dido (2); Zara died soon after, but Caesar and the twins were still going strong in 1774. Caesar ended up, stuffed, in Sir Ashton Lever's museum in Leicester Square.[257] Nero, probably the most long-lived lion in the Tower, is recorded as aged 67 in 1762 and described in 1774 as 'the oldest lion in the Tower, from Gambia, very old and greatly in decline'; parts of a lion's skeleton, grossly distorted by arthritis, which are probably those of Nero, survive in the Hunterian Museum in the Royal College of Surgeons.[258]

In 1775 the *Gentleman's Magazine* noted that

> a fine young lion [Nero 2] was landed at the Tower, as a present to his Majesty, from Senegal. He was taken in the woods out of a snare, by a private soldier, who being set upon by two savages that had laid the snare, he killed them both, and brought away the game. His Majesty for his bravery, has ordered his discharge, and a pension for life of £50 a year.

This appalling attitude was not universal—in a letter to the magazine a few months later Mr D.H. protested, 'in the name of all that is sacred, is a lion's whelp an equivalent with the K. of England for the lives of two human creatures?'[259]

Although the majority of the lions were Barbary lions from North Africa, they were occasionally joined by Asiatic lions—Nero (3), 'said to be the first of its species successfully brought from the East Indies', was brought from India by the Governor of Bombay William Hornby in 1784, and a pair was recorded in 1800 as 'newly arrived from Bussora on the Gulf of Persia'.[260]

The reuse of the names of lions after death is confusing, but there is no mistaking the lioness twins, patriotically named Miss Howe and Fanny Howe, who were 'whelped on the glorious first of June

1794, and named after the gallant Admiral who conquered on that day'. Britain was at war with Revolutionary France, and the fleet commanded by Admiral Howe 'gloriously' destroyed much of the French Atlantic Fleet, which was guarding a convoy carrying relief supplies of food from the United States to the starving French; nevertheless, the convoy reached France safely.

The earliest record of tigers, *sensu stricto*, at the Tower, is a brief mention in the *St James's Chronicle* in 1776 stating that the menagerie contained 'two royal Tygers from Bengal (the only Males that were ever brought into England)'; one of these may be the tiger seen by Thomas Pennant in 1787, which had been in England for some time and whose 'ground-colour had faded into a pale sickly sandiness'.[261] More is known of two male 'Royal tigers' which both arrived in 1785 as presents for the King: the first, 'reckoned to be the most beautiful tiger in all England', arrived from Bengal in August, a gift from Lord Aylesbury; the second, 'the largest ever brought to Europe, very savage', was sent from Madras by Lord McCartney.[262] Pennant saw them both in 1787 and described one as 'young and vigorous, and almost fresh from its native woods'.[263] Thereafter in the Tower and in other menageries true tigers can usually be distinguished from the lesser felids by the epithet 'Royal'. Two more tiger cubs arrived on the *William Pitt* East Indiaman, one in 1790 and the second in 1800. The first, named Harry, a present from Evan Nepean, the Secretary to the Board of Admiralty, was exceptionally tame and survived until about 1809.[264]

'The all-black tiger, which Mr Hastings brought with him from the East Indies', as Sophie von la Roche wrote in 1786, was not a tiger: Pennant correctly recognized it as a leopard:

> In April 1787, there was a leopard, of a quite unknown species, brought from Bengal. It was wholly black, but the hair was marked, on the back, sides, and neck, with round clusters of small spots, of a glossy and the most intense black; the tail hung several inches beyond the length of the legs, and was very full of hair.

The melanistic leopard, named Jack, survived in the menagerie until at least 1800.[265]

The felid whose arrival at the Tower in 1759 attracted the most attention, even that of Horace Walpole, was the 'strange Indian

Plate 1. *The Family of Sir Thomas More* in 1526; note the monkey (lower right) (Rowland Lockey, after Holbein).

Plate 2. Hannah Twynnoy's tombstone, 1703, Malmesbury Abbey.

Plate 3. The Duke of Richmond's 'Java Hare' (agouti) (from Catesby, 1747).

Plate 4. *The Groasbeak from Gambia in Guinea*, belonging to the Duke of Chandos (from Albin, 1738).

Plate 5. Prince Frederick's 'female zebra' (quagga) in 1751 (from George Edwards, 1758–64: i).

Plate 6. One of Pidcock's 'farthing' tokens.

Plate 7. *The Duke of Richmond's First Bull Moose* (George Stubbs, 1770).

Plate 8. 'The Experiment with the Hunting Tyger' (George Stubbs, 1765); originally known as *Portrait of a Hunting Tyger*, currently named *Cheetah and Stag with two Indians* in Manchester Art Gallery).

Plate 9. The main room of Polito's *Royal Menagerie, Exeter Change, Strand* (from Ackermann, 1812).

Plate 10. The Exeter Change with Edward Cross's Royal Menagerie; note the Lyceum several doors to the right (George Cooke, *c.*1820).

Plate 11. George IV's giraffe, with two Arab attendants, a milk cow, and Edward Cross (J.-L. Agasse, 1827).

Plate 12. *The Camelopard, or a New Hobby* (William Heath, 1827: BM 1868,0808.8815).

Plate 13. Polito's Royal Menagerie; the date (*c*.1830) of this Staffordshire figure suggests that it commemorates John Polito's travelling menagerie (V&A 2006BB1940).

Plate 14. *Portrait of Edward Cross*. Cross is standing in the conservatory at the Surrey Gardens; painted by his friend J.-L. Agasse.

beast, the Shah-goest' (caracal cat) sent from Bengal by Robert Clive (Section 4.5).

At least ten bears arrived at the Tower in the reign of George III; the most long-lived was a black bear brought from New York by Captain Lee, who presented it to the Duke of York, who gave it to the King in or shortly before 1762; it was still alive nine years later. All the bears in the Tower were from North America, except for two dancing bears from Archangel and a 'fine brown bear' from Russia. It was probably this bear, recorded in 1782 and 1784, that was slaughtered on the order of the Prince of Wales after it had mauled and very nearly killed the Keeper's wife, who had inadvertently left the door of his cage open. Charles Catton Jr drew a 'yellow bear' 'from nature' at the Tower in about 1776; he described it as being perfectly tame and sociable, with thick silky fur, bright orange, of a reddish cast—it was probably a light-coloured grizzly.[266] A polar bear sent from northernmost Canada in about 1787 by Colonel Clanoe was described, incorrectly, as 'the first ever seen of this colour in England'. It survived for at least five years,[267] and a fine young Greenland bear lived in the menagerie from 1796 to about 1800, when it was replaced by a 'large Greenland Bear, and a small ant bear from Canada'.[268]

A large wolf from Saxony was housed in the Tower from about 1762 to 1778; when John Bewick visited the Tower in 1788, he wrote home to his brother Thomas in Newcastle that there were no wolves left, but a few months later two large wolves (and a large eagle) arrived from St Petersburg, and a very fine wolf belonging to the Prince of Wales, which arrived in December 1790, was probably the 'most ferocious animal, long in the Tower', sold to Stephen Polito for his travelling menagerie in about 1800.[269]

The arrival of a fine striped hyaena from the island of Salset near Bombay in 1786 received much attention from the press. Mr Gooch, first mate of the *Lord Camden* East Indiaman, presented it to the Prince of Wales— 'he is larger than a full grow mastiff, exceedingly tame, and beautifully marked, like a zebra, with stripes of black upon a white ground' and was 'the only beast of the species that has been brought over since those presented to George II about the year 1750'. 'His Highness was pleased to express much satisfaction at receiving so rare a present, and ordered great care to be taken of him in the Tower.'[270] A spotted hyaena, previously kept by John Hunter at Earl's Court, was lodged in the Tower after Hunter's death in 1793; it was so tame it would allow strangers to stroke it, but after it had been sold on to Pidcock's menagerie, it soon 'exhibited symptoms of ferocity equal to those of the most savage of his kind. He was at last killed by a tiger, the partition between whose den and his own he had torn down by the enormous strength of his jaws.' Thomas Bewick saw it in Newcastle in 1799, and it figured in the fourth edition of his *General History of Quadrupeds* (Section 4.2).[271]

One of the oddest animals to be incarcerated in the Tower was a 'Sand or Bear Hog' that arrived from Bengal in about 1788, 'a great curiosity', the first ever seen in England, having 'talons like a bear, tail like a dog, an amazing long head, and his colour dirty white, ears like two holes'. It was figured by Bewick in *Quadrupeds*:

> The Sand bear. We have given the figure of this animal, drawn from one kept at the Tower; of which we have not been able to obtain any further description than its being somewhat less than the badger, almost without hair, extremely sensible of the cold, and burrows in the sand. Its colour is yellowish-white, its eyes are small; and its head thicker than that of a common badger: Its legs are short; and on each foot there are four toes, armed with sharp white claws.

It was a hog badger from South East Asia, now on the Red List of endangered species.[272] Racoons arrived from time to time from North America and sometimes even produced young; a female that came in about 1792 was still there in 1806, though 'blind with age', but her mate had fallen prey to a hungry polar bear. A coati mundi from Honduras and a glutton arrived in 1797.[273]

The 'Eagle of the Sun taken in a French Prize by Admiral Boscawen' appears in most if not all of the editions of the *Historical Description of the Tower of London* between 1762 and 1809. This is a reference to Boscawen's victory in 1759 at the Battle of Lagos off the Mediterranean coast of Portugal, in which he destroyed much of the French fleet and returned to the Spithead bringing with him the *Modeste* and *Téméraire* as prizes, along with many hundreds of French prisoners, and, so it seems, an eagle. A note in John Hunter's manuscript *Of the Anatomy of Birds* allows it to be identified as a bald eagle.[274] A hardy bird indeed, it must have been caught in North America or the Caribbean, caged, and taken across the Atlantic, enduring the noise and chaos of a battle at sea, and finally ended up in the Tower of London, where it survived for nearly fifty years. Other raptors included eagles from Norway,

Orkney, and Russia (a present from the Czarina), as well as a horned owl from Gibraltar.[275]

In Payne's time as Keeper, the 'School of Monkies' included a female 'baboon' 'from the Brazils, or perhaps from the Cape of Good Hope, where monkeys of this kind are more frequent', a 'red-faced monkey' from the Malabar coast, and an 'ape' from the East Indies. Another primate, described as 'a curious animal of the squirrel kind brought from the island of Joanna on the Eastern Coast of Africa, remarkable for his beautiful red eyes', was a maucaco from the island of Anjouan in the channel between Madagascar and the African mainland, now considered as a variant of the mongoose-lemur of Madagascar, which does indeed have red eyes. In 1787 the school was further enlarged by two monkeys from South America—a 'musk monkey having a bag under its chin'—that is, a howler monkey—and a 'red-faced black ring tail monkey which can swing from tree to tree by its tail'; in addition there were various other monkeys 'too tedious to mention'.[276] How long these monkeys survived is uncertain, but, when Jumbo the baboon was added in 1796 or shortly before, they were still numerous enough to constitute a 'school', as depicted in Rowlandson's print *The Monkey Room* published in 1799, in which visitors gawp at monkeys huddled around a fire, with one monkey behind bars on the left and a lion in its cage just visible through the door on the right (Figure 4.13).

In 1776, a year after Payne's appointment, the *St James's Chronicle* reported that the menagerie now contained the largest and most curious collection of wild beasts that had ever been seen in the Tower. All these animals had to be fitted into the limited accommodation available in the Lion Tower and the adjacent buildings, and moreover they had to be fed, so costs must have risen dramatically, soon exceeding Payne's daily allowance of 12*s.* 6*d.* To make matters worse, early in 1783 Payne was sent to Marrakech in a fruitless attempt to formalize trading relations between Britain and Morocco. While lingering in the Sultan's extensive and beautiful palace gardens he came across a menagerie containing two chained leopards, a lion, and a lioness, which were fed for his entertainment. After an absence of about twenty-one months, he arrived back at his country house, 'Old Brooklands' in Surrey, much impoverished, as the considerable expenses of his trip were never reimbursed.[277] To make ends meet Payne sold off some of the more spectacular animals—a 'noble

Figure 4.13. *The Monkey Room at the Tower* (Thomas Rowlandson, 1799) (BM 1894,0417.30)

lion and lioness from the Tower' to William Cross in 1794. The tiger brought in the *William Pitt* in 1800, a lion, 'whelped at the Tower when the late George Payne Esq was shepherd to his Majesty's Menagerie', and the most ferocious wolf were all sold to Polito in 1800.[278]

The first and perhaps only newspaper advertisement publicizing the menagerie during Payne's tenure appeared in March 1788, only a month before the opening of Thomas Clark's rival establishment at the Exeter Change in the Strand (Section 4.2). From then on the menagerie at the Exeter Change, with its wide variety of animals, its numerous visitors and its exuberant advertising, seriously rivalled the Tower collection as a tourist attraction.

The arrival of 'a stupendous male Elephant from Calcutta' in 1795—the first to be housed in the Tower for any length of time since 1255—must have been an added burden for Payne, since it involved the construction of a new 'den', and at the same time it was reported that the menagerie was to be enlarged, 'having been found to be too contracted for the number of animals, particularly in warm weather'.[279] Although the number of animals had burgeoned, there is no indication that visitor numbers increased, so it is no surprise that by 1796 the admission charge had risen from *6d.* to *9d.* per person, and in 1800 it stood at *1s.*—the same price as at the Exeter Change, though the charge there was only *6d.* for children.

Although the Tower was a popular attraction, records of actual visits to the menagerie are few; Mrs Thrale brought her children and a guest, Count Manucci, to see the 'Lyons' in March 1776, before going on to visit Brooke's Original Menagerie in the New Road to buy some birds.[280] Thomas Bewick, who spent most of his life in Northumberland, visited London in 1776–7—even though no such visit has been recorded, it is inconceivable that the future illustrator of *A General History of Quadrupeds* failed to visit the menagerie. However, in 1788 his brother John sent him sketches of two of the lions at the Tower:

> these two sketches are as near as I can draw from nature, the old Lion [Nero?] has very little shag about him, & that of the young one is intirely caked together in flat pieces the size of the palm of your hand, which appears rather ugly, whether they are so in their wilde state I cannot say, there is no long hair at the end of the tail.[281]

In 1786 the German novelist Sophie von la Roche considered the cages to be large and light, so that she was able to get a good view of the lions, leopards, tigers, wolves, and hyaenas. She saw a young polar bear playing with a young black bear, as well as monkeys, which she had always loathed. She admired Warren Hastings's black 'tiger', but thought 'his tigery glance all the more horrible'.[282]

Four cheetahs arrived at the Tower at the end of October 1800; virtually all cheetahs brought to England after 1799 were claimed to have been brought from Tipoo's Court at Seringapatam, but in this case it was probably true, as they were presented to George III by General Harris, who had stormed Seringapatam in May 1799, when Tipu, the Sultan of Mysore, was killed. It was said that there were 300 cheetahs—often referred to as 'tygers' or 'hunting-tygers'—in a courtyard of his palace, 'which were reserved as curiosities by private Gentlemen, or servants of the Company'. The cheetahs were brought to the Tower in the care of three Indian keepers; one was 'Tippo's huntsman, an old man, who seemed so much affected at parting from his favourite beasts, that he cried bitterly'. The cheetahs soon died, and so, a month later, on 7 December 1800, did George Payne, suddenly and intestate, leaving his widow to manage the finances of the menagerie, which were not settled until 1807.[283]

Joseph Bullock was awarded a 'Patent for Life' as Keeper on 23 May 1801. Within two months, ten more of the 'curious animals' from the Tower, including a wolverine, had been sold 'at enormous expense' to Pidcock at Exeter Change, and a few years later Miss Howe the lioness and the second tiger brought to England for the Prince of Wales were sold to Polito for his travelling menagerie.

When Cape Colony was relinquished to the Dutch in 1803, General Dundas, the acting governor of the colony, returned to England, bringing with him 'a most beautiful [female] Zebra from the Cape of Good Hope', which was deposited in the Tower. Although she was very intractable, she sometimes allowed the keeper who had accompanied her from the Cape to ride on her back for 200 yards or so, before 'obliging him to dismount'.[284] This must be the zebra painted by Jacques-Laurent Agasse in 1803, as it was the only one in London at that date. The painting was originally owned by William Herring, who may have been one of the Keepers at the Tower at the time (Sections 5.1 and 5.3).

The most notable animals recorded in 1805 were four cheetahs, the second group of Tipu Sultan's animals to arrive at the Tower; they were brought by the Marquis of Wellesley, who had been General Harris's second-in-command at the capture of Seringapatam; they probably died or were sold soon after their arrival. Wellesley, the future Duke of Wellington, was to be instrumental in the demise of the menagerie later in the century (see Chapter 6). A watercolour in his collection of paintings by Indian artists in the British Library is inscribed 'Drawn from the life from a Cheeta that was found in the Palace of Tipoo Sultan at Seringapatam. 1799', but there is no reason to suppose that it was one of the four he brought to England.[285]

The other animals in the menagerie in 1805 included at least five lions, Harry the Bengal tiger, who now had a little dog as his constant companion and was so docile that his keeper could enter his den with impunity, Duchess the leopardess, a 'curious ring-tail'd Tyger from Bengal, brought by Admiral Reynier' (possibly the first record of a clouded leopard in Britain), a black leopardess named Miss Peggy from Mozambique, and 'Miss Nancy, a bright spotted leopardess'. Several other new leopards arrived, as well as a striped hyaena and a 'young wolf from Mexico, sent in a flag of truce from Admiral Masserano, in Spain, to Lord St Vincent, and by him presented to the King'. This is probably a reference to the capture by the British in October 1803 of three Spanish ships laden with treasure from the New World.[286]

Several newspapers reported the arrival in July 1809 at Plymouth of 'a beautiful male tyger, a tyger-cat, several sheep from the Cape of Good Hope, a land and sea tortoise, together with many other Oriental quadrupedes'. They were brought by Admiral Sir Edward Pellew when he gave up command of the British fleet in the 'East India Station'. However, the only animal of Pellew's recorded at the Tower was a white tiger from India, seen in 1811 by Louis Simond, a French businessman visiting London from the USA; despite the fact that he considered it 'proverbially vulgar' to visit the menagerie, he wrote that 'the tiger was so tame that the sailors had been able to pare its toenails on the voyage, and it was led through the streets of London to the Tower like a dog'. He commented that the menagerie was small, ill-contrived, and dirty and that the animals looked sick and melancholy.[287]

A famous grizzly bear named initially Martin and subsequently Old Martin arrived in 1811, a present for George III from the Hudson's Bay Company. A terrible accident occurred that year—a soldier reached into a cage to stroke a tiger's paw, the tiger somehow managed to drag him into the cage, from where he was 'rescued insensible when a man forced a stick down the animal's throat', and conveyed to a surgeon.

The menagerie at the Tower went into a sad decline during Bullock's tenure, from which it was rescued after his death in 1821 by the most enterprising of all the Keepers of his Majesty's Lions, Alfred Cops (Section 5.7).

5

George IV as Regent and King, *c.*1811–1830

5.1. The 'Original Menagerie' in the New Road: Paul Brookes and William Herring, *c.*1810–1828

Although the 'Original Menagerie' in the New Road was still owned by 'the assignees of Mr. E. Whitehead' (Section 4.1), on his return from South America Joshua Brookes's son Paul took over the management, advertising the fact on 1 May 1810[1] and issuing a charming handbill (Figure 5.1).

In December 1811, on Edward Cross's recommendation, Paul Brookes wrote to Thomas Bewick asking him to make a wood (or pewter) block that could be used to print large posters to advertise the New Road menagerie, but as far as can be ascertained nothing came of the idea.[2] In 1813, when the house, its contents, the land, and a 'choice and valuable collection of animals of the feathered and quadruped kind' were put up for sale, Paul acquired the lot.[3] He carried on with the business until he retired in 1820, when the menagerie with its 'extensive Stock of Foreign Birds, of every description, Westphalia Hogs and Pigs' was again offered for sale.

Since his first journey to South America in 1782–4, Paul had spent most of his life travelling, networking, and acquiring animals from far-flung parts of the world. This is clear from the handbill describing the revived Original Menagerie as well as from his obituary. His apparent friendship with Edward Cross suggests that he may have supplied animals to other menageries, including that of his brother John in Piccadilly, and he certainly provided animals, possibly mostly dead, to his brother Joshua, the surgeon. He died in St Petersburg in 1826 or 1827. According to his obituary he was

Figure 5.1. The 'Original Menagerie' in 1810, revived by Paul Brookes; note the similarity to the menagerie in his father's time (Figure 4.1) (anonymous handbill: Yale Center for British Art, Eph Ent 9).

> much respected by most zoologists as an indefatigable traveller in the pursuit of natural history. For the last thirty years, with the exception of two or more that he resided in the New-road London, he was engaged in zoological researches in France, Holland, Germany, Portugal, and Africa, also in North and South America. Having sold his house, he became an annual voyager to both the capitals of the Russian empire, viz. St Petersburgh and Moscow, as well as occasionally to Sweden, Lithuania, and even Lapland.[4]

The purchaser of the Original Menagerie in 1820 was Edward Cross, the owner of the menagerie at the Exeter Change (Section 5.3), but this was a temporary arrangement, as two or three years later it was being run by his former employee and brother-in-law William Herring, who advertised himself as 'successor to the late Mr P. Brookes, at his old established Menagerie, New-road, Regent's Park' and begged 'to inform the Nobility, Gentry, and Public in general' that he had pheasants and many other birds for sale, as well as foxes, deer, sporting and fancy dogs, gold and silver fish (although retired, Paul was probably abroad, certainly not dead).

Before moving to the New Road, Herring had probably been employed first at the Tower menagerie, as it is known that he owned the picture of a zebra painted there by Jacques-Laurent Agasse in 1803,[5] and then by Cross at the Exeter Change, as it is on record that by the time he moved to the New Road in 1822 the elephant Chunee at the Change had developed a particular dislike for him. Four years later, described as 'Mr Herring of the New-Road', he was involved in the famous episode in which Chunee went berserk (Section 5.3).[6]

Herring continued to supply noblemen and gentlemen with animals needed for shooting and hunting, even occasionally including red deer. His menagerie in the New Road was conveniently near to the newly founded Zoological Gardens, which the Zoological Society's authorities were anxious to stock with animals (Section 6.1). Before the official opening by the Duke of Wellington on 27 April 1828, Herring sold them several pairs of birds, plus four more in exchange for an Eskimo dog, and later in the year he exchanged two monkeys for some of their Chinese geese.[7] At about the same date he had 500 fallow deer for sale, with 'Caravans for removing them to all parts of England with the utmost care'. He imported red and fallow deer for Queen Victoria and was known to Charles Darwin. Herring, or more probably a relative, Thomas Herring, continued in business until 1876.[8]

Although Philip Castang, Paul Brookes's brother-in-law, was declared bankrupt in 1813, he continued to sell pheasants, 'large and small Water Fowl from Holland', and a few Chinese pigs and foxes at his menagerie in the Hampstead Road, a few hundred yards from the Original Menagerie. The sale of foxes is of particular interest, since it implies that, by the early nineteenth century, far from exterminating a predatory pest, the hunting fraternity was running short of foxes for the hunt.[9]

Castang specialized in breeding pheasants. Bonington Moubray, in his *Practical Treatise on Breeding, Rearing, and Fattening All Kinds of Domestic Poultry, Pheasants, Pigeons, and Rabbits*

(1819), referred to the 'intelligent and experienced' Mr Castang; the book includes a chapter written by Castang himself on the rearing of pheasants.[10] One of his sons sold poultry in Leadenhall Market and was, 'By Appointment to Her Majesty the Queen, His Royal Highness Prince Albert, H. H. Abbia Pasha, the Viceroy of Egypt, the King of Portugal, PURVEYOR of ORNAMENTAL WATER-FOWL, POULTRY, and PHEASANTS'. Like Herring, he was known to Darwin.[11] Other scions of the same family were involved in circuses in Europe and the USA until the 1930s, and today, many generations after Joshua Brookes's daughter Sarah married Philip Castang, Philip Castang Ltd sells pet food in Potter's Bar in Hertfordshire.[12]

5.2. The Exeter Change 2: Stephen Polito, 1810–1814

Three days after taking over the menagerie premises at the Exeter Change and acquiring most of the animals auctioned at Gilbert Pidcock's posthumous sale on 21 March 1810 (Section 4.2), Stephen Polito, whom we met in Section 4.3 as the owner of a travelling menagerie, advertised the fact that the menagerie remained open, and that the premises were undergoing a thorough clean, adding that in a short while he would add other curious subjects from his own 'well known and valuable Menagerie'. He also had to dispel the rumour that his entire travelling menagerie had been lost while attempting to cross the Irish Sea, and announced that the animals in the Exeter Change could be viewed, 'in healthy, beautiful and compact order'. Agasse returned to the Exeter Change and painted a huge picture of a tiger (7 ft × 6 ft), perhaps commissioned by Polito to decorate the menagerie.[13]

Once the improvements at the Exeter Change had been completed, Polito was able to concentrate on the acquisition of new stock, including the essential pair of lions, replacements for those that he had failed to secure at Pidcock's sale. In July he acquired a new and most spectacular animal, an infant Indian rhinoceros from the Malabar Coast, as well as a zebra and two new ostriches, probably at the great Mart in Portsmouth. These prize animals were quickly transported back to the Exeter Change, where, along with the animals purchased at Pidcock's sale, they were displayed in 'three compartments, constantly aired with a patent retort stove, that not only affords considerable heat, but admits a circulation of pure air, which totally prevents any unpleasant odour'.[14]

Even though Polito had now acquired a lucrative London showroom, unlike Pidcock he had no intention of abandoning his travelling menagerie, and to a large extent the two operations functioned separately. While the improvements at the Exeter Change were taking place, Polito's original menagerie, including the elephant landed from the *Winchelsea*, the lion from Senegal, the lioness from the Tower (Miss Howe), the ursine sloth, the black swan, and an emu, was travelling in the south of England. In June they were in Oxford, where a snide academic wrote:

> Likewise, the wonderful Signor Polito, who has come here (*h*as he says) in consequence of the vacancy of the Natural Philosophy Chair, to exhibit and lecture on the qualities of a number of inhabitants of Asia, Africa, and America, yclept wild beasts; among which is a beautiful elephant, which has come all the way on purpose, with his travelling trunk.

Some interchange of animals between the two collections did take place, as it seems that, after the travelling menagerie had appeared at Cheltenham in August, the elephant from the *Winchelsea* was transported back to London, and replaced by another, a male landed from the *Walthamstow* East Indiaman, which had arrived in England from Ceylon in June. It was this animal that was taken north to Edinburgh, where it proved so popular that its stay was extended from early January 1811 until March, marred only when 'Mr Polito's private apartment in the menagerie' was broken into and about £100 stolen. The menagerie was not only highly visible on the 'Earthen Mound'; it was also very audible, whenever the lion roared, the male jackal in a neighbouring cage responded with a penetrating howl, and the strange concert could be heard along the whole length of Prince's Street.[15]

While the travelling menagerie was on its the way to Edinburgh, Polito's staff were busy at the Exeter Change. After Bartholomew Fair in September, to which they took a selection of animals, including the prize rhinoceros and the zebra, they had to prepare transparencies for the celebrations marking the fiftieth anniversary of George III's accession. As on similar occasions, the Jubilee on 27 October was to be marked all over town by illuminations of 'flags, transparencies, and devices', but this time there was trouble.

By nine o'clock in the evening the illuminations had still not been lit, and a posse of boys and pickpockets ran down Catherine Street, next to the Exeter Change, into the Strand, breaking every unlit window. The lamps, including Polito's, were hastily lit, but huge crowds had gathered, blocking the road, and throwing fireworks, causing much annoyance to 'passengers, especially those in carriages', frightening the horses and causing general mayhem; several people were badly hurt—those most active in fomenting the disorder were arrested and taken to Bow Street.

In the spring of 1811 a llama and a vicuna arrived from the 'Land of Silver'—that is, Peru—joining 'upwards of three hundred of the most rare and beautiful quadrupeds and birds... assembled at the Exeter Change'. In the summer, having 'at great expense' acquired some new caravans, Polito took a selection of the animals, including the rhinoceros, the zebra, and a newly acquired alpaca, to Portsmouth for the Mart in July. He was probably only too well aware of the fate that had met Pidcock's rhinoceros at the Mart in 1793, but all went well, except for the death of one of the ostriches, for whose carcass William Clift, the Conservator of the Museum at the Royal College of Surgeons, paid 'Polito's man' 5 guineas.[16]

Polito's main travelling menagerie left Edinburgh in March 1811 and, after visiting Glasgow, travelled south to Leeds. The newspaper advertisements stated that the elephant that had arrived in the *Walthamstow* was 'absolutely the only male one at the present time in Europe', which suggests that the *Winchelsea* elephant at the Exeter Change had either died or been sold. The advertisements included a print of the small woodcut of a lion that Polito had commissioned from Bewick in 1808. It was used again when the menagerie reached Castle Hill in Norwich in August. After visiting Yarmouth towards the end of August, the travellers probably came to London, for, unusually, Polito hired two pitches at Bartholomew Fair. Thomas Miles's menagerie (Section 4.2) was also there; a partially illegible account relates how the roars of the inhabitants of Polito's menagerie mingled with those of Miles's 'rugged Russian bear', lending their aid to an infernal cacophony of drums, trumpets, horns, gongs, tambourines, flutes, and clarinets, 'as if all hell, with congregated might, had rushed rebellious to the realms of light'.

After the fair the travelling menagerie was off again, this time with the rhinoceros and the zebra on a long tour of the south-west of England. When they reached Exeter in October, they were not alone; Miles's 'Grand Menagerie of Living Curiosities' was also in town. Both menageries encamped on Southernhay Green, Miles on the higher part, Polito near the hospital. Polito's large advertisement appeared in *Trewman's Exeter Flying Post* next to an even larger one of Miles's that was decorated with symbols of Masonry, while both advertisements were crowned with the royal coat-of-arms. One can imagine much antagonism. Both menageries included elephants, Polito's the 'only male that travels', while Miles's noble elephant (probably the one he had acquired by private contract from Pidcock's sale) was 'the largest ever seen in Great Britain, and drawn by eight horses'. Each had a zebra, but Miles boasted an even rarer equid from Africa, a quagga, which he claimed untruthfully was the first ever seen in the kingdom. Unable to match Polito's rhinoceros or his noble lion and lioness, Miles countered with a 'sea lion taken on a mountain of ice, 300 miles from any land, and brought to North Shields on board the *Isabella*'. Miles's noble panther was the largest ever seen in England, Polito's merely noble. Miles's kangaroos had the edge on Polito's pair with several young, 'which come out and retire into the cavity or pouch at pleasure', and he had a 'tyger-cat'—that is, a cheetah—described as always as 'from Seringapatam'. Miles's Bengal tiger was 'the largest and most stately'; Polito's was 'never excelled for size and beauty' and was, moreover, accompanied by a tigress. Polito outdid Miles in the matter of birds, with a real African male ostrich, an emu, a cassowary, a black swan, and two adjutant cranes to Miles's single condor. But Miles had two 'real Satyrs' (cebid monkeys?), 'six feet high, from the interior of South America'. Miles charged ladies and gentlemen 1*s.* admission, servants and others 6*d.*; Polito, more upmarket, ladies and gentlemen 2*s.*, tradesmen etc. 1*s.*[17]

Polito scored over Miles the following week with a repeat of his advertisement and an extra puff, which nicely evokes the glamour of the menagerie at night:

> On the evenings of Monday and yesterday in particular, we never witnessed such vast crowds going and coming from this truly gratifying collection. The interior as well as the exterior parts of the caravans were fancifully illuminated; in front was a brilliant display of variegated lamps in different forms; and in the space encircled by the carriages, were upwards of one hundred large mould candles, whereby the numerous spectators could distinctly discern the vast

assemblage of the most beautiful animals and birds ever introduced into this part of the country...

Miles responded furiously and issued a large handbill:

T. Miles... notwithstanding the Opposition and groundless reports of other Proprietors of Wild Beasts, is proud to assert with the strictest veracity that his Collection has to boast of the most *extraordinary Productions of Nature*, which has cost him many Thousands of Pounds to procure: and having lately succeeded beyond his most sanguine expectations in obtaining such wonderful admirable Objects, in addition to his original Collection; he has purchased at an enormous Expense, A MOST NOBLE ELEPHANT...[18]

Miles's additional objects were the animals that he had purchased from Pidcock in 1808 and at the auction in 1810, but he no longer mentioned the Change in his publicity, presumably because Polito was now able to advertise himself as the 'Sole Proprietor of the Royal Menagerie at the Exeter Change'.

Having spent the winter touring through Devon and Cornwall as far as Penzance, by 7 May 1812 Polito's menagerie was back in Exeter. At that point at least some of the animals were taken to London, for in June the great African ostrich (11 feet high) died at the Exeter Change. Its corpse was sent to Mr Bullock's Museum in Piccadilly; while the bird was being prepared to be stuffed for display, the viscera were removed and sent to William Clift at the Royal College of Surgeons, who noted that, despite its great size and having 'the finest plumage in nature', the poor bird was tubercular.[19]

While the travelling menagerie was negotiating the tricky roads of south-west England, the precious rhinoceros was on its way back to the Exeter Change. It arrived in December 1811 in time for the 'Season', now 'wonderfully improved in size and condition... full ten feet in length, and secured in a new apartment built for him while he was away'. It was perhaps in this season that the famous actor John Philip Kemble decided that he simply had to have a ride on the rhinoceros. The story goes that Kemble, very drunk, found himself alone in the Strand one morning at 4.15, when the roar of a lion drew his attention towards the menagerie. With some difficulty he managed to ring the doorbell and summon a keeper, who appeared bleary-eyed, dressed in a Beefeater's coat and a Welsh wig. Kemble demanded to be introduced to the rhinoceros; when the keeper demurred, Kemble insisted that he rouse Mr Polito, who duly appeared. The following conversation ensued:

KEMBLE. Mr Polito I presume? [Polito bowed.]You know me I presume?

POLITO. Very well, sir. You are Mr Kemble, of Drury Lane Theatre.

KEMBLE. Right, good Polito! Sir, I am seized with an unaccountable, an uncontrollable fancy. You have a rhinoceros?

POLITO. Yes, sir.

KEMBLE. My desire is to have a ride upon his back.

POLITO. Mr Kemble, you astonish me!

KEMBLE. I mean to astonish the whole world. I intend to ride your rhinoceros up Southampton Street to Covent Garden Market!

KEMBLE. It's next to an impossibility, Mr. Kemble.

Realizing that Kemble was not to be dissuaded, and sweetened by the offer of 10 guineas, Polito finally agreed, and told the keeper to bring the rhinoceros down. Kemble mounted on its back and the keeper gravely led the animal up Southampton Street. Daylight had broken and they were greeted at Covent Garden by a cheering mob. Kemble dismounted, tipped the keeper a crown and staggered off, supported by a friend who happened to be passing.[20]

Another story is that of *Mr Polito's Portmanteau*:

> Mr Erskine was one day pressed in defending the proprietors of a stage coach, in an action on the case for negligence, in losing Mr Polito's portmanteau. The plaintiff had sat in front of the coach, with his luggage on the top. 'Why did he not', said the witty counsel, 'take a lesson from his own sagacious elephant, and travel with his trunk before him?'[21]

That winter Polito acquired his most famous animal, whose name would later become synonymous with the Exeter Change—a male elephant aged about 6 years. Chunee (though he had not yet acquired the name) had arrived from Bengal in July 1811 on board the *Astel* East Indiaman and was purchased by Mr Harris, the manager of Covent Garden Theatre, for £800. He was initially housed at Deptford, where six Indian lascars (seamen employed by the East India Company) were engaged to care and train him. Chunee's uneven temper soon became manifest, for one evening in late November, when he was being walked from Deptford to Covent Garden, the lascar leading him was knocked down by a drunken spectator; Chunee attacked the drunkard, who fled, pursued by Chunee at full gallop. After the drunkard had reached safety in a ham shop, the lascar managed to calm the furious animal, and 'the procession moved on amid the acclamations of the multitude'. Thereafter Chunee's activities became the talk of the town and were reported on almost daily in the London papers, which were soon able to announce that 'the elephant, whose engagement at Covent Garden has excited so much waggery... is to be introduced in a pantomime'.

On 26 December, after some weeks of rehearsals, when he was found to be worryingly restive on stage, Chunee made his debut in the pantomime *Harlequin & Padmanaba, or the Golden Fish*. Led by a lascar, and with Mrs Parker and her father on his back, Chunee advanced downstage, but the shouts and cheers of the over-enthusiastic audience unnerved him, and he fled. Nevertheless the papers reported that, 'upon the whole, his debut was favourable'. On the second night Chunee was induced to kneel briefly at the front of the stage, but he suddenly rose up, gave a hideous roar, and made a rapid exit. Realizing that Chunee was too much of a liability, Parker (or perhaps Harris) sold him to Polito.[22]

Wishing to capitalize on his new and expensive acquisition, Polito quickly hired Chunee out to Astley's 'New Pavilion Theatre' in Newcastle Street, only a short distance from the Exeter Change along the Strand. Chunee was to appear in 'an entirely new Equestrian and Pedestrian Legendary Melo-Dramatic Spectacle' along with a cast that included Circassians, slaves, tartars, Chinese, Mongols, Russian, Turks, etc., as well as horses. 'In the course of the Spectacle various splendid processions, introducing the largest and most sagacious Elephant in this kingdom, the property of Mr Polito, whom Mr Astley, jun. has prevailed on at a very great expense to suffer this favourite and most stupendous animal to appear in a public theatre.' After several very successful performances at the New Pavilion, Chunee reappeared for two nights at Covent Garden, and then returned to the Exeter Change, performing almost every evening to packed houses at the New Pavilion, billed, until the end of the season on 21 March, as 'the largest and most sagacious elephant in this kingdom'.

The text accompanying the famous engraving of the main room of Polito's menagerie on the first floor of the Exeter Change, published by Ackermann in July 1812 (Plate 9), states that the image was drawn with a certain amount of artistic licence. The elephant was actually housed in a different room, through the door on the right between the cages, and the cage that it occupies in the picture was divided horizontally into two, with a hyaena below and a young lion above. Note the keeper behind the safety barrier, explaining the animals, with the tapir in the cage above him. The other animals in this room were a leopard, panther, a 'real jaguar or tiger cat recently imported from Amboyna' (which later joined the travelling menagerie), a male and a female ursine sloth, Pidcock's 'great Egyptian camel', and many kinds of monkeys, parakeets, and so on. In the third apartment, on the floor above the other two, were the large birds, including an ostrich, cassowary, a pelican, and two emus, as well as Pidcock's female elk (that is, a North America red deer), two pairs of kangaroos, including a female with a young one in her pouch, a nilghai, and, of course, the skeleton of the whale, 60 feet long.[23]

In September 1812 Polito took a selection of his animals to Bartholomew Fair; as ever Miles was also there. The animals, or at least the elephants, numerous 'enough to equip a Nabob', were shown for three days instead of the usual one. The travelling menagerie then set off on a tour of East Anglia, leaving Polito, who was seriously ill, in London, so seriously ill that he wrote his will. He was attended by John Hunter's brother-in-law, the surgeon Sir Everard Home, who had overall charge of Hunter's collection at the Museum of the Royal College of Surgeons; in lieu of payment Home received the corpse of one of Polito's adjutant birds for the museum.[24]

Edward Cross, who had left Polito's employment at the Exeter Change in 1811, was now persuaded to return. It seems that he had left in order to work on his own account, probably because of his marriage the previous year: on 14 October 1810, Edward Cross 'late keeper to his Majesty's menagerie in the Tower, but now at "Exeter Change"', married Maria Ann Cross, the eldest daughter of William Cross, of Bow'. William Cross, as we have seen in Section 4.3, had been the proprietor of his own rival travelling menagerie. The coincidence of surnames may have been just that, but it is also possible that they were related in some way.[25] Edward acquired his own 'Exhibition of Peruvian Animals'—that is, a llama and an alpaca, which had been shown in the spring in Piccadilly; in December he exhibited them in Oxford, and the same animals took part in the 'Grand Parade' at the Mart in Portsmouth the following July; when not travelling they could be seen at his house in Brook Street in London. In his insurance policy of 27 July 1812 he is described as 'Edward Cross, 1 Little Brook Street Grosvenor Square, gent.... dealer in birds'. One of his birds, a mysteriously named 'African cassowary', died in November and was sold to the Royal College of Surgeons for 2 guineas.[26]

Polito recovered and in December was able to advertise that crowds were flocking to the menagerie to see the stupendous theatrical elephant, the immense rhinoceros, the majestic lion, a tiger, a nylghai, and a llama, 'combined with every description of rare and curious birds, from all parts of the globe... particularly worthy of the attention of the Juvenile World'. A few months later he published a print of Samuel Howitt's *Taken from Life, amongst the Numerous Assemblage at the Royal Menagerie, Exeter Exchange*, which shows many of the animals, including the tapir and the ursine sloth in an imaginary rural landscape. Polito was certainly in town in April 1813, when he and his daughter Sophia signed the register as witnesses at the marriage of his other daughter Rebecca to John Huntington at the local church, St Mary le Strand. In the autumn the menagerie at the Exeter Change received its most notable visitor, the poet Lord Byron, who wrote in his journal on 14 November 1813:

> Two nights ago I saw the tigers sup at Exeter 'Change. Except Veli Pacha's lion in the Morea,—who followed the Arab keeper like a dog,—the fondness of the hyaena for her keeper amused me most. Such a conversazione!—There was a 'hippopotamus', like Lord L.L in the face; and the 'Ursine Sloth' hath the very voice and manner of my valet—but the tiger talked too much. The elephant took and gave me my money again—took off my hat—opened a door—trunked a whip—and behaved so well, that I wish he was my butler. The handsomest animal on earth is one of the panthers; but the poor antelopes were dead. I should hate to see one *here*:—the sight of the *camel* made me pine again for Asia Minor.[27]

Byron's hippopotamus was the tapir, and Lord L. the Prime Minister Lord Liverpool; from this account it seems that the Exeter Change had perhaps acquired a new sloth bear.

Early in 1813 Polito took, or sent, his travelling menagerie through East Anglia to Lincolnshire, before returning south for a visit to the great Mart in Portsmouth in July, where it was enhanced by the addition of a nylghai. The animals were then toured around Hampshire before being shown to Queen Charlotte, the royal princesses, and Princess Charlotte at Windsor. After visiting Leeds

Fair and Liverpool in the winter of 1813–14, the menagerie reached Manchester, and there, on 18 April 1814, disaster struck:

> Died. On Monday, at Manchester, aged 50 years, Mr Stephen Polito, of Exeter 'Change, London, the worthy and well known proprietor of the celebrated Menageries, that have in this, and various other parts of the kingdom, afforded gratification to the curious, and instruction to the naturalist.

Sarah Polito carried gamely on without her husband. She took the menagerie back to Liverpool, where she donated a pound 'part of the profits arising from the exhibition of wild beasts' to the Liverpool Infirmary and then went on to Northampton, where a large advertisement of 25 June includes the words 'S. Polito has brought the Whole of *her* Travelling Collection . . .'.

While Sarah was in the north, activities at the Exeter Change continued much as before, probably under the management of Edward Cross, though it was still referred to as Polito's. The Napoleonic War had at last come to an apparent end, and on 10 June 1814 the menagerie, like almost every other building in central London, was ablaze with colour and light—the 'Illuminations in Honour of the Peace'. Polito's menagerie 'made a grand and whimsical appearance; it was ornamented with a superb plume of the Prince's feathers, diamonds and festoons of coloured lamps, the word "Peace" in large letters; the whole enclosed within an illuminated arch, ornamented with laurel, while the transparencies of Louis, tigers, elephants, ostriches, etc. gave it a novel appearance'.[28] Foreign statesmen and aristocrats poured into London for the celebrations. Prince Metternich, the Prince of Bavaria, the Duchess of Lagau, and 'a great many other foreigners of distinction', took time off from the round of grand dinners, soirées, balls, and processions to visit the Exeter Change and admired, of course, the 'stupendous rhinoceros', the 'scientific elephant', the 'majestic lion', the 'royal tyger', 'noble panther', nylghau, quagga, and the black swan.[29]

Sarah must have returned to London during the festivities, for only three months after her husband's demise, she became ill; she died on 28 July and was buried at the church of St Andrew in Holborn. It was reported that, after her death, a lion, a tiger, and a wolf died in quick succession. Although Polito had stipulated in his will, proved on 9 May 1814,[30] that the collection was to be sold for her benefit, there is no evidence that this actually happened, even though the insurance policy dated 20 July covering household goods and animals is in her name, for in her own short will Sarah left just £300 to two of her daughters, Sophia and Rebecca.[31]

Edward Cross took charge of the menagerie at the Exeter Change (Section 5.3), but the travelling menagerie was acquired by *John* Polito (Section 5.4).

5.3. The Exeter Change 3: Edward Cross, 1814–1829

As we have seen, Edward Cross was already managing the menagerie at the Exeter Change at the time of Stephen Polito's death in April 1814, but it was not until Sarah Polito's death three months later that he became its owner, while the mysterious John Polito took over the travelling side of the business.

Towards the end of June 1814, while Sarah was in Northampton, the menagerie's most valuable asset, the rhinoceros, was sold. Having failed fourteen years earlier to obtain Pidcock's rhino, Antonio Alpi had reappeared in London and had purchased the rhino, no doubt at vast expense, even though the animal was little more than a calf. It was shipped across the North Sea to Rotterdam, where it was disembarked on 27 June 1814 and then toured around Europe for ten years before being exhibited by Madame Tourniaire in Amterdam. Having survived travelling for quarter of a century, Tourniaire's famous rhino died of the cold in Königsberg in Prussia in 1839.[32]

Activities at the Exeter Change were muted in the early days, with very few advertisements appearing in the newspapers, perhaps because of a shortage of cash; hence the sale of the rhinoceros. Odd cuttings and receipts in various archives confirm that the menagerie was still in business and that Alfred Cops, the future keeper of the animals at the Tower, was working for Cross. However, criticisms prompted by a fire in a pub behind the Exeter Change in May 1816 raised concerns about the very existence of a menagerie in the heart of London. The possibility that the fire might have spread to the menagerie and released the large cats meant that the staff had had to consider slaughtering them, but fortunately it was brought under control and the animals saved. The *Morning Chronicle* voiced the opinion of many: 'Certainly a more improper place as a receptacle for wild beasts could not be chosen. Exeter Change itself is a nuisance, and it is rendered more so by the incessant roar and stench of these animals.' Cross responded:

> Having read a learned opinion of the Editors of the Chronicle, respecting the Exeter Change and its inhabitants, I think it requisite to state my humble opinion, with respect to that concern in the same street, viz, the newspaper called the Morning Chronicle, is a greater nuisance (politically speaking) to England than all the Wild Beasts in the country. I am Sir, with respect, your most obedient, humble servant, Edward Cross.[33]

Nevertheless, business at the menagerie in Exeter Change began to look up. In February 1816 Cross offered two kangaroos, a pair of emus, a black swan, and a bleeding-heart pigeon from the Philippines for sale to the 'Nobility and Gentry'. When Agasse (whose portraits of a tiger and of a group of lions had probably been painted there in 1814) took his friend Adam Wolfgang Töpffer to visit, Töpffer wrote enthusiastically, if somewhat inaccurately to his wife,

> His most intimate friends are those whose job it is to show animals to the public, he is welcome everywhere and these people consider him to be a great authority. He took me to a beautiful menagerie where there were about a dozen lions, lion cubs, some tigers, young elephants, and, believe it or not, a whale. There I saw a lamb in the same cage as a very large lion, and this lion, being one of Agasse's favourite friends, he ran to embrace him. He is very attached to a leopard too but their friendship is slightly tinged with mistrust.[34]

The whale must have been the 60-foot-long skeleton that Pidcock had acquired many years before, still hanging from the ceiling in the upper room. It is one of the attractions in a poster probably issued in 1816 or 1817,[35] in which Cross states that, as successor to the late S. Polito, he has substantially improved the menagerie and adds that he himself has been superintendent of the menagerie for 'upwards of twenty-five years'. New additions include a margay cat, 'two great white polars, generally exhibited under the title of sea lions, the most tremendous animals of the frozen ocean' (polar bears?), a quagga from the Cape of Good Hope, a magot (Barbary ape), a capybara, a famous samboo (deer) from the Coromandel coast, a lyrate antelope from Senegal, a water buffalo from Bombay (presumably domestic), and the mysteriously named 'Newtre, from Buenos Ayres'. Having grown to nearly 10 feet in height and over

4 tons in weight, the elephant Chunee (though the name was not yet in use) is now the single occupant of the apartment previously shared with two camels.

One of the Exeter Change's most famous animals arrived in England in August 1817, a female orang-utan from Java. She was brought by Dr Clarke Abel, the medical officer and naturalist to Lord Amherst's unsuccessful mission to China—Britain's second attempt to establish diplomatic relations with that country. On the return journey the ship carrying the expedition was wrecked in the Sunda Strait, and, although everyone was eventually rescued and taken to Batavia, most of the specimens of plants and minerals collected by the scientists on the expedition were lost. While awaiting the replacement ship, the *Caesar*, in Batavia, Abel acquired a young female orang-utan, which made a nest by intertwining small branches and covering them with leaves in a tamarind tree near his lodging. Once on board the *Caesar* she was given the freedom of the ship, sometimes being chased by the sailors through the rigging. She usually slept at the masthead wrapped in a sail, but off the Cape of Good Hope she began to suffer from the cold, especially early in the morning, when she would descend from the mast, shuddering with cold, and running to any of her friends, would climb into their arms, clasping them closely to warm herself. In Batavia she had lived mainly on fruit, but on board ship she ate all kinds of meat, especially raw, and was fond of bread, although she always preferred fruit if available. She preferred tea and coffee to water and would even quaff wine, sometimes stealing the Captain's brandy bottle.[36] On arrival the orang was taken to the Exeter Change, and Cross was able to announce the good news, adding a genuflection to the 'Great Chain of Being':

> A gentleman in the suite of Lord Amherst has brought home with him a fine individual of that wonderful species of animal the Ourang Outang, or what is vulgarly called the 'wild man of the wood'. The formation, habits, and motions are so much the apparent result of thought, that it may truly be said to be the connecting link in the chain of nature uniting the human species with the brute creation.[37]

The orang was sometimes taken on outings in a coach with her keeper, dressed in a beribboned smock and a hat. On one occasion they called at an inn, where, without noticing that the keeper's 'friend' was a biped of a different race, the landlord served each with a glass of liquor and was astonished when he noticed the orang's 'rufous hairiness' and the length of her fingers. The poor orang did not end well; a rich diet, beer to drink and lack of exercise combined to make her 'daily more corpulent, and on Sunday [she] displayed great symptoms of plethora and inclination towards apoplexy'. Sir Everard Home administered the most common treatment for almost all human ills at that time—being bled, copiously, from the temporal artery. The orang's recovery was duly noted in the press, where it was stated that she had been visited 'by nearly all the faculty and fashionable in town'. England's first live orang-utan is said to have died on 1 April 1819. However, Agasse's cousin André Gosse, while on holiday in England, recorded a visit in 18 June to the menagerie of Mr Kross [*sic*] where Agasse was drawing an orang-utan, and, on 9 September, wrote that the little orang utan 'went better' implying that even she had been ill she was still alive; these dates accord with the statement by Edward Donovan that she died early in 1820. Notwithstanding the confusion about dates, a life-size painting by Agasse in Geneva is entitled *L'Orang Outang Joko* (Londres 1819).[38] Her body was taken to the Museum of the Royal College of Surgeons, where her stomach was preserved, her skeleton mounted, and her skin stuffed.[39] Although Abel was the first Western scientist to report the presence of the orang-utan on the island of Sumatra—*Pongo abelii* (Lesson 1827) is named for him—the orang he brought to England was a different species, *P. pygmaeus* from the highlands of Borneo.

In July 1819 another rare anthropoid ape arrived at the Exeter Change, a small black chimpanzee from the Gold Coast. Described as extremely mild and tractable, it must have been ill, for its body soon followed that of the orang-utan to the Royal College of Surgeons, where its skeleton was mounted and its skin stuffed.[40]

A report in the *Morning Chronicle* in 1819, written, one suspects, in Cross's own words, is of interest, not only because it records the arrival of an immense snake, but also because it suggests that the menagerie was subject to regular inspections by officials from the parish:

> MONSTROUS JAVA SERPENT—Wednesday, agreeable to custom, the Gentlemen composing the Parochial Court Leet Jury, with their Beadles in full dress, paid their quarterly visit to the Royal Menagerie Exeter Change, when the Foreman, Mr Burgess and Jury, were pleased to signify their entire approbation with the security, cleanliness and order, with which the different apartments are kept. During their stay (and in the presence of Sir Wm Scott) they were

spectators of that wonderful of the Snake Tribe, the Boa Constrictor, swallowing a whole fowl, and several fowls' heads by way of dessert.

In the autumn of 1819 Cross expanded his business by purchasing the livestock and premises of Kendrick's menagerie in Piccadilly opposite St James's Church, which specialized in large reptiles and birds. But two years later he transferred the birds to the Original Menagerie in the New Road (modern-day Euston Road), which he had purchased from Paul Brookes and which was soon managed by Cross's brother-in-law, William Herring (Section 5.1).[41]

Cross wrote a detailed guide, *The Companion to the Royal Menagerie*, in 1820. The elephant in the Great Room was now 10 feet high and weighed about 5 tons: 'familiarly speaking, he may be called an animated mountain.' A lioness had given birth to eleven cubs in four litters, but the only cub to have survived teething was now 'perfectly tame and fondled in the arms of visitors'.

> The lion alluded to, which is the larger one, was found, when a cub, wandering near the sea coast of the Cape of Good Hope, by a Hottentot woman, who immediately conveyed the helpless little stranger to her hut, cherished it, and actually suckled the animal from her own breast, until with her care and attention, the animal got over the weaknesses of the first months, and finally grew large and strong...

The cheetah, as so often, was said to be from Seringapatam, though Cross does not go as far as to claim that it had belonged Tipoo Sultan. One of the lynxes arrived from 'Lebida, sea-port of Tripoli, whence it was shipped at the same time, and in the same vessel, as the head of Memnon, now in the British Museum'. Cross makes surprisingly little of the polar bear or his pair of beavers, whereas 'the hunch of the Zebu eats like ox tongue'.[42]

The animals at the Exeter Change came from many different parts of the world—those that are recorded are Holland, France, Denmark, Siberia, India, Persia, Egypt, Libya(?), Algeria, Senegal, Ghana, 'Negro Land', Abyssinia, Madagascar, southern Africa, Brazil, Peru, South America, North America, Canada, the Arctic, Australia, the East Indies (usually meaning India), Java, 'islands in the South Seas', and even 'Lakes in Tartary'. Native people brought large numbers of animals to seaports for sale to foreigners, who shipped them home in the hope of making a quick buck when they landed in

England. For example, when Mrs Bowdich returned to England from the Gambia in 1818, there were 300 caged parrots on board, presumably intended for sale, although in this instance all the birds had died on the voyage. Some of Cross's staff were employed to scout along the quaysides lining the Thames and its estuary to bargain with sailors and others for those animals hardy enough to have survived the passage from the tropics to the cold of England, so new acquisitions to the menagerie were acquired on an ad hoc basis. Cross may also have made arrangements with particular sea captains to bring back animals for him. He himself made at least two lengthy visits to France to purchase birds from the menagerie at the Jardin des Plantes in Paris. It is likely that he also made trips across the North Sea to purchase animals from the dealers whose shops lined the quays in Amsterdam and Rotterdam, and from where small merchant ships plied to London almost every day.[43]

There was much buying, selling, and exchange of animals between the travelling menageries and those in London. For example, in August 1815 Cross purchased a 'carriage and apparatus for a lioness, tigress, &' from William Samwell, who, although better known for his 'Equestrian Circus', also had a small menagerie of wild beasts. The money, £120, was delivered to Samwell in Colchester by Alfred Cops. Cross also acquired the famous lioness that had escaped from George Ballard's menagerie when it was on its way to Salisbury Fair—it had attacked a dog and one of the horses pulling the mail coach from Exeter to London across Salisbury Plain. Ballard recaptured the lioness and, having acquired the injured horse and the dog, showed all three as his prize animals, finally exhibiting them at Bartholomew Fair in the autumn of 1825, where Cross purchased the lioness. On 23 September1822 Davis's Royal Amphitheatre (late Astley and Davis's) presented an equestrian event, *Alexander and Thalestris*; taking part was 'a stupendous real Camel from Alexandria'. This was a Bactrian camel, with two humps, not, as one would expect given its origin, a dromedary. It was purchased by Cross and noted in a doggerel poem on a handbill for the Royal Menagerie, Exeter Change, which listed most of the animals on display.[44]

Some of the most valuable animals at the Exeter Change were purchased by agents from France and Germany. As we have seen, Antonio Alpi, who visited Britain in 1814 in order to buy Polito's rhinoceros and other animals for his travelling menagerie, had previously acted as an agent for the Emperor of Germany obtaining animals from Gilbert Pidcock. In 1827 an unnamed merchant made considerable purchases of animals from the Exeter Change and other collections for the King of Spain (Ferdinand VII). Cross sold him a young elephant, which was put on board a ship bound for Gibraltar—'Mr Cross, having, it seems, previously taken the precaution of securing payment of £700 *en argent comptan*'.

An invoice of 1824 shows that Cross advertised himself, not only as an 'importer of foreign birds and beasts', but also as a 'dealer in all sorts of pheasants, fancy poultry, swans, water fowl, every description of fancy pigeons, gold and silver fish and British birds of every description', so he must have had extensive premises elsewhere in addition to the cramped quarters over the Exeter Change. A small piece of evidence suggests that he had some land at St George's Field in Kennington, south of the river, where the larger grazing animals and others not required for show could be kept awaiting sale. William Clift's diary records a visit he made there in 1820 to inspect a dead gnu, which he purchased for five guineas for the Royal College of Surgeons. Cross must also have had stables to house the horses that pulled his carts, and the wagons and caravans used to transport his animals, probably also in Kennington.[45]

Cross had surprising success with his lions, not only in terms of longevity, but also in breeding. On 24 May 1817 he was able to announce that his lioness had produced two very beautiful whelps, male and female, which were being suckled by a spaniel. Seven months later she produced another litter, but this time 'her savage Majesty has graciously and unexpectedly condescended to suckle them herself'. In the course of time he bred twenty-one lions, six tigers, four jaguars, and four leopards. Cross also produced two hybrids between a black leopard and one of normal colouration, but as they are the same species (black leopards being merely the melanistic form) this hardly justifies a claim for inter-specific cross-breeding.[46]

The menagerie must have employed a very large staff: animal keepers, doormen in their famous Beefeater costumes to tout for business, ticket-sellers, book-keepers, cage-makers, carpenters, blacksmiths, farriers, carters, stablemen, scouts, taxidermists, porters, errand boys, domestic servants, and so on. Early every morning, before the menagerie opened, the cages would have needed mucking out, vast amounts of stinking bedding carried down the stairs from the first and second floors of the Exeter Change to waiting carts to be disposed of, perhaps to burgeoning dung heaps at St George's Fields, to be replaced with fresh hay and straw. Large quantities of food—grass, hay, and beans for the vegetarians, meat and offal for the carnivores, and live fish for some of the birds—had to be carried up the stairs. Dead animals had to be removed or fed to the carnivores; new arrivals accommodated.

Only a few names of Cross's staff have been recorded; most are those mentioned in the various accounts of the life and death of Chunee and relate only to the elephant-keepers employed between 1815 and 1826. One man who is said to have worked at the Exeter Change for thirty years was Harry Richardson, who later became Head Keeper of the short-lived Manchester Zoological Gardens, which opened in 1838, only to close four years later.[47]

The management of the menageries in London was in the hands of a small group of people, many of whom were connected by marriage, with Edward Cross at the centre. While there was undoubted rivalry between them, there was also much cooperation, exchange of expertise, and it seems mutual support. In 1810, when employed at the Exeter Change, Edward Cross married Mary Ann Cross, the daughter of the late William Cross of Bow, who had previously kept a menagerie of his own (Section 4.3). Whether they were related to each other is uncertain, but one or both may have also been related to John Cross, who, in partnership with Joshua Brookes senior, managed the menagerie at the corner of the Haymarket and Piccadilly in the 1770s. Stephen Polito married Mrs Sarah White, a widow who may have been the sister of William Herring (Section 5.1) at St Andrew Church in Holborn in 1790. As we shall see, Stephen's relationship with John Polito is uncertain; but John married Edward Cross's sister Elizabeth in 1814 (Section 5.4). Edward's other sister Ann's marriage in October 1817 to William Herring at St Andrew in Holborn was witnessed by 'E. Cross and Mary Wolf', and Cross referred to Herring as his brother-in-law. Three of Herring's four sons were baptized at St Clement Dane's, which suggests that he was employed at, or living near, the Exeter Change from at least 1819, even though by 1822 he was running the Original Menagerie in the New Road, which Cross had acquired in 1820.

Edward and Mary Ann Cross had no children of their own, but Edward was clearly devoted to his extended family and in his will of 1854 made bequests to William Herring and Herring's four sons, as well as to his sister Elizabeth Polito and her son John jr. Edward

Cross may have been related, perhaps through his wife's family, to another William Cross who kept a menagerie in Liverpool in the late nineteenth century, but there is no evidence for this.[48]

The menagerie at the Exeter Change (Plate 10) was one of the most popular sights of London. It was open every day, except Sunday, from 9 a.m. to 9 p.m. The admission fee was usually 1*s.* 6*d.* for the elephant, 1*s.* for the Great Room, or 2*s* for both; 6*d.* extra for feeding time.[49] Cross issued numerous handbills and placed endless advertisements in the newspapers to publicize the menagerie, always drawing attention to the latest wondrous arrivals and capitalizing on visits made by the rich, famous, or aristocratic. He was not given to false modesty; a typical advertisement reads:

> A beautiful Black Tyger and an immense serpent, originally intended for the King of the Netherlands, have just been presented to Mr Cross, Proprietor of the Royal Menagerie Exeter Change, as a great mark of esteem for his unremitting exertions in procuring (regardless of expense) every living production of foreign climes—such as a continual source of instruction to the rising generation, as well as to Artists, as is afforded by this truly grand Depôt, far out-rivals every Establishment of this kind upon earth, and justly deserves every patronage that a great Nation is capable of bestowing.

Children were much encouraged:

> As the Study of Natural History forms one the principal Branches of Education nothing can be more desirable for young Ladies and Gentlemen, during the Vacation of the approaching Holidays, then a visit to the ROYAL MENAGERIE EXETER CHANGE where the living productions of first celebrity of the extensive forests of Asia, Africa, and America, abounds in number and variety, in perfect health and condition. The sight of so many rare and beautiful inhabitants of foreign regions must strike every beholder with wonder and admiration, and agreeably fill the mind with contemplation on the wonderful works of the Supreme Author of nature.[50]

In 1823 Cross changed the feeding time of the big cats from 9 p.m. to 8 p.m. for the benefit of visiting children. He liked to present himself as a kindly patriotic man and on two occasions he invited no fewer than 300 boys of the Royal Military School in Chelsea to visit the menagerie; not doing anything by halves, he provided refreshments, a bun for each boy and a glass of wine to drink the King's health, while a band struck up 'God Save the King'. The police were on hand to ensure that the 'little heroes' behaved with proper decorum.

To encourage the more knowledgeable visitors, Cross would show off his familiarity with the world of science—for example, his 'Tongarabarra wombat alive ... is the least known of any of the animals found in Australasia, neither Buffon, Goldsmith, Shaw &c had in their time the least knowledge of its existence'. The menagerie was undeniably an invaluable artistic and scientific resource, much appreciated by visiting artists and scientists: 'Mr Cross, whose extensive menagerie at Exeter Change, and whose urbanity and readiness to assist all persons interested in the zoological department of natural history in a considerable degree supplies the want of a national collection ...'[51]

The menagerie at the Exeter Change was much valued by artists and zoologists. Samuel Howitt's drawings record some of the rarer animals shown there, including the capybara listed in the handbill of about 1816–17, which must have been one of the earliest depictions of this species, Dr Abel's orang-utan, and an African hunting dog, now included in the genus *Lycaon* established in 1827 by Dr Joshua Brookes (Section 4.7).[52]

Both Thomas Landseer and his younger brother Edwin drew animals at the Exeter Change, first in Polito's time and then in Cross's. Edwin went on to become one of the most famous animal painters of Victorian England, indeed of all time. Although Thomas produced hundreds of drawings and engravings of animals that were used to illustrate books on natural history, he is less well known, and his work is often attributed to his brother. Both boys contributed to their father's *Twenty Engravings of Lions, Tigers, Panthers and Leopards* ... first published in 1823. All the plates, of superb quality, were engraved by Thomas and some, including the *Lioness and Bitch, from Cross's Menagerie*, were drawn by Edwin. The plate of the *Black-Maned Lion of Africa* was 'corrected by a reference to Nature'—that is, to the lion donated by Lady Castlereagh (Section 5.5). It was probably this animal's corpse that Cross gave to Edwin and his friend Thomas Christmas in 1820, which they trundled away to their studio in order to dissect and study its musculature. Edwin must have used the knowledge he had gained from his dissections and observations at the Tower and the Exeter Change when he was commissioned in 1857 to sculpt the four great lions, whose casts by Carlo Marochetti are still recumbent at the foot of Nelson's column in Trafalgar Square. When Edwin Landseer died in 1873, black wreaths were placed around their necks.[53]

John Frederick Lewis, a talented painter and draughtsman, was another friend of Edwin's, and the two boys are said to have played truant in order to draw animals at the Exeter Change. Lewis's drawings include a sketch of the elephant *At the Old Exeter Change*, *Puma and Pug at Exeter Change* drawn in about 1820, and *Nero at Exeter Change* (not be confused with Nero at the Tower, as the name Nero seems to have been a standard appellation for elderly male lions). Lewis is said to have rented rooms 'off the Paddington Canal, where he was wont to paint from dead animals procured at the Exeter Change'.[54]

Edward Griffith described many of the mammals shown at the Exeter Change in his *General and Particular Descriptions of the Vertebrated Animals* (1821), and in the first five volumes of his edition of Georges Cuvier's *Le Règne animal distributé d'après son organisation: The Animal Kingdom: Arranged in Conformity with its Organization* (1827), published in quarterly parts between 1824 and 1827, with new illustrations and much additional text. The contributions by Major Charles Hamilton Smith and the drawings and engravings by Thomas Landseer were used as selling points in numerous newspaper advertisements.[55] A particularly attractive plate entitled *Marmot* (Figure 5.2), which is actually a viscacha from Argentina, was seen at the Exeter Change by two French naturalists Henri de Blainville and Frederic Cuvier (the keeper of the menagerie at the Jardin des Plantes), taking advantage of the temporary peace in 1814, who identified it, incorrectly, as a jerboa. After its death its remains were delivered to Joshua Brookes, who described it as a new species of rodent, *Lagostomus trichodactylus* (now known as *Lagostomus maximus* (Brookes 1828), the plains viscacha).[56]

Griffith described the first clouded leopard known to science, said to have arrived from Canton in about 1817 and kept at the Exeter Change; he named it the 'clouded tiger' *Neofelis nebulosa* (Griffith 1821); the rather poor illustration by Howitt was replaced in 1827 in the *The Animal Kingdom* by an improved version, engraved and probably drawn by Griffith himself (Figure 5.3). However, it has recently been pointed out that the lack of a type specimen renders the attribution to Griffith invalid and a skin in the Natural History Museum has been chosen as the 'neotype'. There is a simple

Figure 5.2. The 'Marmot', described by Joshua Brookes as a new species of rodent, *Lagostomus trichodactylus* (the plains viscacha) (from Griffith et al. 1827: iii).

Figure 5.3. The 'Clouded Tiger' (clouded leopard) whose skin was cut up to make caps for the keepers in a travelling menagerie (from Griffith et al. 1827: iii).

explanation for the lack of a type specimen: Griffith wrote forlornly that the clouded leopard was 'taken into the country with an itinerant exhibition and died there, and so little attention did Zoology, at that time, receive here, that as far as appears, its skin was cut up to make caps for the keepers'.[57]

The text of the Ungulate volume was largely, if not completely, the work of Major Smith, and most of the plates engraved by Landseer were based on Smith's drawings. The volume includes illustrations and descriptions of many of the antelopes and deer in Cross's collection, including a 'musk deer' (more probably a muntjac, as it was said to have come from Sumatra) drawn from life in 1821 and a Malayan rusa deer drawn in 1818 when Georges Cuvier himself was in London, the implication being that Smith, who was a friend of Cuvier's, had taken the great zoologist to visit the Exeter Change. Smith wrote that 'the sailors and people at Exeter Change named it Jamboe and Great Water Stag'. Cross advertised it as 'a famous Samboo, the most elegant quadruped of Hindustan'.[58]

The ever inquisitive William Clift frequently visited the Exeter Change to view the animals and to purchase any that had died, so they could be dissected and preserved in the Museum of the Royal College of Surgeons, among them a llama, a kangaroo, the Great Lion (died 1815), a margay cat, a toucan, a gnu, the trunk of a young elephant, a coypu, various monkeys, an oryx, an Indian zebu, a wombat, a ring-tailed coati, several ostriches, a cassowary, a Cyrus crane, a vulture, a secretary bird, a golden eagle, one of England's first dingoes, Dr Abel's orang-utan, and a small black chimpanzee.[59]

Sir Everard Home carried out a rather bizarre experiment at the Exeter Change in 1822. He had received the head of a young Sumatran elephant preserved in a keg of spirits from Stamford Raffles and, having examined the ear drum, was keen to discover whether its structure could explain any aspects of the elephant's ability to hear. So, with the cooperation of Edward Cross, he put the matter to the test. The results are best described from his own text read to the Royal Society in December 1822 and published the following year with plates of the dissection drawn by Clift:

> To see the effect of high and low notes upon the elephant in Exeter Change, Mr Broadwood kindly sent one of his tuners with a pianoforte to make the experiment: the higher notes hardly attracted notice, but the low ones called up the elephant's attention. He

> brought his broad ears forward, remained evidently listening, and he made use of sounds rather expressive of satisfaction than otherwise. The full sound of the French horn produced the same effect.
>
> The effect of the high notes of the piano-forte upon the great lion in Exeter Change, only called his attention, which was very great. He remained silent and motionless; but no sooner were the flat notes sounded, than he sprung up, endeavoured to break loose, lashed his tail, and appeared to be enraged and furious, so much so as to alarm the female spectators. This was accompanied with the deepest yells, which ceased with the music.[60]

One of Cross's most long-lived, and best loved animals, was a striped hyaena, which had actually been acquired in order to provide a skull for William Buckland, the geologist. When excavating animal bones from Pleistocene deposits at Kirkdale Cavern in Yorkshire, Buckland had found a portion of a fossil skull that he believed to be that of a young hyaena. Having no comparative material to hand, he requested William Burchell the explorer to send him a young hyaena from the Cape of Good Hope. In 1821 a baby hyaena arrived at the docks; it had been a great favourite with the sailors, who had named it Billy. Cross acted as agent and undertook to deliver poor Billy to the Dean to be slaughtered for the sake of science. However, moved by the little hyaena's good temper and playfulness, Cross begged for its life, and he was allowed to keep it on condition he obtained a skull of another young hyaena for Buckland, which he eventually managed to do. Billy became a much loved pet and made an unusual contribution to science. Buckland had excavated coprolites from Kirkdale Cavern, delicately referred to as '*album græcum*', of uncertain origin; after Billy had dined on bones, his droppings were found to be identical to those found in the cavern, proving it to have been a hyaena den. Buckland's son, the equally entertaining Frank Buckland, claimed that Billy served the cause of science in another way, for when Billy cracked beef marrow bones, the resulting fractures matched exactly those on bones excavated from Kirkdale. In addition, Billy polished the sides and floor of his wooden den in the same manner as his ancestors had done to the sides and floor of their den of stalactite in the Yorkshire hills.[61]

Despite the fame and success of the menagerie at the Exeter Change as an exhibition of animals, Cross considered himself a wild beast merchant, not a mere showman—animals came and animals went, some dying before they could be sold, others kept on for their appeal to the public. Most of the owners of private menageries were his customers, and Cross sold small animals, especially parrots and monkeys, to be kept as pets by anyone who could afford them. At the foot of almost all his handbills are the words, 'The utmost value given for all kinds of FOREIGN BIRDS and BEASTS, by E. CROSS. Noblemen and Gentlemen supplied with Gold, Silver, Tame Bred Pheasants, Foxes, Deer, &c &c'—that is, he was supplying the landed gentry and aristocrats with animals to hunt on their country estates. Animals that died in the menagerie, and there must have been many, were also of value; their bodies could be stuffed for sale, or sold, or given to museums, especially, as we have seen, the Museum of the Royal College of Surgeons, Joshua Brooke's Museum and Bullock's Museum in Piccadilly.[62]

By 1829 Cross was able to claim that many noble and illustrious personages were among his customers, and that the names of no fewer than five monarchs were to be found in his books. But his most important client by far was George IV; in September 1824 he supplied him with two wapitis and a chevrotain, and from 1825 until the King's death in June 1830, 'by Special Command of His Majesty' he acted as his agent, purchasing animals, supervising their transport from the point of arrival to the menagerie in Windsor Great Park, and managing them after arrival. Cross was responsible for landing the King's most celebrated animal, the giraffe, in 1827 and the delicate business of transporting it to Windsor, as well as its subsequent care, so it is not surprising that he is one of the figures depicted by Agasse in his famous painting of the giraffe (Section 5.6). Cross's invoices in the Royal Archives at Windsor form one of the most detailed records of the activities of any of the early menagerists, as well as indicating the high prices that exotic animals might command; between 1825 and 1830 Cross provided animals to Windsor to the tune of nearly £1,000.[63]

The spectacular and awful death of Chunee the elephant in 1826 is the single most reported event in the history of the menagerie at the Exeter Change, but it was the climax of many years of mistreatment, in particular the total lack of exercise and confinement of the ever-burgeoning behemoth in a cage in which he could barely turn round. The erratic behaviour of such a powerful animal with huge tusks became so alarming that attempts to control him became ever more barbaric, although one of his keepers did manage to calm him with kindness for a while.

Chunee was the most popular attraction at the Exeter Change in Polito's time, having been taught various tricks by his keeper Alfred Cops, but, after Cross had taken over the menagerie, Chunee became increasingly restive. He had grown so much that he could scarcely lie down in his cage. In 1815, after he had attacked Cops, his management was passed to George Dyer, who mistreated him. Dyer slept in an apartment above the Chunee's cage, and one day in 1819 Chunee, perhaps out of boredom, or perhaps intentionally, took his revenge by breaking through the ceiling, purloining Dyer's clothes, and consuming some of them. On another occasion, in Dyer's absence, Chunee pulled down the ceiling, completely covering himself in plaster. He was then 'physicked in order to reduce him, with 97 lbs of salts, 12 lbs of treacle and a quantity of camomile', none of which had any effect. It was decided that two men were needed to control him, and John Taylor, an experienced animal keeper who had had his right arm torn off by a lion and eaten in front of his eyes, was employed to assist Dyer, but, when Dyer was attacked by Chunee, Taylor took over the care of the elephant, realizing that what the animal needed was loving kindness.[64]

By 1822 Chunee was 10 feet tall and his weight had reached 5 tons. Cross claimed, probably correctly, that the 'stupendous elephant' was 'unquestionably the largest Animal ever seen in Europe'. A Mr Harrison was called in to construct a more commodious den for him in the Great Room. At first Chunee seemed pleased, but he soon became unruly again, so Cross decided to have his tusks cut off—thus allowing George Wombwell, the owner of a travelling menagerie, to take advantage of the situation by advertising the fact that the elephant in *his* menagerie was the only one in England with tusks! Cross must have been greatly relieved when a man from America offered to buy Chunee for £500, but, as 'no ship trading to that country could be found capable of transporting him', the plan had to be abandoned.

Relative peace reigned for the next few years while the elephant was cared for by Taylor, who taught him to ring a bell at 8 p.m. each evening to announce feeding time, but in 1825 Taylor, the one person who could manage Chunee, left the menagerie, after what he claimed was a misunderstanding. Ill-treated yet again. Chunee became unmanageable and killed one of his keepers, John Tietjen—the coroner's verdict was accidental death and the elephant was fined 1s.[65]

Early in 1826, after more mistreatment, which included being deprived of food, Chunee, in a terrible state of neglect and dirt, and covered with wounds where he had been goaded by his keepers, became extremely savage and badly damaged the bars of his cage. Harrison was called in to repair it, but refused, fearing for his life. Cross resorted to the usual punishment and administered vast amounts of 'opening medicine', which only made Chunee even more furious.[66] Added to that, he was in a state of musth or, as the *Morning Post* delicately stated, 'at certain seasons of the year was ungovernable', and became so alarming that Cross decided that the only solution was to kill him.

Several conflicting accounts of Chunee's death on 1 March 1826 were published, three of them in the same edition of the *Morning Chronicle*; another by his former keeper, John Taylor; another, said to have been related by Cross himself, in Hone's *Every Day Book*. A recent account based on these sources has been related in entertaining if somewhat embroidered detail by Jan Bondesen.[67]

Cross's brother-in-law William Herring 'of the New-Road', for whom Chunee had previously developed a particular dislike, was called in. He rushed off to a gunsmith in Holborn and returned with several rifles. Meanwhile Cross hurried to Great Marlborough Street to Joshua Brookes the surgeon for advice. On his return he learnt that Herring had fired numerous shots at Chunee, which had only succeeded in infuriating him further. Fearing that the building might collapse and release the big cats, the keepers were managing to hold the elephant at bay with long pikes, while one of them, named Cartmel, repeatedly stuck a sword into Chunee's side, to no avail. The proprietor of the entire building, Thomas Clark (the son of the original owner of the menagerie), cleared all the shoppers out, who assembled in a large, riotous crowd in the Strand. Soldiers were summoned from Somerset House, but even after 152 'musquet balls' had been fired at the unfortunate animal, they only succeeded in wounding him. The situation was finally resolved when Dr Joshua Brookes killed Chunee with a poisoned harpoon, or, in another account, when Herring finally managed to administer a *coup de grâce* by firing a bullet under one of his ears. Silence. Chunee lay in a monstrous bloody heap, surrounded by the fallen timbers of his cage, his head, as in life, still upright.

After several days lying-in-state, when hundreds of visitors came to 'pay their respects', the stench of Chunee's enormous cadaver became so overpowering that even the shops below were deserted. A Bow

Street magistrate, Sir R. Birnie, sent a message that, unless the body had been removed by the following Monday morning, 'Mr Cross would hear from Sir Richard, in a way he would not like'. It took nine butchers twelve hours to flay the hide, which was sold to a Mr Davis for £50, and the skinned corpse was then dissected, or rather hacked to bits, by a Mr Ryals, under the direction of Dr Brookes, assisted by several other surgeons and watched by medical students. Cross denied feeding the putrid flesh to his animals, and it was reported that four tons of it were carted off to purveyors of cat meat; Brookes did not escape censure, he too was reputed to having dressed and eaten part of the meat, though he denied the charge. The *Caledonian Mercury* reported: 'Mr Deville's son on Friday took a cast in Plaster of Paris, of the elephant's head at Exeter 'Change, to place among his phrenological collection of murderers, at his museum in the Strand. The elephant, our readers are aware, had killed his man.' Chunee's hide, having been tanned at Greenwich, was on sale six years later in Leadenhall Market for £32 12*s.* 6*d.* His mounted skeleton was displayed in his ruined cage; it ended up in the Museum of the Royal College of Surgeons. The last vestiges of Chunee met their end when the college was bombed in 1941.

The existence of a menagerie, filled with dangerous animals, on the first floor of an ancient building in the heart of London had long been a matter of complaint and anxiety. In 1823 the *Morning Chronicle* published a letter that called for a national menagerie along the lines of that in the Jardin des Plantes in Paris and stated that it would require only a small sum of money to release many of the wretched animals pining away in Exeter Change. Although signed C.T., the letter may have been written by Cross himself, who had recently returned from Paris. Within a few months Cross changed the name of his collection from 'Royal Menagerie' to 'Royal Grand National Menagerie', presumably with this in mind.

Matters came to a head only a few days after Chunee's death, when they were debated in the House of Commons. On 21 March, when Charles Arbuthnot moved leave to introduce the 'Improvement to Charing Cross Bill' and Sir M. W. Ridley added that after what had recently happened at the Exeter Change there were the strongest reasons for removing the menagerie. Arbuthnot said no one was more alive than he to the inconvenience to which the public were exposed in the vicinity of the Exeter Change and that he understood that Lord Exeter, the owner of the buildings, proposed to make great improvements when the leases were up.

It seemed that a solution might be at hand, for when Arbuthnot, in his position as First Commissioner of Woods and Forests, received a request from Lord Auckland and Sir Stamford Raffles, on behalf of the as yet unconstituted Zoological Society of London, for a grant of land for the establishment of a zoological garden similar to that in the Jardin des Plantes, they reinforced their request by adding, 'we are happy to state that Mr Cross, of Exeter Change, has offered his lamas and birds and such part of his collection as we may choose, to the Society, with a tender of his services in promoting our views'. In the event Cross offered the Society his collection for a mere £3,000, but was turned down on grounds of cost (Section 6.1).[68]

The writing was on the wall, but, undeterred, Cross continued much as usual. Late in April a large consignment of animals, brought from India on an East India Company ship, disembarked at Blackwall. Cross purchased the lot, including 'a most superb Royal Bengal Tygress, two remarkably beautiful leopards, a most elegant antelope, two Axis deer, and a noble Samboo deer'. They were followed by a male elephant, which arrived from Rotterdam as a replacement for Chunee.[69]

Only a few days after the arrival of George IV's giraffe in August 1827, Cross was back at the dockside to purchase a second 'most beautiful' elephant,

> landed from the Hon. Company's ship, the Thames . . . added at enormous expense to the Menagerie . . . it is by far the most sagacious and docile animal that was ever imported, being the same that had, by his numerous sagacious tricks and extreme docility, rendered himself a favourite and almost constant inmate at the Court of His Highness the RAJAH MEER, Achand Ali Sing.

The possession of two elephants lasted only a month or so, for in November Chunee's replacement was sold to the King of Spain. The following January, John James Audubon was in London soliciting subscribers for his magnificent book on the birds of America,[70] and wrote in his journal: 'We went to Mr Cross at the Exeter 'Change, and I had the honor of riding on a very fine and gentle elephant.'[71]

In an advertisement issued in April 1828 Cross added a postscript, saying that, although the menagerie was 'to be pulled down . . . we trust only to rise Phoenix like, more glorious from its ruins'. Three handbills, also dated to 1828, show that the menagerie was far from moribund. In September Rajah Meer's elephant was hired out to Astley's Amphitheatre to appear in the melodrama *Blue Beard*—

after the performance it showed off its 'surprising and sagacious tricks, Mr Cross, of Exeter Change, having sent the Keeper, who arrived with the Elephant in England'. The most notable animals listed were two Persian lions, a caracal, a sloth bear, some emus, and kangaroos bred at Windsor, a short-lived 'long-tailed Ouran Outang from Negroland' (presumably an African monkey), three 'extraordinary compound animals the Gnus', and two boa constrictors (probably pythons), which 'swallowed six large fowls'. Six crocodiles were shown for a short time and then taken on tour in Devon in 'Richardson's Menagerie of Reptiles'.[72]

It was not to last. The menagerie founded by old Thomas Clark in 1788, owned successively by Gilbert Pidcock, Stephen Polito, and Edward Cross, was about to come to an end, threatened not only by the impending demolition of the building, but, as we will see in Chapter 6, by the Zoological Society's menagerie in Regent's Park.

5.4. Three Early Nineteenth-Century Travelling Menageries: John Polito, George Wombwell, and Thomas Atkins

When Edward Cross acquired the menagerie at the Exeter Change after the deaths of Stephen and Sarah Polito in 1814, John Polito took over the travelling side of the business. He seems to have appeared out of nowhere, but it is likely that he had previously been employed by Stephen. Their relationship is uncertain, for, even though John referred to Stephen as his late brother, he is not mentioned in Stephen's will. I suspect that John was an Englishman who changed his name to Polito for convenience. His association with Edward Cross was cemented by his marriage to Edward's sister Elizabeth in 1814 at St Clement Dane's in the Strand,[73] only a few yards from the Exeter Change, and thereafter he referred to Cross as his brother, clearly meaning brother-in-law.

In October John Polito took the travelling menagerie to Oxford, advertising it as 'Polito's Royal Menagerie from Exeter Change', stating that he had 'succeeded to that inimitable Collection so lately possessed by his late brother, Mr. S. Polito', and adding that he had also acquired many new animals. It is clear from the advertisements issued in Hereford in August 1815 and in Plymouth in October that most of the animals in his collection had previously been in Stephen and Sarah Polito's travelling menagerie—the stupendous sagacious male elephant, the zebra, the nylghai, the lion that had 'been in the collection for upwards of 20 years', the lioness 'the only survivor of the original breed of lions that have been in the Tower of London' (i.e. Miss Howe), the ursine sloth, and so on.[74] Another of the animals that had been included in Sarah's advertisement issued in Northampton in 1814 was a 'tortoise-shell tiger'. It was probably the animal brought from Sumatra, described by Dr Horsfield as a 'rimau-dahan' (now the Sunda clouded leopard, *Felis diardi*), which was exhibited at the Exeter Change and which later died at Axminster in an un-named travelling menagerie, presumably John Polito's. The skin was tanned and transferred to Bullock's Museum in Piccadilly.[75]

John Polito probably spent the winter of 1815–16 travelling slowly northwards from Plymouth to Edinburgh, where he remained from March until August, and where he announced that he had 'made considerable additions to the inimitable collection of his late brother Mr. S. Polito'. He wrote that he had prevailed upon his 'brother' Edward Cross to bring two boa constrictors (pythons?) all the way from London to be shown with the other travelling animals. In October Polito's menagerie was in Aberdeen in the north-east of Scotland for the races, and then toured round Scotland, before returning to Edinburgh in March 1817.

Several Staffordshire pottery figures depict Polito's travelling menagerie, with models of figures, including one of a woman, perhaps Mrs Polito (Plate 13).

Newspaper advertisements show that from March 1816 to November 1819 Polito's travels were confined to the north of England and Scotland. His menagerie was so often on show in Liverpool, especially in the winter months, that one suspects it was based there, the port being a convenient point of departure of shipping to Ireland and more importantly a port where exotic animals arrived and might be purchased.

In November 1819 Polito shipped the menagerie across the Irish Sea to the Isle of Man, where it included

> a noble full-grown royal lion ... intended as a present for the Prince Regent; but the proprietor, at an enormous expense, purchased it for the intended purpose to exhibit it through every Island of Great

> Britain. Also the singular curiosity—the great baboon or, wild man of the woods from the coast of Malabar. When full-grown stands six feet high, and walks erect like a man.

The menagerie sailed on to Dublin late in 1819 and encamped in Abbey Street, where it was visited by boys from the Claremont Deaf and Dumb Institute. Letters written by two of the boys paint an unpleasant picture; they saw a woman beating a panther with a stick and a man doing the same to a monkey and a stork. The other animals they described were a water buffalo from Bombay, an elephant, a camel, a zebra, a kangaroo, a hyaena, a sleeping lion ('his tail was down pendulous, through rail my hands were touching tail'), and a porcupine whose 'quills are long, and black and white. I felt his quills. We were afraid.'[76] In January 1820 the menagerie was still in Dublin, where, on 15 March, 'the Great Serpent at Polito's menagerie took a rabbit at three o'clock on Saturday, to the great gratification of all present'. At some point, perhaps having toured in Ireland in the meantime, the menagerie moved to Ormond Quay in Dublin, where it was visited in August 1821 by Prince and Princess Esterhazy. It is also recorded at Donnybrook Fair near Dublin in the summers of 1821 and 1822.

In March 1823 Polito's menagerie was again on the Isle of Man, presumably when en route from Ireland. There had been a heavy snowfall in which the caravans overturned, releasing 'tygers, bears, hyænas, and other terrific animals, but their ferocity was totally paralysed, either by the cold, or the appearance of the snow all around, and happily they were again got into secure confinement, and thus the island was saved from being possessed by these unwelcome colonists'.[77]

A few weeks later John Polito's menagerie sailed for France, probably from Liverpool, and landed at Bordeaux on 22 April 1823, touring thereafter through much of Europe. Although it seems that the menagerie itself never returned to England, Polito himself visited London from time to time to purchase animals. Initially his advertising material stated in various languages that the 'La Grande Ménagerie de M. Jean Polito' was 'd'Exeter Change de Londres', but in later years merely 'London'. In Marseilles in March 1827 he offered his services to the professors at the Musée d'Histoire Naturelle in Paris to transport the famous giraffe, Zarafa, which had been captured with George IV's giraffe, from Marseilles to Paris. His offer was declined because they surmised that he wanted to do it only for the money. He is last heard of in 1852.[78]

Much has been written about Wombwell's, the most famous and notorious travelling menagerie of the nineteenth century, though much of the information is unreliable.[79] He is said to have started in a small way in 1805 by showing some boa constrictors that he had purchased at the London Docks,[80] and by 1808 he had a menagerie at 207 Piccadilly, showing two crocodiles. Two years later, when he was preparing his animals to show at Bartholomew Fair, the horse pulling the caravan containing his 'fine Bengal Royal Tyger' took fright and bolted, the caravan overturned, and the tiger escaped, terrorizing Piccadilly until at 4 a.m., when the keepers managed to capture it. Wombwell was soon touring a collection of animals around the country, including an Indian rhinoceros, which he seems to have had on loan before buying it at auction in Norwich in 1817. It probably died soon after February 1819, when it is last recorded.

Handbills and newspaper advertisements show that Wombwell amassed an extraordinary collection, not only in size, but also in the wide range of species it contained; he took it all around Britain as far north as Glasgow. In 1821 he claimed to have bred nine lions, two leopards, and a panther. He had zebras, kangaroos, camels, a nilghai, a 'sea cow' (manatee?), various small cats, guinea pigs, jackals, a pelican, a variety of porcupines and serpents, rattlesnakes and crocodiles, and many smaller animals and birds, 'too numerous to mention'. The whole of this collection was to be seen in one booth, which must have been enormous. A handbill issued when the menagerie was touring East Anglia at about this time adds a male elephant, two diminutive elephant calves, a sloth (sloth bear?), a lion, a tiger, and a leopard in the same cage, seven other tigers, 'a desert of panthers', and 'a wilderness of apes, baboons, monkeys, opossums etc', as well as a quagga, and an onager.[81]

Wombwell's most notorious exploit was to stage a lion-baiting—perhaps the first since the days of James I—in Warwick in the late evening of 26 July 1825. A large cage containing the lion Nero was placed in the yard of an old factory before a huge crowd of expectant spectators. Six dogs were thrown in, but Nero, a docile animal, appeared not to understand that the dogs meant to attack him, but attack him they did and he was badly mauled. However, when Wombwell realized that Nero's life was in danger, he called the fight off, much to the spectators' disgust. But, having whetted their thirst for bloody combat, he rearranged the fight for the following Saturday night; this time he used a lion named Wallace, who, despite having been born in captivity (in 1812 in Scotland), was of a much fiercer nature. Three pairs of dogs were thrown in, and Wallace put paid to each in less than a minute. Using his lions for such a vile purpose was no sudden whim of Wombwell's; he had posted handbills in all the neighbouring towns as well Birmingham, Coventry, and Manchester further afield, promising potential punters that 'the battle was to be for £5000'; tickets cost 2 guineas (about £100 in today's money). The event attracted such bad publicity and Wombwell made so little money from it that it was not repeated, which did not prevent him adding such phrases as 'THE CONQUERING LIONS, NERO AND WALLACE; THE SAME LIONS THAT FOUGHT AT WARWICK' to his banners and other publicity material.[82] After Nero's death two years later, his body was sent to 'Mr Martin Noble Shipton, surgeon of Birmingham, for dissection'. Wallace died in 1838; his body was stuffed and may still be seen in Saffron Walden Museum.

One September when Wombwell, having left his menagerie in Newcastle, was in London, he heard that his rival Thomas Atkins was preparing to show his animals at Bartholomew Fair. Determined to not to be outdone, Wombwell hurried north and then force-marched his menagerie down the Great North Road to London. Unfortunately the pace was too fast for his elephant, which collapsed and died on the first morning of the fair. Atkins delightedly announced that his show included the only live elephant in the Fair. Wombwell responded with a huge banner, 'the only dead elephant in the show'; which proved a greater draw for spectators than Atkins's live one.[83]

Thomas Atkins is best known for his success in breeding lions and for having crossed a tigress with a lion on five occasions—the first in 1824, when he showed the resulting cubs to George IV at Windsor. He toured a large collection of animals around southern England from about 1814, initially with a partner named Gillman. The descriptions in their handbills of some of the animals, including their 'stupendous elephant', are identical to those used by Stephen Samwell, who also ran a menagerie for a while, but gave it up to concentrate on his 'Equestrian Circus' (Section 5.3). Newspaper advertisements suggest that Atkins's tours were mainly in the west of England, though he did reach Edinburgh in 1826; Wombwell toured mainly in East Anglia and the Midlands. Both menageries had bands, with the musicians and the front-line staff dressed, like those at the Exeter Change, as Beefeaters. Atkins's most notable animals were a gnu, described as 'the truly singular and wonderful animal the aurochos', a Persian onager, a 'Nepaul Bison' (probably a gaur, the first in Britain), a moose, an alpaca, a quagga, and an 'ostrich which comes from Brazil' (meaning a rhea). Many of his advertisements add the phrase 'birds and beasts bought, sold and exchanged by the proprietor' or words to that effect.[84]

Atkins's menagerie grew and grew, though not to the same extent as Wombwell's. In September 1829, when both menageries were at Bartholomew Fair, Atkins's occupied nine caravans, whereas Wombwell's needed fourteen or fifteen.[85] A few years later Atkins ceased travelling and founded instead the Liverpool Zoological Garden, whereas Wombwell's collection grew and grew until by 1831 he had three menageries touring simultaneously.

5.5. Some Early Nineteenth-Century Country-House Menageries

The Duchess of York kept a collection of animals at Oatlands Park in Surrey. She was Princess Frederica, the daughter of the King of Prussia, married in 1791 to George III's second son, Frederick Duke of York. In the early years of their marriage the Duke was rarely at home, and the Duchess spent her days helping the poor and needy and lavishing her affections on her collection of animals, which at that time included parrots, horses that were never ridden, and dogs.[86]

In 1801 the Duchess received a pair of kangaroos,[87] probably from the Queen at Kew, although a print of her farmyard shows two dark-coloured animals, which look more like wallabies than kangaroos. Other animals in the yard are an ostrich, a female antelope(?), several goats, a white zebu, a partially visible dark-coloured bull or cow, some Somali sheep and several other sheep, chickens, some peacocks (one white), and a black swan.[88] The Duke sometimes visited Oatlands with members of his fast set, including Beau Brummel, who gave the Duchess a little dog, which they named Fidelité, and Thomas Raikes, who mentioned more animals in his reminiscences of life at Oatlands: 'There was a large menagerie in her flower-garden filled with eagles, macaws, and various creatures; a little colony of monkeys on the lawn before the windows of her boudoir; a herd of kangaroos, ostriches, &c, in the paddock . . .'[89]

The Duchess's menagerie included a ring-tailed lemur, which gave birth to a young one, though probably stillborn. The preserved infant was acquired by Henry Bernard Chalon, who had been appointed Animal Painter to the Duchess in 1795; he presented it to the Museum of the Royal College of Surgeons.[90] She also owned a monkey, probably a black and white colobus from Africa, depicted on her lap in her portrait by J.-F.-M. Hüet-Villiers.[91]

The Duchess was given a leopard by the redoubtable naturalist and biographer of Georges Cuvier, Mrs Sarah Bowdich.[92] When she was in The Gambia with her first husband, Mrs Bowdich acquired a tame leopard (referred to as a panther) named Sai from the King of the Ashanti. On the voyage home in 1818 the ship was boarded by pirates, and there was consequently very little to eat, especially for such an extravagance as a tame leopard. However, there were 300 parrots on board, all of which died en route, and the leopard was allowed to eat one corpse daily; the diet did not improve his health, and in her anxiety Mrs Bowdich dosed him with calomel pills. Sai survived and on arrival at the London Docks was presented to the Duchess, who lodged him temporarily at the Exeter Change; on the morning before her return home she visited Sai, playing with him and admiring his healthy appearance and gentle deportment; but alas all was not well, for in the evening, when her coachman came to collect him, Sai was found to have died.[93]

The Duchess owned a llama or perhaps a guanaco from South America, which was seen at Oatlands by Charles Hamilton Smith and described and figured in the fourth volume of Griffith et al., *The Animal Kingdom* (1827). The crowning glory of her menagerie was to have been an infant bull elephant, which was displayed at the Exeter Change in June 1820 alongside Chunee, where it 'affords the greatest treat imaginable to see the wonderful difference of size between it and the enormous old one that has been there so many years'. After the Duchess's death in August 1820, it seems that the Duke gave the little elephant to the Exeter Change, along with a 'beautiful wild cat from Bangalore',[94] and presented two of her macaws to Raikes in her memory.

Lady Castlereagh, the wife of the great, unpopular Foreign Secretary Lord Castlereagh (Marquis of Londonderry from 1821), kept a menagerie at North Cray Cottage, their surprisingly modest country home in Kent. The house still exists; traffic roars past on a dual carriageway on one side, and, on the other, tranquil lawns and meadows stretch down towards the River Cray. An inconspicuous blue plaque reads 'Robert Stewart Viscount Castlereagh (1769–1822), statesman lived and died here'.[95] Near the house is a range of stables, which once housed his wife's remarkable collection.

Lady Amelia Anne Hobart, usually known as Emily, married Lord Castlereagh in 1794. They were a devoted couple; Lady Emily, plump, pretty, and gregarious, provided a light-hearted, or perhaps frivolous and indiscreet, foil to her austere, handsome husband. They lived in St James's Square in London, and in 1810 acquired the lease of North Cray Cottage, with its 40-acre farm and trout stream. Emily immediately started assembling a menagerie; one of the first animals to arrive was a zebra. Castlereagh wrote to his brother in Spain: 'Lady C. is very much obliged to you for executing her commission about the ass, which will be an excellent cross for the zebra.'[96]

The Comtesse de Boigne remembered life at North Cray:

> The visitor's carriage drew up at a little gate and a walk between two beds of ordinary flowers led to a six-roomed house. Though the entrance to the house was exceedingly mean it was situated in charming country, and enjoyed a magnificent view; behind it there was a considerable enclosure, with rare plants, a menagerie, and a kennel, which, with the green houses, divided the attention of Lady Castlereagh... somewhat unintellectual, she was most benevolent, while her social manners were entirely ordinary and displayed no great knowledge of the world... She had the good taste to lay aside her finery at Cray, and was to be found in a muslin dress, with a straw hat on her head, an apron round her waist, a pair of scissors in her hand, cutting away dead flowers.[97]

Emily's eccentricities were not confined to her garden and menagerie. At a grand banquet during the Congress of Vienna in 1814, she shocked some of the illustrious guests by appearing dressed in white as a vestal virgin, and a few days later, at the grandest of all the grand parties of the Congress, she famously and ostentatiously wore her husband's ribbon of the Order of the Garter in her hair.[98]

Richard Rush, the gentlemanly American Ambassador, who was in the process of negotiating the Anglo-American Convention of

1818 with Castlereagh at North Cray, was not altogether impressed by what he saw:

> there was something that I had not expected; it was a menagerie. Taste, in England, appears to take every form. In this receptacle, were lions, ostriches, kangaroos, and I know not what variety of strange animals... And here, amidst lawns and gardens, amidst all that denoted cultivation and art, I beheld wild beasts and outlandish birds the tenants of uncivilized forests and skies set down as if for contrast![99]

At dinner on a visit by Rush the following year, Emily sounded him out about the possibility of acquiring a humming bird:

> [she] said that she had now two of my countrymen in her collection, a mocking bird, and a flying squirrel; but the bird, vexed perhaps at being stolen from its native woods, would neither mock anything, nor sing a note of its own; and as to the squirrel, none of her efforts had been able to make it fly; still, there was one other thing she wanted from the United States a humming bird, having never seen one. I said it would make me most happy to procure one for her if possible. Thank you, she said, but will it hum in England? I said I would disown it as a countryman, if it did not...[100]

In August 1821 the menagerie included an antelope, several kangaroos, emus, a pair of ostriches from Barbary (a present from the Prince Regent), and a tiger.[101] The 'tiger' must have been an American felid, perhaps a puma or a jaguar, as it had been brought from the West Indies by Lord Combermere, the Governor of Barbados, as a present for the Duke of Wellington, who gave it to Emily.[102] The male ostrich was injured, first by a playful llama, which kicked it in the neck, and then one night when a fox stole into the premises and seized it by the wounded part of its neck; for the rest of its life the unfortunate bird carried its head and neck in a crooked position. Castlereagh liked to boast that he was the only person at the Coronation of George IV in 1821 who wore a plume of ostrich feathers from his own birds.

Figure 5.4 depicts the 'black-maned' lion from the Cape that Emily gave to Edward Cross at the Exeter Change, which may have proved too dangerous an animal to be kept at Cray. She was well acquainted with the keepers there, including Alfred Cops—it was due to her influence that his application in 1821 to be Keeper of the menagerie at the Tower of London was successful

Figure 5.4. Lady Londonderry's 'black-maned' lion, drawn and engraved by Thomas Landseer (from Griffith et al. 1827: iii).

(Section 5.7).[103] Emily's faith in him was more than justified, for eight months later in a letter to the *Examiner*, Cops was able to report the presence in the menagerie of numerous animals ranging from lions to 'nutmeg birds'. Her support was followed by the gift in December of 'a most beautiful Tiger, and the Ostriches which, added to the great improvements within the last six months, form an assemblage at once beautiful and interesting'.[104]

Another of the animals exhibited by Cops at the Tower was a jaguar, which had been given to Emily by Lord Exmouth. It was said to have been obtained by Exmouth 'while on the America Station' and to have accompanied him when he commanded the fleet that bombarded Algiers in 1816, almost destroying the city and resulting in the release of about 3,000 slaves, including 1,000 Christians, who had been captured by Algerian pirates. One problem with this story is that the only American posting that the Lord Exmouth had was in 1786–9, when, as Admiral Edward Pellew, he was 'Commander of the Newfoundland Station'. It is also not at all clear when he gave it to Emily, but it is recorded that she gave it to George IV, who ordered it to be placed in the menagerie at the Tower, where it was exhibited as a panther. Cops's handbills of 1826–8 include references to a 'beautiful panther presented by the

Marchioness of Londonderry: the most exquisite animal of the kind ever seen in Europe'.[105] She also gave him a great sea eagle and a golden eagle, probably obtained from Norway.

In 1822 the idyll at North Cray Cottage was shattered. Castlereagh's mind became deranged, and he was advised to leave London and go to Cray for much needed peace and quiet, which he did. Although Emily removed all his pistols and razors from his dressing room, he managed to secrete a little knife in his pocket book and severed his carotid artery with anatomical accuracy.[106] At Emily's insistence, Castlereagh was buried in Westminster Abbey, next to William Pitt. Emily died suddenly and painfully on 12 February 1829 at St James's Square. The *Morning Chronicle* reported: 'The widow of a deceased statesman had, at his death, the most extensive menagerie of beasts and birds in the kingdom. This was a very costly establishment, and her Ladyship has chosen to diminish it. The collections at the Tower and at Exeter Change have been enriched by the dispersal of the collection.'[107]

The most detailed descriptions of Emily's animals are of those presented to the London Zoo, either by her during the last few years of her life, or by her sister-in-law (the new Marchioness of Londonderry) when the menagerie was being dispersed before or soon after her death; among Emily's animals given to the Zoo were a Russian wolf, three turkeys, two wigeons, a crested curassow, and a red curassow.[108] After her death a pair of ostriches was presented to the Zoo by her nephew the Marquis of Lothian.[109] Like one of the pair she had given to the Tower menagerie in 1822, one of these birds was unwell:

> The fine ostrich presented lately by Lady Londonderry to the Zoological Society having, in the course of its rambles, approached too near the cage of one of the wild beasts, received a severe wound in the neck by a bite from the latter. A great deal of attention was paid to the bird, not only on account of the noble donor, but also for the sake of the ostrich itself, it being a remarkably beautiful bird, and a great favourite at the gardens, eating every thing that was presented to it. Notwithstanding the best surgical treatment, after a few days of disconsolate wanderings, it died. Agreeably to the scientific character of the society, Mr Joshua Brookes, the anatomist, in the presence of 50 or 60 gentlemen, three days gave what may be termed an extempore lecture, with a demonstration of the anatomy of the ostrich . . .

The identity of the remainder of the collection given to the zoo seems not to be recorded, but several of the stuffed animals in the Zoological Society's Museum in Bruton Street had belonged to Emily, including a crowned crane 'remarkable for the showy wattle under its throat', a pine marten from Siberia, and a jerboa from Egypt.[110]

The sixth Duke of Devonshire kept a small collection of animals at his Palladian villa at Chiswick on the western edge of London, designed and built by the third Earl of Burlington, and completed in 1729, with one of the earliest landscape gardens, designed by Burlington and William Kent 'in the Italian style'.[111] The Duke inherited Chiswick House in 1811, as well as the family seat at Chatsworth in Derbyshire, various other houses, and nearly 200,000 acres of land. Deaf and a serious collector of minerals, coins, and books, the 'Bachelor Duke', as he was known, spent enormous sums of money altering and improving Chiswick House and at about the same time acquired his first, and most spectacular animal, an elephant.

The Duke was 'asked by a lady of rank what she should send him from India'; he answered, 'nothing smaller than an elephant'.

> He was surprised to find, at the expiration of some months, a very handsome female of the species consigned to his care. The Duke of Devonshire's elephant was kept at his Grace's villa at Chiswick, under circumstances peculiarly favourable to its health and docility. The house in which she was shut up was of large dimensions, well ventilated, and arranged in every particular with a proper regard for the comfort of the animal. But she often had the range of a spacious paddock; and the exhibition of her sagacity was therefore doubly pleasing, for it was evidently not affected by rigid confinement. At the voice of her keeper she came out of her house, and immediately took up a broom, ready to perform his bidding in sweeping the paths or the grass. She would follow him around the enclosure with a pail or a watering-pot, shewing her readiness to take that share of labour which the elephants of the East are so willing to perform. Her reward was a carrot and some water; but previously to satisfy her thirst by an ample draught, she would exhibit her ingenuity in emptying the contents of a soda-water bottle, which was tightly corked. This she effected in a singularly adroit manner . . . the man who had the charge of her in 1828, when we saw her, had attended her for five years . .[112]

It has recently been discovered that the elephant's keeper, Walter Elliot, received the first instalment of his salary on 25 December 1811, and, from a newspaper report, that the 'lady of rank' was the Marchioness of Hastings:

> The Duke of Devonshire's *dejeune* on Thursday to the Grand Duke Michael, at Chiswick, was attended by nearly 100 fashionables. The beautiful elephant which the Marquis of Hastings sent to his Grace from India, was promenaded over the pleasure grounds by the keeper, who came over with him. The Band of the Guards performed on the lawn.[113]

When a new keeper, John Leggrove, arrived in about 1820, Sadi (as she had been named) became intractable, although she gradually adjusted and 'would cry after him whenever he was absent for more than a few hours'.[114]

Gradually other animals arrived. In 1820 the Duke's sister Lady Harriet Cavendish wrote: '

> He is improving Chiswick most amazingly, opening it and airing it and a delightful walk is made round the paddock, open and dry, with a view of Kew Palace—and a few kangaroos (who if affronted will rip up a body as soon as look at him), elks, emus, and other pretty sportive death-dealers playing about near it. The lawn was beautifully variegated by an Indian bull and his spouse, and goats of all colours and dimensions. I own I think it a mercy that one of the kangaroos died in labour, vu that they hug one to death.[115]

Two years later he had a pet mongoose:

> Hart [the Duke] is in a most amiable humour. You will sympathize with me as to a new acquisition he had made—a sweet little pet called an ichneumon—the size of a large rat, with a nose like a weasel's, so tame that it springs up into one's face, gets into ones plate at dinner, and when one drinks tea runs rapidly up ones back, over ones shoulder and puts its dite nose into ones cup. Its peculiarity is I believe, a delight in sucking human blood.[116]

When Prince Pückler-Muskau visited Chiswick in 1826, he was much impressed:

> There is a menagerie attached to the garden, in which a tame elephant performs all sorts of feats, and very quietly suffers anybody to ride him about a large grass plat. His neighbour is a lama, of a

> much less gentle nature; his weapon is a most offensive saliva, which he spits out to a distance of some yards at any one who irritates him, and fires so suddenly at his antagonist, that it is extremely difficult to avoid his charge.[117]

Two years later Walter Scott wrote admiringly:

> A numerous and gay party were assembled to walk and enjoy the beauties of that Palladian dome. The place and highly ornamented gardens belonging to it resemble a picture of Watteau. There is some affectation in the picture, but in the ensemble the original looked very well. The Duke of Devonshire received everyone with the best possible manners. The scene was dignified by the presence of an elephant, who, under the charge of a groom, wandered up and down, giving an air of Asiatic pageantry to the entertainment.[118]

Sadi died in 1828 or 1829 aged about 21, of pulmonary consumption, and her corpse was presented to the newly constituted Museum of the Zoological Society.[119]

Other members of the menagerie at Chiswick included kangaroos, a Neapolitan pig, emus, ostriches,[120] many aquatic birds, gold and silver pheasants, a cockatoo, several species of monkey, and a bear.[121] There was also an Arctic fox, whose coat changed from dirty blue in September to white in January (many Arctic mammals do the same).[122] At one time when there was no male emu at Chiswick, a female 'collected her eggs together and sat upon them' (normally male cassowaries do most of the incubation).[123] Although it has been claimed that the Duke acquired a giraffe in 1827,[124] the giraffe that arrived in England in 1827 actually belonged to George IV (Section 5.6). The confusion arose because the Duke, having been invited to the King's birthday banquet, was staying at Windsor at that time, and recorded its arrival in his diary.[125]

In 1831 the Duke agreed to be a patron of the Surrey Zoological Gardens (Section 6.3), but by 1836 the menagerie at Chiswick had come to an end, the surviving animals having been moved to Chatsworth.[126] The Duke's patronage of the Surrey Zoological Gardens paid off when, many years later, he borrowed a giraffe to entertain another Russian grandee, Tsar Nicholas I.

John Bligh, fourth Earl of Darnley, an original subscriber to the fledgling Zoological Society in 1824 and later a member of Council and Vice-President,[127] created a menagerie at Cobham Hall in Kent, described by J. C. Loudon as

> the finest [menagerie] we know in England ... the kangaroo, the opossum, the zebra, the quagga, several kinds of goats, sheep, and deer, the ostrich, the emu, the cassowary and many other birds and beasts, live in harmony together on a lawn of several acres, finely ornamented by foreign trees and shrubs, and surrounded by a wire fence fifteen feet high. There are suitable ornamental structures for the animals to retire into, and constant attendants to see that they are properly provided with food, and that they do not injure one another.[128]

When the Earl's llama died, he gave the corpse to Joshua Brookes, who displayed its skeleton in his museum in London.[129] Some of the animals kept at Cobham before 1830 were illustrated in *Wonders of the Animal Kingdom*: a roebuck, an antelope, a 'Ganges Stag', and three kangaroos brought from 'New Holland' by Captain Bligh, which did well for a while 'with every expectation of their being naturalized; [but] they died a few months after their portrait was taken'.[130] The Captain may well have been Bligh of *Bounty* fame, who had named Darnley Island in the Torres Straits in honour of the Earl in 1792.[131] If so, the kangaroos must have arrived in 1810, when Bligh returned under a cloud from Australia. It seems that the collection was dispersed in or shortly before 1836, when the fifth Earl died, for a lemon-crested cockatoo in the Surrey Zoological Garden was then said to have been 80 years old, 'having spent nearly fifty years in the family of the late Lord Darnley'.[132]

Robert Heron, for many years MP for Peterborough, kept a large number of animals at Stubton Hall in Lincolnshire from about 1809 until his death in 1854. A volume of his *Notes* printed in 1851 deals mainly with political and social economy, but also includes observations on some of the animals in his menagerie. He was a serious zoologist and one of the first 'subscribers' to the Zoological Society in 1825,[133] the year before it was formally constituted, and he continued to exchange animals with the society after the gardens were set up. Two of Heron's suppliers were Kendrick, the knowledgeable dealer in Piccadilly who provided fish and reptiles, and Edward Cross, with whom he exchanged kangaroos and various birds, and to whom in 1823 he gave a panther, which had arrived as a present on HMS *Conway*. He was a member of the Africa Association, which may have facilitated the acquisition of animals from Africa. He also owned slaves on the island of Grenada

Heron bred goldfish, so successfully that from 1809 to 1815 they increased from about a dozen to about 1,100 in a paved pond in his flower garden—'of my original stock, six came from Burleigh, and six from Kendrick's, in Piccadilly. I have since had a few which Kendrick calls Brazil fish, but which do not differ from the others'. He also obtained some 'black gold fish', which he said were found only in lakes on a volcano in China.

In 1818 Kendrick supplied Heron with, 'three Brazil tortoises, 2¾ inches in the greatest length of the shell...they delight in the warm water of the aquarium... ', which suggests that as well as the fishpond he had a heated aquarium. Heron also received a chameleon, but three months after its arrival in 1820, 'a stupid under gardener destroyed him by hastily closing one of the lights on which he had climbed. He was brought from Brazil. During a journey of six months no food was given him, and it was a month longer before he recovered his appetite. His brother who travelled with him is at Exeter Change.'

By the 1820s Heron's birds were doing so well that he was able to send four pairs of golden pheasants, and a pair of black swans, to the Exeter Change, from where he later acquired Balearic cranes, and a pair of young emus in exchange for an old male; clearly he had breeding in mind. In July 1829 he was rewarded:

> In the winter, my emus had laid some eggs: we then confined them to their house, with a small temporary yard. The cock soon formed a nest, and arranging eleven eggs which were left him, began sitting so perseveringly, that he was never seen off the nest, and was obliged to be fed by hand. He never suffered the female to interfere in the daytime, yet, in the first week, the eggs were increased to fifteen; in sixty-seven days five were successfully hatched towards the end of March, and are now all grown and healthy.

Heron's attempts at breeding black swans were so successful that he was able to send cygnets to his friends, as well as the Exeter Change and the London Zoo; and many years later he was the first person in England to breed rheas.[134]

In 1815 Heron acquired a pair of half-grown kangaroos from the Marquis of Bath at Longleat in Wiltshire, who had bred a substantial number from a pair presented to his mother by the King and Queen in 1801 or 1802,[135] followed by a larger male from Sir Joseph Banks, and in 1820 by a male from Exeter Change. He must have acquired more females in the interim and had some success in breeding, since in 1821 he wrote that his three females had all produced young. In July 1832 he wrote: 'Three weeks ago, an infant kangaroo was found abandoned and nearly dead. The mother was caught, and the young one I put in her pouch; there it has completely recovered.' He received 'a pair of kangaroos from New Holland, of a species rarely seen here,—small and very dark' in 1832, which he identified as 'k. enfumé of Cuvier' (they were probably swamp wallabies), which soon produced an offspring. Heron's menagerie went from strength to strength over the next twenty years until his death in 1854.

Most of the meticulous records of Lord Fitzwilliam's menagerie at Wentworth Woodhouse in Yorkshire date from later in the century, but in 1803 he was already interested in importing black swans and by 1806 was attempting to breed (black?) swans and Greenland geese. Prince Pückler-Muskau visited Wentworth Woodhouse in 1827 and noted 'an inclosure made of wire fence, running along the gay parterres, peopled with foreign birds, a clear brook running through it, and planted with evergreens, on which the feathered inhabitants could sport at pleasure'.[136] A bear pit at still survives at Wentworth, but it is uncertain whether it was ever occupied.

5.6. George IV and his Adored Giraffe

As Prince of Wales, the future George IV had been involved with the importation of both red and fallow deer from Europe to improve the stocks of deer that were hunted on his various estates, and he had shown some interest in the animals presented in his name to the menagerie at the Tower. After having achieved his long-wished-for position as Prince Regent, with the final incarceration of his poor mad father in 1811, he received various gifts of animals, including a zebu cow 'of the Braminico or true Bengal breed' from Lord Minto the Governor-General of India, which was kept at Carlton House, and a 'very fine zebra a present from Sir John Cradock, the Governor of the Cape'.

Following the death of his father in 1820, George IV was able to indulge himself to any even greater degree than he had as Prince Regent, and he soon built up the most notable private menagerie in Britain. From 1824 most of his animals were housed in a series of sheds and paddocks beside the Lodge at Sandpit Gate on the edge of Windsor Great Park, some of the other large herbivores were kept at another entrance, Bears Rails, where they were cared for by a 'Keeper of the Wapeti Deer and other animals', while aquatic birds lived on the banks of Virginia Water. The first keeper of the menagerie, John Clarke, lived in the Lodge; he had previously been one of the 'Deer Park Keepers', then 'Senior Keeper', and then, 'From Age and Infirmity, his duties were, by His Majesty's Gracious Command, to be confined, more particularly to the care of the Animals and Birds at Sandpit Gate'.[137] John Frederick Lewis painted a portrait of Clarke in 1826 at Sandpit Gate with some of the animals in his care, including two small deer, a gazelle, a kangaroo, a pony, several peacocks, an emu, a macaw, a cockatoo pecking at Clarke's top hat in the foreground, and in the distance more peacocks and a zebu. Clarke stands slightly hunched holding a dish of food in one hand, while the other caresses the pony. Clarke and most of the animals are illuminated by a sudden shaft of sunlight, while the trees in the dark background merge into a black stormy sky—peace and tranquillity in an otherwise hostile environment, which is perhaps how the King perceived his menagerie.[138]

In 1825 Edward Cross, who had previously supplied several animals to Sandpit Gate, was charged 'by Special Command of His Majesty' with responsibility for superintending the animals at Windsor. In July he and Alfred Cops, the Keeper of the Tower menagerie, were involved in the transfer of some animals brought to England by Major Dixon Denham and Captain Clapperton, when they returned 'from the most successful expedition that has perhaps ever been made into the evil-omened regions of Central Africa'. Cops invoiced Sir William Knighton, Keeper of the Privy Purse, for the 'Cartage of 6 Horse Boxes from the Menagerie Royal to the London Docks' and 'for Lighterage, Landing &c &c &c on 2 Mandara Horses, 1 Sheep, 1 Monkey, 3 Dogs, 1 Ichneumon, 1 Sashe, 4 Ostriches ["a present from the King of Sokatoo, in Central Africa"], 5 Parrots', all presumably intended for Windsor, apart from the mysteriously named sashe and the ichneumon, which remained at the Tower. The accounts preserved in the Royal Archives at Windsor Castle show that, between May 1825 and George IV's death in 1830, Cross provided animals to the tune of nearly £1,000, as well as overseeing the transport of animals given to the King by various ships' captains, returning soldiers, and landowners.[139]

Almost all the mammals at Sandpit Gate and Bears Rails were herbivores; the most numerous being wapiti (American red deer, also sometimes termed 'elk'), some fine specimens being turned into the park. They may have been acquired with a view to improving the red deer required for stag-hunting. Some or perhaps all of the Indian deer—three chitals and three rusas (sambars)—were allowed the freedom of the Great Park; however, the two rusa stags not only 'disdained to congregate with the other deer', but fought each other, so it was thought prudent to transfer one of them to the Tower. The two other species of deer taken to Windsor were a roebuck and a 'South American roe deer' (South America has several species of small deer, but no roe deer). Cross charged £26 5*s*. for a pair of 'Corine antelopes', which were not antelopes in modern terminology, but Edmi gazelles. A dorcas gazelle was figured by Rennie, whose description reads: 'The most elegant of antelopes is the gazelle, the animal here represented was in the King's menagerie at Windsor in 1827, since dead.' An albino blackbuck brought from Bombay by Captain Dalrymple of the *Vansittart* was transferred to the Tower from Sandpit Gate, where it had been 'an especial favourite of His Majesty, on account of its gentleness and beauty'. There were also several llamas at Windsor, which 'were allowed to range in a paddock, but they did not long endure the climate'; several chamois suffered the same fate.[140]

On 8 June 1827 Cross charged £10 for 'disembarking and conveying to Windsor from the ship Slaney, a Zebra and two Gnus, plus an extra 12*s*. for a farrier and for three men to dress their feet', with £1 extra to the labourers who unloaded them. The transport of a third gnu from the West India Docks on 4 January 1828 cost a mere £4 5*s*. The gnus were said to have been among the King's favourites, which is probably why in 1828 Cross's friend Agasse was commissioned to paint them; he was paid 100 guineas 'for a picture of the Gnus by command' and £18 17*s*. for the frame.[141]

By 1830 the menagerie contained two pairs each of mountain and Burchell's zebras; one of the zebras was probably the 'perfect male zebra' supplied by Cross in the previous year for the large sum of £240; the epithet 'perfect' perhaps indicated that it had not been castrated, so it may have been the sire of the two zebra/donkey hybrids that were in the collection in 1830. Two quaggas supplied by Cross in 1826 were joined by a third, a gift from Sir Archibald Campbell, brought to England on the *Caesar* and landed at Gravesend in July 1829. Cross supplied at least two kangaroos, and others may have been received as gifts; at any rate, like their compatriot emus, they bred so freely in the park that by 1830 they numbered thirty or forty. Only one primate is recorded in the menagerie, 'a large Satyr' (probably a baboon) conveyed to Windsor in 1828; its absence from later records suggests that it did not survive long.

Picturesque aquatic birds were one of the attractions of Virginia Water, which the King had had deepened and cleared of weeds and where he built a picturesque Chinese fishing temple. He liked to entertain his mistress Lady Conyngham and other guests with 'picnics' in his royal barge, constructed of mahogany and teak, with green silk hangings and ornamented with carved dolphins and the royal coat of arms. Near the temple was an aviary; and gold and silver carp supplied by Cross swam in the lake. Most of the birds were supplied in pairs; they included crowned cranes (probably both the grey-crowned and black-crowned from southern Africa), demoiselle cranes, spoonbills, a scarlet ibis, and seven swans, including a black swan from Australia. Many kinds of geese graced the lake and its banks—Canada geese, barnacle geese, a Greenland (snow) goose, and four white-fronted geese (probably from Holland, where they breed in large numbers), and six shelduck. The 'Cereopsis geese' bred particularly well. The four 'Carolina summer ducks' (wood ducks from North America) were supplemented by native species, and Cross supplied a brace each of Chinese partridges and quails.[142]

The birds kept at Sandpit Gate were a king vulture, two sea eagles, a peregrine falcon, two great-eared owls from India, four macaws, two cockatoos, a king parrot, a scarlet lory, two golden parakeets (the national bird of Brazil), two roselle parakeets from Australia, a hornbill, five widow-birds from Africa, two curassows (donated by Sir Astley Cooper, 'Sergeant Surgeon' to the King, from whose head he had removed a cyst), and more than forty peacocks and peahens.

One of the ostriches presented by the King of Sokatoo died in 1827:

> The bird was in the finest plumage and condition; and to render the skin and internal structure of so interesting a specimen available to the purpose of science, his Majesty signified his pleasure that it should be presented to the Zoological Society. The skin has been accordingly preserved; and with preparations of several of the internal parts, will be placed in the museum in Bruton Street. On Friday, the body of the ostrich was dissected, in the presence of a

> number of distinguished professors and lovers of science, when a most interesting demonstration of its anatomy took place. Large pieces of wood, Iron nails, eggs, &c, were found in the stomach. The cause of death was pronounced to be obesity. After the operation the gentlemen partook of a portion of the flesh, which was declared to be excellent, and much resembling beef.

George IV sometimes showed interest in animals belonging to other people. In October 1824 the Marquis of Conyngham, the complaisant husband of the King's mistress, brought the King's attention to an extraordinary event in Thomas Atkins's travelling menagerie, which was then showing at Windsor Fair—a tigress had given birth to three cubs fathered by a lion.

> The King commanded [they] should be brought for his inspection to the Royal College [*sic*], Windsor Park. Mr A. accordingly took them on 1st November, taking with him also the terrier bitch, which acted in the capacity of wet nurse to the heterogeneous brood, as the Tigress had not evinced any disposition of acting the part of a mother to her family. His Majesty appeared highly pleased with this novel exhibition, and asked a great number of questions respecting them . . . and gave the name Lion-Tigers to the cubs . . .[143]

When the menagerie reached Bath two months later, Atkins claimed that, when the cubs had been shown to the King at Windsor, the King had taken them in his arms saying, 'long may you live and prosper, and be beneficial to your master'. In consequence of such royal approval, hundreds of people, 'from the peer to the peasant', flocked 'to review those extraordinary productions of Nature, an entirely new species of animal, between the Small Lion and Bengal Tigress'. Six months later Agasse completed a painting, *A Group of Whelps Bred between a Lion and a Tigress at Windsor*, which shows the three cubs and the terrier, with the parents looking on in the background from behind bars. It is inscribed: 'Bred Windsor Octr 1824. Painted when six months old.' It has been assumed that he painted the cubs in the royal menagerie at Windsor for George IV, but this was not the case. After their birth in Atkins's travelling menagerie at Windsor *Fair*, they were shown in various parts of the country, including Clapham, a south London suburb, which was probably where they were painted by Agasse.[144]

George IV spent the last few years of his life in seclusion at Windsor Castle, where one of his main enjoyments was driving briskly around the park in an open carriage with a parrot on his wrist and Lady Conyngham seated beside him; another was visiting his menagerie, where a glass of cherry gin always awaited him.[145] In August 1827 he received the most remarkable present, England's first living giraffe.

In the autumn of 1826 two young Nubian giraffes, little more than calves, were grazing with their mother on the plains bordering the Blue Nile in the Sudan. They were captured by two Arabs, who sold them to Mukar Bey, the Governor of Sennar, who sent them north on camel-back to Khartoum, where they were loaded on to a boat and shipped down the Nile to Cairo, to the Viceroy of Egypt, Mehmet Ali Pasha. Pasha was persuaded that the giraffes would make suitably prestigious gifts to influence the policies of Charles X of France and George IV, so they were sent on to the French and English consuls in Alexandria to arrange export. The English consul Henry Salt and Bernardino Drovetti, the French consul-general, were rival Egyptologists who sold and exported their excavated plunder. Drovetti moreover trafficked in animals; he had supplied Arabian stallions to Vienna, Nubian sheep to Moscow, gazelles to Princess Caroline of Naples, and a superb horse from Dongola to a prince in Turin. The problem arose as to who should have which giraffe, and it was decided to draw lots. Drovetti won, he chose the larger, healthier animal, and arranged for it to be shipped across the Mediterranean to Marseilles accompanied by three antelopes, as a gift from himself to Charles X.[146]

The French giraffe, known as Zarafa (the Arabic word for giraffe), spent the winter in Marseilles; the problem then was how to transport her to Paris. John Polito, whose menagerie 'La Grande Menagerie de M. Jean Polito d'Exeter Change de Londres' was travelling on the Continent, offered his services to the professors at the Musée d'histoire naturelle in Paris.[147] The professors declined his offer, because they thought that, as he only wanted to do it for the money, he might charge for people to view the animal en route. It was decided that the least hazardous way for Zarafa to travel the long road to Paris was on foot, under the supervision of the naturalist Geoffroy St Hilaire. She walked, accompanied by a cavalcade of guards, milk cows, her Arab keepers, and sundry attendants, covered 550 miles in 41 days, and arrived at the Jardin des Plantes in perfect health. Paris went Zarafa-mad. Parisians in their thousands flocked to see her, songs and poems were composed in her honour, and her image proliferated on china, furniture, knick-knacks, jewellery, textiles, and topiary. In women's fashions, giraffe-yellow was all the rage; women styled their hair à la giraffe and men wore giraffe hats and cravats.

It was a very different story with the sickly English giraffe. After being quarantined on Malta for six months, in June 1827 she was taken on board the *Penelope* Malta trader, for the long voyage to London. Two months later news of her impending arrival in the Pool of London was brought to George IV, who he ordered Edward Cross to attend to her care and landing.[148] The *Literary Gazette* described her arrival. The *Penelope* docked at a wharf under Waterloo Bridge at about six in the evening on 11 August 1827:

> A large craft, with a suitable awning of tarpaulins, was provided in which the camelopard, with two Egyptian cows (we believe in the character of wet nurses), two Arab Keepers, and an interpreter were brought from the vessel. They were immediately lodged in a roomy warehouse under the Duchy of Lancaster office. Here they remained until Monday morning, about 5 o'clock, when Richardson's spacious caravan, with four horses, was ready to transport them to Windsor. In this vehicle they were all safely stowed, and by it conveyed to Windsor the same evening. Having been lodged in security, the King himself hastened to inspect his extraordinary acquisition, and was greatly pleased with the care that had been taken to bring it to his presence in fine order...[149]

The King was quick to commission Richard Barrett Davis to record his prized new possession in a portrait, probably painted in the stables at Cumberland Gate in Windsor Great Park, where the giraffe was initially housed. Another painting by Abraham Van Worrell shows Edward Cross leading the giraffe by a halter, with the two cows in the background; an amulet is hung around her neck to protect her from the evil eye.[150] By October, on the King's orders, a warm, commodious stable, with a large paddock, had been built for his precious animal at Sandpit Gate. It was here that a third portrait was painted, this time by Agasse (Plate 11). It shows the giraffe in an enclosure, bending its long neck towards a bowl of milk held by two Arab keepers; Edward Cross is standing beside them. Agasse referred to the painting in his record book as *For His Majesty. The Zarifa and Portraits of his three Keepers*. In the painting Cross is looking down at the bowl with a slightly quizzical smile, perhaps unsure about the diet of milk, which had been ordered by the King; however, the order must have been rescinded, for records show that she was fed on barley, oats, and split beans, as well as leaves and branches of acacia, mimosa, and ash.[151]

By November arrangements had been made for the giraffe to be viewed by the general public at Sandpit Gate. The *Morning Post* announced the good news:

> His MAJESTY has granted permission for it to be seen every Saturday and Monday—on the former day before one o'clock, and at any time on the latter. In this menagerie they are not pent up in miserable dens, but have huge open sheds and spacious paddocks, to range in; water in plenty; and spreading trees to shade them from the noon-day sun . . . Here may be seen the giraffe, various species of antelope and deer, kangaroos in great numbers, zebras, quaggas, ostriches and emeus rearing their young as fearlessly as the barn-door fowl.[152]

Although the English giraffe did not provoke the same 'rage' in London as her sister in Paris, she was commemorated on china and various other objects, and a Monsieur J. J. Vallotin placed almost daily advertisements in the press begging 'most respectfully to acquaint the Nobility and Gentry that he has just returned from Paris, with a large stock of French Goods . . . ', including giraffe dresses, French ribbons à la giraffe, and giraffe reticules. The fashion pages of the newspapers suggested that a lady might dress her hair resplendently with 'puffs of giraffe-crape, crossed by plaits of hair, and crowned by a full plume of white feathers'.

The giraffe's knees became so weak that she was soon unable to rise without assistance. It took two men to hoist her up into a standing position by means of slings under her body, a pulley, and a windlass. The same procedure in reverse allowed the poor animal to recline. The construction of the hoist and other necessities for raising her to her feet was organized by Cross. He charged £100, not only for 'attending the Arrival of the Giraffe many journeys down the river for the purpose of preparing every thing necessary for landing it from on Board the Penelope and conveying it to Cumberland Lodge & remaining with it many days afterwards in attending and assisting in removal to its habitation at Sandpit Lodge', but also for 'very many journeys to construct a steel support in consequence of its weakness & attending the application of it—and also to construct a triangle to remove it from the slings by which it was suspended in order to remove it into the great Paddock, continuing with it near Six weeks during the period it remained out of doors'.[153]

Sir Everard Home, by now the King's Sergeant-Surgeon and Vice-President of the Royal Society, surveyed the external anatomy of the giraffe and then presented the results in a pompous lecture at the Royal Society; one of his conclusions, that the giraffe's prehensile tongue was coloured black in order to protect it from sunburn, proved highly risible to the *London Magazine*. Home's other contention that 'the giraffe preferred licking the hand of a lady to that of a man' was also much enjoyed.[154]

George IV's love of his giraffe provided a heaven-sent target for the cartoonists of the day; within a month of its arrival, William Heath's engraving, *The Camelopard, or a New Hobby* (Plate 12), was published.[155] It shows the King astride a prancing giraffe, with the distinctly curvaceous Lady Conyngham seated sideways behind him. The giraffe appears furious, as well it might, given the size of its overweight riders; several Arabs bow obsequiously in the background. The hoist provided yet more fuel for caricature; a print *The State of the Giraffe* shows the King and Lady Conyngham desperately turning a windlass to raise the drooping giraffe, its forefeet dressed in shoes and stockings and hanging over a chest full of medicine bottles.[156]

Various newspapers reported that the giraffe at Windsor was 'in a very drooping state', and in the summer of 1828 the King's physicians, Sir William Knighton and Sir Henry Halford, were brought in to advise on the state of its health—they pronounced it very precarious. They were said to have reported to the Privy Council 'that the indisposition of the Giraffe at Windsor has arisen from the animal's loyal sympathy to his Majesty's twinges in his toe, in his late fit of gout'. It was the general opinion that it would not survive the following winter.[157] Remarkably the giraffe did survive the pain and indignity of its treatment at Sandpit Gate, and continued to be an attraction throughout the following spring and summer. But on 11 October 1829 'His Majesty's giraffe' died.

The King could not bear to be parted from his beloved, so in death as in life no expense was to be spared in preparing the giraffe for its afterlife in the galleries of Windsor Castle.[158] James de Ville was brought in to make plaster casts of the body. The King's veterinary surgeon dissected the body, discovering in the process that the giraffe had suffered a dislocated shoulder in addition to its other maladies. Home was allowed to take away the stomach for detailed study, and on Christmas Eve he presented his findings on its anatomy to the Royal Society. Then the bones were cleaned and mounted as a skeleton by the articulator from Guy's Hospital. Meanwhile the Zoological Society's 'stuffer' John Gould used De Ville's casts to shape a wooden framework over which the skin was stretched and then stuffed. Once completed, it stood more than 11 feet from head to toe.[159]

The King did not live to see the transfiguration of his giraffe, for on the 26 June 1830 he died. His brother William IV lost no time in donating the stuffed giraffe and its mounted skeleton to the Museum of the Zoological Society in Bruton Street in London, and almost all the animals from Sandpit Gate and Bears Rails to the Society's Zoo in Regent's Park. Great efforts were being made to reduce expenditure at Windsor, and many staff were laid off, so the chance of finding a suitably prestigious institution to take on his brother's animals must have proved a godsend. The animals were sent in several consignments to London in 1830; the only animals left behind were three gnus, a roebuck, an axis deer, two eagles, and the majority of the birds at Virginia Water.[160]

Just over a year later the carnivores from the royal collection at the Tower menagerie followed George IV's 'graminivorous' animals to the Zoo. The Tower menagerie survived, *sans* Royal animals, in the capable hands of Alfred Cops.

5.7. The Tower Menagerie Revivified: The Excellent Management of Mr Cops

By 1815 the menagerie was in a bad way—*An Improved History and Description of the Tower of London* stated laconically of the animals, 'you view them through large iron grates, like those before the windows of a prison, so that you may see them with the utmost safety, be they ever so savage. Sometimes the dens are all occupied by beasts and birds; at other times there are many vacancies occasioned by that cruel enemy *Death*'—the menagerie had been reduced to a single lion, two lionesses, a panther, a hyaena, a tigress, a jackal, a mountain cow(?), and Old Martin, the grizzly bear.[161] The lion was probably Nero (2), drawn in 1814 by Edwin Landseer when he was only 12 years old; his drawing was engraved by his brother Thomas and published as *Nero, a Lion from Senegal, Now Exhibiting at the Tower of London*.

The decline of public interest in the Tower menagerie coincided with the ever-increasing popularity of the menagerie at the Exeter Change. Nevertheless a few animals continued to arrive, among them an elephant—a newspaper reported in 1820 that 'His Majesty has sent to his Royal Menagerie in the Tower the Elephant upon which the King of Candy, in the Island of Ceylon, was accustomed to ride', which, with a startling ignorance of zoogeography, was reported to have been accompanied by 'two emus from Africa and three large and peculiarly handsome ostriches, recently brought from India'.[162]

On 12 November 1821 it was announced: 'Died . . . Mr Joseph Bullock in the Tower of London, Keeper of his Majesty's Royal menagerie, formerly house-steward to the late Hon. Wm Pitt.'[163] Alfred Cops lost no time in applying for the post. A manuscript note in the Fillinham Collection in the British Library notes that 'Alfred Cops made application to the Marchioness of Londonderry & through her interference with Lord Liverpool [the Prime Minister] obtained him the situation . . . ', adding sourly that Cops was 'an American by birth, illiterate and of low origin, formerly a showman to the Company visiting the Menagerie at the Exeter Change';[164] as Cops was born in Whitechapel, it seems unlikely that he was American. As we have seen, the Marchioness—the wife of the Foreign Minister usually referred to as Castlereagh—had a considerable menagerie of her own in Kent and must have encountered Cops at the Exeter Change.

On 2 April 1822 the *Morning Post* announced that 'Alfred Cops, many years Superintendent of the Menagerie, Exeter Change, is appointed Keeper to his Majesty's Royal Menagerie in the Tower of London, and we have no doubt that under his judicious management that national establishment will soon become one of the first in Europe'. For the first time in its history the menagerie was to be managed as a commercial enterprise by a professional animal-keeper, one who was anything but illiterate, and who brought expertise and a flair for publicity reminiscent of the Exeter Change. He soon moved into the Lion House with his wife Sarah (née Willoughby or Willoughway) and their two daughters.

The King of Candy's elephant died, but dozens of new animals arrived, including a crocodile and an immense 'boa constrictor' (probably a python). Lord Liverpool gave Cops a mouflon (actually an African hair-sheep), and, as we have seen, the Marchioness of Londonderry presented a leopard, a pair of ostriches, a golden eagle, and a 'great sea-eagle' (white-tailed eagle). In August 1824 Cops advertised the menagerie as 'the daily resort of the Rank and Fashion of the Metropolis'. If his figures are to be believed, visitor numbers were soaring; upwards of 2,000 were claimed to have visited in a single week in June 1825. As tickets cost 1s per person, he was raking in over £100 a week in addition to his annual allowance as Keeper of the Lions of £206. An advertisement issued in July 1825 listed seven species of reptiles, thirty-two of mammals (including twelve primates), eighteen species of birds, and 'a great variety of other birds of most splendid plumage', amounting to over a hundred individual animals.[165] Cops issued at least five handbills between 1826 and 1829 listing the animals present in the menagerie, which, together with some of the guidebooks to the Tower, allow many individual animals to be tracked over time. The animals present in 1828 were described in some detail by the zoologist Edward Bennett in his well-known book *The Tower Menagerie Comprising the Natural History of the Animals Contained in that Establishment* (1829).

Although many of the animals that arrived in the Tower were gifts to the King, Cops was definitely in the business of buying and selling. The statement 'the highest prices given for every kind of foreign beasts and birds' appears at the foot of the handbills and in some advertisements. He sold, or hired out, some animals to 'Earl James and Son's Royal Collection of wild beasts from the Tower of London' in 1826; a lion from the Cape and two cheetahs from Senegal were removed 'owing to the spirit of commerce, and passed into foreign hands and are now [1828] on the Continent of Europe', and a young Asiatic bull elephant that arrived at the Tower in 1828 was shipped off very promptly to New York; eleven years later, having acquired the name Bolivar, he recrossed the Atlantic to perform at the Adelphi Theatre.[166]

In 1825 a newspaper reported that, 'from the dilapidated state of that part of the Tower called the Lions Office, Mr Wyatt, the architect, and clerk to the Board of Works, received directions to pull it down, and erect another building on the site of the old one, for the security of the royal collection of wild beasts'.[167] The ancient Lion Tower was about to be demolished. No doubt the improvements were also needed to accommodate Cops's ever-burgeoning number of animals. The cages were 'arranged around a large area, [and] with the benefit of the open air, it has the advantage over other exhibitions of the kind'—clearly a dig at the Exeter Change, where animals were crowded indoors.[168] The cages in a second yard, which also was open to the sky, were arranged in a double tier, and housed the carnivores. Some of the stabling was replaced by a room 70 feet long for the herbivores and birds. 'Within the interior, and supported by an artificial atmosphere, a numerous collection of living serpents of gigantic length and circumference, repose in folds, or twine in various ways their mottled bodies,' while the floor of their spacious cage was covered with red baize and filled with warm water.[169] Monkeys were housed in 'the fifth division'. A print entitled *The Menagerie in the Tower* (Figure 5.5) is said to date from about 1820, but probably depicts the cages in the second yard after Wyatt's improvements.

In 1825 there were four new lions in the Tower, all African: a Cape lion, a 'black-maned Caffrarian lion', and a 'full grown majestic lion and a lioness from the coast of Barbary'.[170] The pair of Asiatic lion cubs from Bengal presented by General Watson in 1823 had matured, and on 20 October 1827, the day of the Battle of Navarino, the majestic silver-maned male named George fathered three cubs, probably by an African lioness; they were the first lion cubs to be born in the Tower since the Glorious 1 June 1794. The lion's daily diet of 8–9 lb of beef 'exclusive of

Figure 5.5. *The Menagerie in the Tower* (*c.*1820; probably later) (anonymous engraving: Art Archive AA 352420).

bone' was clearly adequate. One of the cubs died, but two more were born in about 1829 and were sometimes displayed in the same cage as their parents.[171]

One of the few tigers to reach Britain in the early nineteenth century was a young male that arrived in April 1828; it had lived for a year in a paddock with a pony and a dog in Penang, and was so tame that the sailors played with him on board ship. Seven months later a female arrived from Calcutta to share his cage. The tigers soon outgrew their tameness, and two years later they were involved in a famous combat with a lion when a keeper inadvertently raised the partition between their cages. The lion sprang through the gap and landed on the tiger with a tremendous roar; the tiger responded in a paroxysm of fury, fiercely seconded by the tigress.

> The roaring and yelling of the combatants resounded through the yards, and excited in all the various animals the most lively demonstrations of fear or rage. The timid tribes shivered with dread, and ran round their cages shrieking with terror, while the other lions and tigers with the bears, leopards, panthers, wolves, and hyenas, flew around their dens, shaking the bars with their utmost strength, and uttering the most terrific cries.

The battle raged on for a good half hour until the keepers managed to separate the warring animals with red-hot iron rods and to drive the lion back into his own cage. He died a few days later.[172]

The handbill of 1826 notes a 'Malayan bear from Bencoolen: presented by Sir Stamford Raffles. The only one in England.' Shortly after its arrival in 1824, it was described as 'very full of action, though its movements may be called slow and measured; Its favourite position, however, is here represented by Mr [Thomas] Landseer—sitting on its haunches and thrusting out its long narrow tongue... it eats about two pounds of bread and milk a day' (Figure 5.6). According to Raffles's friend and colleague Thomas Horsfield, it was 'one of the most attractive and interesting specimens among the animals confined in the Royal Menagerie... the *Helarctos* readily distinguishes the keeper, and evinces an attachment to him. Our animal is excessively voracious and appears to be disposed to eat almost without cessation'. Its greed was its undoing, for one day in the summer of 1828 it 'over-gorged itself one day at breakfast and died. This was a severe loss to Mr Cops, who prized it highly, and to whom, in return, it was greatly attached. It delighted

Figure 5.6. Raffles's sun bear at the Tower; drawn and engraved by Thomas Landseer in 1824 (from Griffith et al. 1827: ii).

in being patted and rubbed.' Its stuffed skin ended up, appropriately, in the Museum of the Zoological Society.[173]

Only a few herbivores were kept at the Tower, as there was no space for them to graze, and those donated to the King were usually kept in his menagerie in Windsor Great Park, but a 'beautiful male nylghau from the Coromandel Coast' is recorded from 1827 to 1829. An albino blackbuck—the King's 'especial favourite'—and a rusa stag were transferred from Windsor. The blackbuck 'bore its confinement in the Menagerie with perfect resignation, and is remarkable for the mildness and tranquillity of its deportment'. The rusa stag, which was anything but gentle, 'exchanged the fresh and wholesome herbage and the unbounded liberty of Windsor Park for a small square of enclosure in the Menagery in the Tower, having for its companion on one side, the snapping pelican, and on the other, the spitting llama'.[174] In 1826 the East India Company's ship *Atlas* brought two zebras from Ethiopia, which were added to the menagerie. One of them was allowed to run through the Tower, with a man by her side, 'her only attempts to leave him were to run to the canteen for a draught of ale, of which she was exceedingly fond'—a very different animal from the aggressive zebra of 1803.[175]

A pair of kangaroos born at Windsor was transferred to the Tower in 1827, and in due course the female gave birth, 'a circumstance by no means infrequent in this country among those which are less restricted of their liberty'. Cops also acquired a more unusual marsupial, advertised as a kangaroo rat, 'the smallest of that species from Botany Bay', probably an eastern bettong, extinct on the Australian mainland since the early twentieth century. It survived in the menagerie for about four years.[176]

Although 'an infinite variety of Simia-Monkey Tribe' was housed in the menagerie, both Cops and Bennett had difficulty with their taxonomy, and Bennett was very dismissive of what he called 'their much vaunted intelligence'. A pig-tailed macaque died in 1828, and no wonder, for 'it was an excessive drinker of porter and sunk under a confirmed dropsy'.[177] A Bactrian camel, a seal pup, a tapir, and a ratel appeared in the menagerie in 1829, remaining there long enough to be illustrated in *Delineations of the Most Distinguished Wild Animals in the Various Menageries of this Country* (1829) and to be described by Robert Huish the following year.

Between 1826 and 1829 the Tower menagerie contained no fewer than four pairs of emus, some obtained from Windsor, where they bred successfully. Sir Joseph Banks's aspiration when the first pair arrived in 1820 that emus might become acclimatized was temporarily fulfilled, but, as Huish wrote: 'they might be found to answer domestic purposes, like the hen, or turkey. It must, however, be acknowledged that a full grown emeu would cut rather a preposterous figure on an English dinner-table.'[178]

By 1827 there were three pelicans from Hungary in the Tower. 'They are allowed daily to take an airing in the court-yard, where they are refreshed by copious bathing, from which they seem to derive great pleasure. Towards the close of the year 1828 the female laid three eggs, building for herself a very comfortable nest in one of the corners of their cage.' The eggs did not hatch, but, while the female was incubating them, the male fed her with some of the three dozen small live plaice that each bird was allowed per day.[179] Two pairs of curassows from Trinidad are listed in all the handbills between 1826 and 1829, along with an adjutant, three crowned cranes, a 'Cyrus or Polish crane', a pair of spoonbills from Holland, a pair of storks from Denmark, four vultures of different species, a horned owl from Virginia, and the usual pheasants, macaws, cockatoos, parrots, parakeets, and 'other birds of the most splendid plumage'. An eagle illustrated in a plate entitled *The Eagle of the Andes*, published in 1829, made a brief appearance in 1828 or 1829.[180]

The six 'boa constrictors' from Java, and the 'anaconda serpents' listed from 1825 onwards must have been pythons from India, Sri Lanka, or South East Asia, as boa constrictors and anacondas are confined to the New World; in 1826 one of these huge snakes nearly cost Cops his life:

> THE BOAS CONSTRICTOR. The following accident which occurred a few weeks since... Mr Cops, the keeper of the Menagerie at the Tower was a few weeks ago holding a fowl to the head of the largest of the five snakes there, when the snake darted at the bird, missed it, but seized the keeper by the left thumb, and was coiled round his arm and neck in a moment. Mr Cops, who was alone, did not lose his presence of mind, and immediately attempted to relieve himself from the powerful constrictor, by getting at its head; but it had so knotted itself upon its own head, that Mr Cops could not reach it, and he threw himself on the floor, in order to grapple with a better chance of success, when two other keepers coming in, they broke the teeth of the serpent, and with some difficulty saved Mr Cops from the fate of Laocoen...[181]

The other reptiles Cops acquired were a pair of rattlesnakes and a harlequin snake said to have come from Ceylon, which was shown to the King, who was 'highly pleased with its beauty and vivacity'; actually there are no harlequin snakes in Ceylon, so this was probably from South Africa. He also had a young 'alligator' from the Nile [*sic*]; another alligator captured on the banks of the Mississippi arrived in 1825, only to be killed by one of the two mongooses.[182] In July 1828 no fewer than a hundred live rattlesnakes in a single consignment arrived from the North America; the majority either died or were sold, for only one is recorded in subsequent handbills. Even so, it was reported in 1830 that a rattlesnake had unexpectedly produced 100 young ones, 'When first discovered they were moving in all directions on the floor of the room in which they were confined. They are extremely lively and incessantly in search of food.' Unlike most snakes, rattlesnakes do not lay eggs, but give birth to live young, so the fact that the birth was unexpected is understandable.

Bennett wrote: 'So excellent is the management of Mr Cops, especially as regards cleanliness, that not a single death has occurred from disease, and one only from an accidental cause: the secretary bird, having incautiously introduced its long neck into the den of the hyaena, was deprived of it and its head at one bite.' Many of the other animals that disappeared from the records probably died, but the longevity of some, including Old Martin, who was still going strong in 1828 after seventeen years in the Tower, must be attributed to Cops's good management. As we have seen, he also had some success with breeding.[183]

On 22 December 1828 the following paragraph appeared in the London newspapers: 'A statement which has appeared in most of the public prints, announcing the intention of his Majesty to present the Zoological Society with a number of beasts from the Royal Menagerie Tower of London, is altogether void of foundation.' However, the statement was *not* void of foundation. Change was in the air, and it would not be long before the menagerie at the Tower would be divested of the King's animals; Cops's spirit of commerce was about to be tested.

6

William IV, *c.*1830–1837

6.1. Sir Stamford Raffles and the Foundation of the London Zoo

The foundation of the Zoological Society and the setting-up of its menagerie, mentioned in several of the preceding chapters, contributed to the demise of the royal menageries at Windsor and in the Tower, or at least was a convenient receptacle for their animals when moves were afoot to close them both down. The society's relationship with the menagerie at the Exeter Change, soon under threat of closure, was more complex—it transpired that, although the zoo authorities hoped to acquire the animals, they had no wish to employ their owner, Edward Cross.

The menagerie of the Zoological Society of London owes its origin to Stamford Raffles, who had conceived the idea after visiting the menagerie at the Jardin des Plantes in Paris in 1817.[1] He had returned to England from the Far East in order to clear his name after having been demoted by the East India Company from the post of Lieutenant-Governor of Java to that of Resident at Bencoolen, a small, ailing company outpost on the coast of Dutch-controlled Sumatra. Things went well in London. He wrote and published a two-volume *History of Java*, was exonerated by the company, married, was elected FRS, and knighted by the Prince Regent.

Raffles was no mean zoologist himself. He was well aware of the pre-eminence of the French and well acquainted with the works of the great French zoologists, especially those of Georges Cuvier, so it is no surprise that, after resuming his post in Bencoolen later in 1817, he employed Cuvier's stepson Alfred Duvaucel and pupil Pierre-Médard Diard to collect animals for him and to catalogue his collection of preserved animals, an arrangement that lasted barely a year, after which he catalogued the collection himself and sent it to London—it was 'communicated' to the Linnean Society by

Sir Everard Home in 1822.[2] Raffles settled his growing family in a country house near Bencoolen, where he kept a menagerie and amassed collections of manuscripts, natural history drawings and paintings, as well as botanical and zoological specimens, some of which were shipped to London. Raffles's activities were not confined to Bencoolen. Anxious to extend British influence *contra* the Dutch, he was instrumental in the establishment of the hugely successful British port on the island of Singapore off the tip of the Malay peninsular, an achievement to which he owes his perpetual fame.

Having decided to retire, Raffles left Bencoolen in February 1824, but on the first night out his ship the *Fame* caught fire—all the passengers and crew escaped in lifeboats and returned to port—but he lost thousands of drawings and papers, more than £20,000, and a 'second Noah's Ark' of live animals—'Tigers Bears Monkeys &c'. He finally arrived in Plymouth on board the *Mariner* in August.[3] While awaiting the *Mariner* in Bencoolen, he had managed to acquire several live animals, including a Sunda clouded leopard, a tiger-cat, and several monkeys,[4] as well as a bear of the species that he had first made known to science in 1822, the sun bear (*Helarctos malayanus* Raffles).[5] The clouded leopard ended its days in the Exeter Change, and, as we have seen, the bear was sent to the Tower.

Back in England, in March 1825 Raffles wrote to his cousin:

> I am much interested at present in establishing a grand Zoological collection in the metropolis, with a Society for the introduction of living animals, bearing the same relations to Zoology as a science, that the Horticultural Society does to Botany... Sir Humphrey Davy [the President of the Royal Society] and myself are the projectors; and while he looks more to the practical and immediate utility to the country gentlemen, my attention is more directed to the scientific department...

The support of the 'country gentlemen' was vital, since it was from them that subscriptions were to be sought to fund the whole enterprise, and to this end the *Prospectus* stated that animals were to be collected with the intention of acclimatizing them as new, useful, domestic animals. It has been argued that the motivation for the establishment of the Zoological Garden was that it should be a showpiece of the British Empire, but it is clear from Raffles's letter to his cousin and from the two versions of the *Prospectus* he drafted that he was primarily concerned with the promotion of the 'scientific aspect' in competition with the French.[6]

Following the success of Raffles's and Lord Auckland's application to the Commission of Woods and Forests for 20 acres in Regent's Park for the establishment of a zoological garden, the Zoological Society of London was formally set up on 29 April 1826 at a meeting in the Horticultural Hall, with Raffles as President. Cross had offered the animals from the Exeter Change to the Society for £3,000, but had been turned down on grounds of cost. Nevertheless it was resolved that 'an arrangement be made with the Keepers at Exeter Change and the Tower for taking charge of such animals as might be presented to the Society till their own establishment is completed'.[7]

The museum and the society's administrative offices were soon established at 33 Bruton Street in fashionable Mayfair, and scientific meetings for the members were held there. Its collections grew at an extraordinary rate—among the first specimens were Raffles's sun bear and several animals from Windsor, including one of the ostriches given to George IV by the King of Sokatoo, and the stuffed body of the King's famous giraffe. By January 1829 the Museum contained no fewer than 600 specimens of mammals, 4,000 birds, 1,000 reptiles, and fishes, 1,000 'testacea and crustacea', and 30,000 insects, many having been sent to England by Raffles. Some of the smaller live animals were also kept in Bruton Street.[8]

Needless to say the society had to appoint a keeper to take charge of the live animals that were beginning to arrive. The two people with the most obvious expertise were Cross of the Exeter Change and Alfred Cops of the Tower. Cross offered his services, but was turned down. It is not known whether Cops was offered the position, but in the event *James* Cops, who must surely have been relative, was appointed in July 1827 with the princely salary of 1 guinea a week; seven months later he was sacked for 'several acts of misconduct'.[9] Alfred Cops was much better off at the Tower.

On 27 April 1828 the gardens were officially opened by the Duke of Wellington, an early member of the society. By January the following year they contained over 430 animals. The society received a Royal Charter from George IV in March 1829. Fashionable London flooded in; although members of the public were admitted only if introduced by a member, in fact tickets could often be purchased from the gate-keepers or in nearby pubs. Nevertheless, care was taken 'to prevent the contamination of the Zoological Garden by the admission of the poorer classes of Society'.[10]

In August 1830, two months after the death of George IV, William IV presented almost all the animals from Windsor to the Zoological Society: 'The inhabitants of Sand-pit-gate Menagerie, so highly prized by his late Majesty, are to be presented to the Zoological Society—the King says he has no taste for birds and beasts.' It would not be long before they were joined by the animals from the other royal collection—the menagerie at the Tower (Section 6.4).

Another person who had no taste for birds and beasts, at least those in the Tower, was the powerful Duke of Wellington. He had been appointed Constable of the Tower early in 1827 and by 1830 had begun an extensive programme of reorganization and improvement and must have seen the menagerie and its buildings as impediments, nor was he happy about the security arrangements.

6.2. Edward Cross: The End of the Menagerie at the Exeter Change and the Move to the King's Mews

Meanwhile time was running out for the menagerie at the Exeter Change. The scheme for the rebuilding of the Strand was coming to fruition, and the entire building was about to be demolished. Once again Cross offered his animals to the Zoological Society with himself as keeper, but, as they only offered to purchase part of his collection, he turned them down. It has been suggested that the reason the society would not accept Cross was that he was essentially a showman, and they doubted if he would have been a suitable manager—an extraordinary decision given that he was the most experienced menagerist in Britain. Even his supervision of George IV's menagerie at Windsor seems to have counted for nothing.

On 22 January 1829 Cross received notice from the Commissioner of Woods and Forests to quit the Exeter Change within a month; it was assumed that he would now have to agree to the terms of the Zoological Society. But Cross did not agree, and negotiations came to an end. His application for an extension of time was turned down, and it was reported that he was gradually disposing of his stock and intended to 'retire upon his fortune'. A hastily printed handbill reads: 'In consequence of the proposed alterations to the Strand, the Royal Menagerie Exeter Change will shortly be pulled down!! Delay not to pay it a visit, or the opportunity will be lost and *You will never see its like again.*'[11] A print depicts the eastern end of the Exeter Change plastered with notices, stating 'LAST WEEK' and 'NO PART COLLECTION REMOVED' (Figure 6.1).

At the last very last minute the government offered alternative, temporary accommodation for the animals in one of the wings of the King's Mews at Charing Cross. In April 1829 Cross trundled the last of his animals along the Strand to the Mews. He arranged a private view of the menagerie on 18 April, which was attended by 'several persons of distinction', 'who regretted only the absence of an elephant'; two days later the *Royal National Menagerie* opened to the public, from nine in the morning until nine at night, admission 2*s*. By June it was 'quite a fashionable lounge... Daily thronged with visitors, feeding time 4 o'clock.'

Figure 6.1. The final week of the Menagerie at the Exeter Change, April 1829 (anonymous engraving: Bridgeman XJF377216).

The star attraction at the Mews was a mandrill named Happy Jerry. He had been taught by a previous owner to smoke a pipe, which he appears not to have relished, although he could be persuaded with a glass of grog. His cage was furnished with an armchair, into which, when ordered, he would seat himself with great gravity every day at four o'clock.

> His keeper having lighted his pipe, presented it to him; he inspected it minutely, sometimes feeling it with his finger, as if to know if lighted, before inserting it in his mouth . . . He was fed chiefly on vegetables, and preferred them cooked; but when he visited Windsor, where he was exhibited to his late Majesty, he is said to have dined upon hashed venison with no ordinary degree of avidity.[12]

Early in November 1829 Cross acquired a very fine, docile elephant aged 7 years, and was able to claim 'this Magnificent exhibition now boasts of two elephants (male and female) an occurrence not witnessed in London for many years . . . '. Rajah Meer's elephant had been joined by a rather nervous female named Lutchm. Perhaps something had been learnt from the death of Chunee, for everything now was to be done with kindness. Later in November another elephant, Mademoiselle Djeck, who had been performing in Paris, landed at Blackwall; for most of December she was on stage at the Adelphi as the star of the show *The Elephant of Siam*, before leaving London to tour the provinces.[13] Not to be outdone and unable to resist the earning power of a performing elephant, Cross arranged for Lutchm, after three weeks' training, to appear at the Coburg Theatre in a satire on the *The Elephant of Siam* written by Thomas Dibdin. It was entitled *Siamoraindianaboo, Princess of Siam, or The Royal Elephant*, and ran for the whole of January. A playbill reads:

> for the 19th time, an entirely New Terrific interesting
> Nondescript translated from nothing, and
> founded on undoubtedly
> questionable facts, which never took place either
> in the BURMESE WAR,
> or an other, with Horrid Music, Dreadful Scenery and
> Dilapidated Ruins, the Incidents perfectly unnatural,—the
> Combats unlike any thing ever attempted in *this* Life, and the
> *Spectres* resembling nothing in each other, their appearance
> being more tremendously ridiculous than the worst
> ever attempted at any

Contemporary Theatre, whether Major, Medium, or Minor, combining all the necessary requisites for a Ghostly Equestrian, Assinian, Monkeyan, Pedestrian, Vocal, Rhetorical, Allegorical, Pantomime Tissue of Nonsense equal to any thing that graced or *disgraced* a Theatre Royal as a melange of Banditti, Soldiery, Peasantry, Nobles, Princes, Princesses, monks and MURDER, under the unassuming Denomination of SIAMORAINDIANABOO, PRINCESS OF SIAM; or the ROYAL ELEPHANT![14]

Lutchm then returned to the King's Mews. Miss Djeck eventually came to a bad end and was slaughtered in Geneva in much the same manner as Chunee in 1837.[15]

The handbills issued from the menagerie at the King's Mews have rather fewer animals than those from the final years at the Exeter Change, but visitors flocked there notwithstanding, and, as before, animals came and went. Charles Knight was able to write 'in the well-arranged menagerie of Mr Cross, removed from the Exeter Change to the King's stables', and Thomas Kelly issued engravings of some of the animals: a gnu, a moose, the American bison, and two 'Condors Minor, or Ash Colored Vultures'.[16] Lord Prudhoe brought him a 'Leucoryx, or bull of the desert from the banks of the Nile'—that is, an oryx; a cassowary, 'the only one in England', arrived in October, along with a 'beautiful young lion kindly deposited by Captain Fitzclarence'.

Although the menagerie at the Royal Mews carried on much as normal throughout 1830, it was not to last, as the demolition of the mews was imminent. Once again Cross had to decide whether to disband the collection and retire, or whether to find alternative accommodation. Cross was a survivor, and it would not be long before he was able to open his very own zoo, rivalling the Zoological Gardens in Regents Park: the Surrey Zoological Gardens.

6.3. Edward Cross and the Foundation of the Surrey Zoological Garden

After, one presumes, much activity behind the scenes, a group of gentlemen met at Horns Tavern in Kennington on 26 January 1831 and agreed to establish the 'Surrey Zoological Garden' at Walworth Manor, south of the river, between Walworth Road and Kennington—a 15-acre site consisting mainly of grazing land, but with a fine landscaped garden and a lake. The expenses were reckoned to be £3,500 to be paid to Cross for the animals, £3,000 for the buildings, £1,000 for fencing, £1,200 for the laying-out of the grounds etc., £500 for lodges and other buildings, and £800 for incidental expenses. It was proposed to raise the money by debentures of £25 each. All agreed that the new establishment would not be a rival to the Zoological Society Gardens and that duplicates might be exchanged between them. Although not explicitly stated, Cross was to be the proprietor and to reside in the Manor House.[17]

Meanwhile the animals remained at the King's Mews, including Lutchm the elephant, still doing 'most of the tricks of the elephant at the Adelphi', but on 4 April Cross announced that 'the 'grandest display is always reserved for the concluding act... the grandest display of Foreign Animals and Birds at the Kings Mews ...This exhibition will close in a few days time. A great variety of Foreign Animals and Birds, &c for sale.'[18]

Many of the animals were taken from the mews to the Surrey Gardens, among them Happy Jerry the mandrill, although he died soon after his arrival—his body was embalmed 'in, we hope, his favourite spirit' and put on display. Another was Dean Buckland's pet hyaena, Billy, which had arrived at the Exeter Change in 1821; when Billy died twenty-five years later, his skeleton was mounted and displayed in the Museum of the Royal College of Surgeons, but his stuffed carcass was returned to the Surrey Zoological Gardens, where it remained on show until 1855. His skin, no longer stuffed, survives to this day in the Natural History Museum in London.[19]

The building works were soon underway. Tenders for the provision of a tearoom were invited, and zebras and alpacas could be seen grazing there; 150 people had become subscribers, among them the Duke of Devonshire, the Archbishop of Canterbury, and Cross's good customer Sir Robert Heron. After a private view for a select few, the gardens opened to the public on 16 August 1831, and by November the number of subscribers had risen to 600.

By January 1832 the Surrey Gardens were rapidly advancing towards completion. A vast circular conservatory, 300 feet in circumference, covered with 6,000 square feet of glass—the largest continuous expanse of glass in England—contained cages for the big cats and other carnivores. Its supporting columns, hung with bird cages, formed a colonnade through which visitors could perambulate, shaded by 'choice handsome exotics', or, according to another report, by 'the greatest number of distinct species of climbing plants ever seen together'. Scientific respectability was reinforced by the appointment of William Swainson as Honorary Zoologist to the menagerie, assisted by the ornithologist Edward Gray of the British Museum. *Kidd's Guide to London*, published in 1832, contains pretty vignettes of some of the buildings, and mentions paddocks for the antelopes, zebras, camels, llamas, the ostrich, and a cassowary. A 'beautiful, picturesque and ruinous pile', 'formerly the vivarium of Mr Joshua Brookes', was erected in the centre of a pond, with chambers for beavers at its base and chained eagles on the ledges. A semi-circular glazed building had been erected for

> those amusing prototypes, the monkeys, where they have abundance of room . . . a long range of aviaries . . . a grand refectory and other refreshment rooms, hermitage &c . . . in various parts of the ground . . . principal walks and avenues planted with every description of native and exotic forest trees . . . each labelled with its common and scientific name . . .

The success of the Surrey Zoological Gardens seemed assured.[20]

Although the Surrey Gardens did not have quite the same fashionable *elan* of the Zoological Gardens in Regent's Park, it was generally reckoned that the grounds, with their mature trees and lake, were more attractive, the animals and plants were properly labelled, and the buildings were both more imposing and more suitable than those at the London Zoo. There were many similarities—both charged 1*s*. for admission; the public were admitted to the Surrey Gardens only on production of an order signed by a subscriber, and at the Zoological Gardens only if introduced by a Fellow; and, while William IV was patron of the Zoological Society, his wife Queen Adelaide was patron of the Surrey Gardens.[21]

Gifts of animals continued to pour into the London Zoo from its aristocratic supporters, and many more were purchased, but the staff had little success with carnivores; many died and few of the young lions reached maturity. Meanwhile, at the Surrey Gardens, Cross with his more experienced keepers still had many of the animals from the Exeter Change and the Mews, including the 'emasculated lion' that had arrived in October 1827, which was drawn by Edward Lear in 1834. Among the many animals at the Surrey Gardens were a common seal, a pair of Indian elephants, a pair of Humboldt's woolly monkeys (one drawn by Lear), various other monkeys, including a pair of Barbary apes, a gnu, an onager 'taken in a pitfall in Astrachan', antelopes, zebras, camels, llamas, ostriches, cassowaries, pelicans, black swans, gold and silver pheasants, a condor, boa constrictors, and gold and silver fish. Cross acquired a young Indian rhino from Burma in 1834—quite an investment as its 'food and conveyance to England' cost him £1,000.

Rivalry with the Zoological Society intensified in 1834 when Cross, having heard that the society had commissioned a M. Thibaut to acquire giraffes from the Sudan for Regent's Park, offered to purchase one of them. The society, of course, refused; nothing daunted, Cross sent John Edington Warwick, 'a travelling naturalist of great merit', to Egypt to acquire some for the Surrey Gardens. Thibaut obtained three males and a female, but, while they were quarantined at Malta, awaiting shipment to London on a naval vessel, the Zoo authorities received the news that Warwick's venture had also met with success. In their anxiety to be the first to exhibit giraffes in England, the Zoo authorities cancelled the booking and chartered a steamship instead. Thibault and his four giraffes left Malta on 4 May 1836; five days later Warwick's giraffes embarked at Alexandria on a converted Austrian brig. Thibaut's giraffes arrived at Blackwall on 24 May and were escorted early next morning to Regent's Park by their Nubian keepers, Thibaut himself, Edward Bennett, the Secretary of the Society, and some other Council members.

Warwick, his giraffes, and their three Abyssinian keepers arrived at Stangate Creek on the Medway estuary on 30 June and, after a week's quarantine, reached the Surrey Gardens on 6 July, along with five ostriches, eighteen Numidian cranes, one camel, five jerboas, and a number of other animals.[22] A week later Cross and Warwick displayed them to Queen Adelaide.

The fifth anniversary of the Surrey Gardens was celebrated in August, with a 'Grand Ridotto al Fresco, and Instrumental Concert', provided by the full Military Bands of the Coldstream and Fusilier Guards. The French Heruclean Gladiators M. Rozet and M. Fleury exhibited their surprising feats, and Herr Werner gave a 'Zoological Concert, introducing the principal animals and birds'; the evening concluded with 'an Imitation of Fireworks, never before attempted'. Admittance was 1*s*. The giraffes were the star exhibits.

A few weeks later one of the giraffes from the Surrey Gardens was taken north and shown in Liverpool and Manchester, and the two others were exhibited at the Cosmorama Rooms in Regent Street in London. Travelling may have undermined their health, for they all died; two were dissected by Richard Owen 'at a full meeting of the medical profession' in January 1837. One of the giraffes at the Zoo also died, but the others settled in well, becoming, as the authorities had hoped, one of the major sights of London.

Despite the large and varied collection of animals, the Surrey Gardens had begun to go downhill—to make ends meet they were sometimes hired out for 'fancy fairs', flower shows, archery competitions, balloon ascents, and the like, and as the years progressed Cross brought in an extraordinary array of attractions. The first of these spectacular pyrotechnic displays was a night-time re-creation of the Eruption of Vesuvius shown on various dates in the summer of 1837, shortly after the death of William IV.[23]

A year later Agasse, many of whose later pictures of animals were probably painted in the Surrey Gardens, painted a portrait of Edward Cross, standing proudly in the great glass dome and holding a lion cub (Plate 14). Cross retired in 1844 and died in 1854 at the age of 80. The following year, when all the animals were put up for auction, Billy was one of the last lots to be sold: 'A stuffed Hyæna on stand, twenty-four years in the Surrey Zoological Gardens.'[24]

6.4. The Demise of the Royal Menagerie at the Tower of London and a New Beginning

By 1830 an extensive programme of reorganization and improvement was underway at the Tower of London, instituted by the powerful Duke of Wellington, who had been appointed Constable of the Tower three years before. He must have seen the menagerie and its buildings as impediments, nor was he happy about the security arrangements. Accidents were not uncommon—one, in May 1831, particularly fuelled his ire. A private in the Foot Guards put his hand on, or more probably into, a cage containing a lion 'couchant', the lion sprang up and seized the soldier's hand in one of its paws, and, although one of the keepers soon managed to beat the lion off, the hand was badly lacerated. The Duke was a formidable adversary and, no doubt with William IV's approval, insisted that the royal animals be given notice to quit.

In November 1831 *The Times* announced that 'the King has presented to the Zoological Society the entire contents of his menagerie, now deposited in the Tower ... Mr Cox [*sic*] still retains his own private collection of wild beasts, and extraordinary and rare birds, the whole of which will continue to be shown as usual'. Cops could not easily be removed. He had been appointed Keeper of the Lions for life, as well as the occupancy of the Lion House, and he had no intention of either leaving or dying.

After the departure of the royal animals in 1831, Cops carried on advertising and managing the menagerie, buying and selling animals as before; he even sold a lion to the Zoological Society for 100 guineas and may have taken grim satisfaction from the fact that the lion, the pride of the society, died in agony in Regent's Park. Duke or no Duke, neither he nor his animals were about to be dislodged. Two handbills issued in 1832 show that he still had a respectable collection of his own at the Tower—the most spectacular being a 'great male elephant with large ivory tusks', which was soon joined by a female. The nylghai acquired in 1827 had been joined by a gnu from southern Africa and a deer from Virginia, and a new zebra arrived in the summer of 1832. As well as the usual 'variety of other birds of splendid plumage', Cops had gold and silver pheasants, a pair of emus, a griffin vulture, a horned owl, and a pair of pelicans. The reptiles included a crocodile and a python, which suggest that his heated accommodation was still in use. As usual, Cops was not up to specifying 'the variety of the Monkey Tribe', apart from a pair of white-headed lemurs, but in the summer of 1832 he added an 'Orang Outang!!':

> That astounding production of nature, OURANG-OUTANG, WILD MAN.—Grand exhibition, Tower of London.—Mr Cops has just returned from a tour through Scotland, with the above rare and extraordinary animal, which has made such an impression on the minds of those who have seen it not easily to be effaced. This great living curiosity, added to the present magnificent Collection, viz., stupendous Male and Female Elephants, in ONE DEN, Lions, Royal Tigers, Panthers, Leopards, Hyenas, Wolves, Great Boa Constrictor Serpents, Crocodiles, Chamelions, and an infinite variety of other Beasts and Birds, is now on view, at 1*s*. each person, and will afford to the reflective and curious a rich intellectual treat.[25]

Sir William Jardine's description and figure of Cops's 'ourang-outang' show that it really was an orang-utan. Dressed in a flannel shirt, this young, very tame great ape lived with Cops in the house in which he was lodging in Edinburgh in August 1832, but probably died soon after reaching London.[26] Quite why Cops was in Edinburgh in the summer of 1832 has not been explained, but he must have had a competent assistant caring for the menagerie in his absence, probably a Mr Roberts, who is said to have been employed at the Tower for at least ten years.

Early in 1833 Cops sold some or all of his lions, his tiger, and some leopards to an American named Lewis B. Titus, a partner in the menagerie firm of June, Titus, and Angevine (also known as the Zoological Institution) based in New York State. Titus took the animals, in the charge of Roberts, to New York on the ship *Julian*. It seems that Roberts had already been training the tiger, for, when the menagerie went on tour in New England, he entertained the crowds of spectators by entering its den. One day, when the menagerie was showing in a small town in Connecticut, the tiger turned on him and mauled him so severely that he died. So ended one of the first and least known of the 'lion tamers'. Roberts's assistant stepped into the breach; his name was Isaac Van Amburgh, soon to became famous, first in the United States and then in Britain and France; his most alarming performance involved placing a boy with a lamb in his arms into a cage containing a lion, a lioness, and a tiger.[27]

Cops's menagerie was still extant in 1834, but with fewer animals and at the much reduced admission charge of *6d*. The Duke of Wellington was becoming increasingly irritated by the presence of the menagerie. He wrote to Major Elrington, the Governor of the Tower: 'I shall be much obliged to you, if you will let me know whether there be any and what wild beasts in the Tower and whether they could be removed.' Major Elrington responded with, 'herewith a Return of the Animals which are the property of the Keeper, A. Cops, who holds the situation by Patent from the Crown'. A serious accident occurred in August 1835, when a guardsman was badly bitten by a monkey, and this time the news reached the King. It was the excuse the Duke needed. He wrote again to Major Elrington: 'The King has spoken to me respecting the Wild Beasts in the Tower since I last addressed you. It appears that his Majesty has been informed that one of them has injured an Officer and a Soldier. The King is anxious therefore that they should be removed from the Tower altogether.' A fortnight later he sent another peremptory note: 'The King is determined that Wild Beasts shall not be kept there. Mr Cops had better dispose of his.' On 27 August 1835 Cops advised Major Elrington: 'I have arranged that his Majesty's wishes should be obeyed and that the Exhibition should be closed tomorrow.'[28]

The *Morning Post* was not amused:

> THE LIONS. The spirit of political economy has at last found its way to the Tower of London. The Reformers have determined to suppress the Royal establishment so long provided for the king of beasts. All the mighty lions and ferocious monsters which so long constituted one of the grand exhibitions of the metropolis have been tried and condemned to suffer secondary punishment or transportation. The dens and all the apparatus for muzzling bears and taming the genus felis are to be sold forthwith to the best bidder; in short, the Tower menagerie is abolished...[29]

More than 600 years after its foundation, the world's oldest menagerie had come to an end. But, menagerie or no menagerie, Iron Duke or no Iron Duke, Cops was not finished, he was still able to draw his salary and to continue to live with his family, rent free, in the comfort of the Lion House. He was involved with Lewis B. Titus and his protégé 'Mr Van Amburgh, the Conqueror of the Brute Creation', after their arrival in London in 1838, and with the acquisition of America's first giraffes in 1840—but those are stories for another day.

Alfred Cops died on 21 March 1853.[30] Six months later the Lion House was demolished. Nothing now remains. The roar of the royal lions has been replaced by the noise of traffic as it grinds its way along Tower Hill.

7 Conclusions

The reasons for keeping exotic animals were many, varied, and overlapping. From the earliest times they were, as they still are, considered to be suitable gifts for royalty, the lion as King of the Beasts being particularly appropriate as an acknowledgement of the royal status of kings of the human kind. Following the example of most of the British monarchs (or their queens) from James I onwards, many aristocrats acquired collections of exotic animals to show off on their country estates. Yet prestige was not the only motive, as John Caius had written as early as 1570: 'for we English men are marvailous greedy gaping gluttons after novelties, and covetous coruorauntes of things that be seldom, rare, straunge, and hard to get.' This greed for novelty underlay the curiosity about the natural world that developed throughout succeeding centuries, spurred on by the discovery of new lands across the seas, resulting in the creation of 'cabinets of curiosity'. This collecting mania is epitomized by the great museum collections of Sir Hans Sloane and the Duchess of Portland, both of whom also owned many exotic animals, and both of whom were serious natural historians. The epithet 'curious' did not mean odd as in modern usage; a curious person was an inquisitive person. It was to the 'curious' as well as to the 'nobility and gentlemen' that showmen and animal-dealers were to appeal for custom.

Although many of the early owners, both men and women, valued their animals because they were exotic, exciting, strange, beautiful, or obviously expensive, they might also obtain much simple enjoyment from possessing them. This is, of course, true of individual pet animals on whom much affection might be lavished, but there are many other examples—when the Duke of Richmond's beloved lioness died, he erected a statue of her in his garden at Goodwood in Sussex and the Duke of Chandos trained his crowned cranes to feed out of his own hand.

In the eighteenth and nineteenth centuries some aristocratic owners expressed an interest in the acclimatization of various animals, believing that they might become useful domesticates; indeed this was one of the stated objectives of the London Zoo when it was first set up. None of the attempts succeeded, probably because the animals were not imported in large enough numbers to avoid the genetic bottleneck that occurs when only a few individuals are used in an attempt to start a breeding population. Cross-breeding was also tried in the hope of creating useful animals, following the example of the improvement of English racehorses when bred with Arabs and Barbs from the Levant and North Africa. As we saw with Lord Clive's zebra, inter-specific crosses were also attempted, but the offspring, if any, were almost invariably infertile, and, unlike mules, not obviously useful.

We should not forget that animal-dealers, showmen, and their wives or widows, valued their animals as a means of earning a living—hard cash. But exotic animals were not simply expensive commodities to be bought, sold, or exchanged. They had to be acquired, fed, watered, housed in secure cages or caravans, mucked out, and transported, necessitating a massive investment in terms of staffing and equipment. Publicity had to be handled, advertisements and handbills drafted, printed, and distributed. Those who managed the day-to-day care of the animals must have acquired an intimate knowledge of their behaviour and needs, far outweighing that of the landowners and zoologists who denigrated them as ignorant showmen.

Needless to say, natural historians were very interested in studying exotic animals, picturing and describing them in attempts to impose order on the apparent chaos of the natural world by devising various schemes of taxonomic classification. The science of comparative anatomy was already quite advanced in the seventeenth century, when a surprisingly large number of animals, both familiar and exotic, were dissected by physicians, including William Harvey and Sir Thomas Browne. Dr Edward Tyson concluded from his dissection of a chimpanzee that arrived in 1698 that in many respects 'our pygmie' was intermediate in form between humans and Barbary apes, thereby adding an extra step to what he called 'this Chain of the Creation'. Nearly 100 years later George Bailey, a London showman, described his baboon—named 'Child of the Sun'—as the 'long-lost link between the human and brute creation'.

Until well into the nineteenth century new or spectacular animals were designated in published books and advertisements as examples of the 'Wonders of the Creation'. For example, the title page of the booklet describing Stephen Polito's travelling *Collection of Living Beasts and Birds* in 1803 quotes Psalm 104: 'O Lord, how manifold are thy works! In wisdom hast thou made them all'; and even the *Prospectus* for the nascent Zoological Society in 1825 defined zoology as a 'branch of Natural Theology, teaching... the wisdom and power of the Creator'. These genuflections to the Creator seem to have been made in order to avoid offending the prevailing morality of the time, for, although scientists such as the great eighteenth-century anatomist John Hunter (who had his own menagerie at Earl's Court) had presented their findings in a somewhat oblique manner, ideas of evolution were gaining ground. It would not be long before zoologists would have to face up to the obvious similarities between ourselves and the great apes and realize that, like apes, human beings are members of the 'brute creation'.

Artists too made use of the living animals in the menageries, as subjects for prestigious oil paintings and for illustrations in the many books on zoology, both scholarly and popular, that appeared in the later eighteenth century, and became a flood in the early nineteenth century when public curiosity—that is, the search for knowledge—was at its peak.

Animals might be valued, not only as living, breathing creatures, nor merely as specimens to be dissected, stuffed, or skeletonized for display in private collections and in museums, but also as symbols—lions as emblems of the power of kings, a camel or a leopard to illustrate the veracity of the Bible, Amazonian parrots to prove the existence of the New World, and collections of exotic animals to exemplify the wealth, status, and international connections of their owners. It has been argued that the London Zoo was established to symbolize the power and prestige of the expanding British Empire. This was not the case: although Raffles was obliged to mollify the landowning aristocracy with promises of useful animals, his intention was to provide a scientific resource in London that would surpass that of the French. The acquisition of animals was to be facilitated by the wide range of Britain's commercial contacts, not only with the colonies, but also by its 'varied and constant intercourse with every quarter of the globe.' Although we can, with hindsight, attribute symbolic values to such exhibitions, it is more pertinent to establish whether they were perceived as such by contemporary eyes—while an elephant, an ape, or a giraffe might excite the curiosity and wonder of a visitor to the Zoo, it is doubtful whether they would impress him or her with visions of Empire.

While we may justifiably condemn the undoubted, if unintended, cruelty to animals in the early menageries, where, deprived of liberty and having endured long sea voyages, animals were confined in appallingly squalid and cramped conditions, fed unsuitable food, teased or goaded into fury—despite much legislation, cruelty to animals is still with us. Factory farming, illegal badger-baiting, and dog fights still take place in Britain and in many other countries, not to mention bull-fighting in Spain, and bear-baiting in Pakistan and in South Carolina in the United States.

We might also pause to admire the sheer ingenuity and organizational skill that characterized the travelling menageries. Despite short northern days and in every sort of weather, in the winter of 1818 John Polito trundled his lions and leopards, a tiger, a nilghai, a zebra, two hyaenas, a porcupine, a wolf, a kangaroo, and a sloth bear, as well as an emu, a cassowary, a pelican, some large cranes, a black swan, and a condor, in a procession of horse-drawn caravans northwards from Edinburgh to Inverness and Aberdeen. The heaviest caravan containing his most spectacular animal, the 'sagacious bull elephant', required no fewer then eight stout horses to drag it along the appalling roads.

In 1829 Charles Knight in his book *Menageries* maintained that, although the animals were confined in miserable dens and their keepers were lamentably ignorant, people of all classes saw *real* animals, not imaginary ones such as griffins or centaurs, and were thus able to acquire a body of *facts*, adding that the travelling menageries were 'among the most rational gratifications of the curiosity of the multitude'. And, when George Wombwell died in 1850, his obituary in *The Times* stated that 'no one has done so much to forward practically the study of natural history among the masses, for his menageries visited every fair and town in the kingdom and were everywhere popular'. This was undoubtedly true, but the sheer scale of Wombwell's menageries and of those it spawned, as well as the demands of the London Zoo and the numerous circuses and municipal zoological gardens set up later in the nineteenth century, must surely have contributed to the decimation of the native fauna in many parts of the British Empire and beyond.

GLOSSARY

GENERAL

aviary an enclosure or cage for housing birds.

Chain of the Creation Great Chain of Being. The scheme in which animals, plants, and minerals were ranked in an immutable order descending from God, through angels, monarchs, European men, European women, non-Europeans, mammals, other vertebrates, invertebrates, and then plants to minerals; current in various forms since classical times.

curious, the used as a noun from the seventeenth century (and perhaps earlier) onwards for people curious about the world around them; no derogatory overtone.

faisanterie, pheasantry an enclosure for pheasants.

Indian (sometimes) foreign.

menagerie from the French *menagerie*, first used in English in 1712, a collection of animals, or a building housing animals, sometimes only quadrupeds in contrast to aviary (birds); sometimes synonymous with pheasantry.

vivary from the French *vivarie*, a place for keeping living animals; sometimes synonymous with aviary; sometimes as a place where quadrupeds were kept, in contrast to an aviary for keeping birds.

volarie, volary from the French *volarie*, a place for keeping flying birds; often synonymous with aviary.

ANIMALS

amedevat (anadavad, averduwtas) red avadavat (*Amandava amandava*).

ape tailless monkey (i.e. Barbary ape), but often used loosely for other species of monkey; see also great ape.

barasingha swamp deer (*Rucervus duvauceli*).

buffalo American bison, African buffalo, or the Indian water buffalo (wild or domestic), but widely used in the past for any exotic cattle, especially zebu.

cassowary—viz. cassawarrens, cassawarwa, cossawarway, cassowar, casheward bird cassowary.

East India goose or Chinese goose swan-goose(?) (*Anser cygnoides*).

Egyptian night-walker monkey of unknown species, sometimes a baboon from Egypt, but also noted as from the 'Coast of Guiney' or 'the island of Borneo'.

elk usually European elk (*Alces alces alces*), sometimes other large deer; the North American 'elk' is a variety of red deer (*Cervus elaphus*). Also = wild swan *OED*.

emu (sometimes) cassowary.

gambo goose spur-winged goose.

gavey bird unknown

gobinite monkey?

great apes chimpanzees, gorillas, orang-utans, and gibbons in contrast to Barbary apes; gibbons are sometimes referred to as 'lesser apes'.

Greenland deer caribou.

Greenland dove guillemot.

hog in armour armadillo?

hunting-tyger cheetah.

ichneumon Egyptian mongoose (*Herpestes ichneumon*), loosely used for similar mammals; also a type of wasp.

jack-an-apes monkey.

kaama tapir.

King of the Vavvous or Vawows (see warwovven).

lyon lion.

magot (French) Barbary ape, sometimes baboon.

man-tiger usually a male mandrill, drill, or baboon.

marmoset originally a small monkey, later used only for several genera of small monkeys from South America.

minor or mino bird mynah.

musk cat civet cat.

panther initially used for large leopards; later for jaguars and pumas from the New World, especially if melanistic, as indeed sometimes to this day.

parrett parrot *sensu late*.

pewett peewit.

possum with a false belly opossum from America.

puette probably polecat, also fart (French slang).

sambar (sambhur, also rusa) deer, usually *Cervus unicolor*.

Swedish owls great grey owl?.

Tartarian pheasant the identity of the Chinese and Tartarian pheasants is uncertain; both golden pheasants (*Chrysolophus pictus*) and Chinese ring-necked pheasants (*Phasianus colchicus torquatus*) are sometimes referred to as Chinese pheasants; silver pheasants (*Lophura nycthemera*) are also found in China and other parts of south-east Asia.

tiger or tyger initially a generic term for any large cat. The epithets tiger, leopard, and panther were used interchangeably not only for leopards, but also for most of the other large and medium-sized cats. Although felid terminology was to a large extent sorted out by the Comte de Buffon, it was not until the first translation of his *Natural History* was published in 1775 that the nomenclature began to be standardized in English, with the term 'tiger' used only for the large, striped (Royal) Bengal tiger (*Panthera tigris*) found in India and the Far East and probably first imported into England in the early eighteenth century.

turtle (sometimes) turtle dove.

Virginian nightingall, or nightingale cardinal.

warwovven king vulture.

white bear polar bear (*Ursus maritimus*); arguably sometimes referred to pale-coloured individuals of the brown bear (*U. arctos*).

zebu humped domestic cattle; originally from India or South East Asia, but introduced to Africa, Sometimes also known as Brahmin cattle.

zibett civet cat, or similar.

PLACES

Banda Banda Aceh on the northern tip of Sumatra.

Barbary Morocco, sometimes including other North African countries that border the Mediterranean.

Barbary Coast the Atlantic coast of Morocco, sometimes also the Mediterranean coast of North Africa.

Bear-Bishes Berbice, Guyana.

Calicut Kozhikode.

Cape, the the Cape of Good Hope.

Capo Verde Cape Verde Islands = Cabo Verde.

East Indies India (most importantly in the eighteenth century and earlier, as in 'East India Company'); sometimes includes South East Asia and the East Indies in the modern sense.

Greenland used by early whalers to denote Spitzbergen (Svalbard).

Guinea, Guiney a country on the coast of West Africa; sometimes used in the past as a general term for areas of Africa situated along the Gulf of Guinea and further south on the Atlantic coast. The English term Guinea comes directly from the Portuguese word *Guiné*, which emerged in the mid-fifteenth century to refer to the lands inhabited by the 'Guineus', a generic term for Africans living south of the Senegal River.

Guyana, Guiana, in South America derived from an Amerindian word meaning 'land of many waters' (*OED*).

Mauritania those parts of Algeria and Morocco that border the Mediterranean, extending west through the Straits of Gibraltar to the Atlantic coast of Morocco and through the Atlas Mountains towards the Sahara.

Muscovy (usually) Russia.

West Indies islands in the Caribbean, but sometimes also used for lands bordering the Caribbean.

NOTES

ABBREVIATIONS

Agasse (1988)	*Jacques-Laurent Agasse 1767–1849* (London: Tate Gallery, 1988)
BL	British Library
BM	British Museum
Clift, *Diary*	Diary of William Clift (Royal College of Surgeons MS0007/1/4/2)
CSP	*Calendar of State Papers*
CTB	*Calendar of Treasury Books*
CTP	*Calendar of Treasury Papers*
Exeter Change File	Exeter Change File, Enthoven Collection, Theatre Museum (Victoria and Albert Museum)
Fillinham Collection	Fillinham Collection, a collection of cuttings from newspapers, advertisements, playbills, etc. (BL 1889. b.10)
Historical Description	*An Historical Description of the Tower of London*, issued nearly annually (1754–1809)
Issues	*Issues of the Exchequer: Being Payments Made out of His Majesty's Revenue during the Reign of King James I.* London: John Rodwell
Lysons, *Collectanea*	Lysons, *Collectania* (BL C.103.k.11)
Menagerie memorabilia box	Menagerie memorabilia box (Rothschild Library, Natural History Museum)
Osborne Scrapbook	Osborne Scrapbook (London Metropolitan Archive SC/GL/BFS/001)
Pie Powder Court Book	Pie Powder Court Book, 1790–1854 (London Metropolitan Archive CLC/308/MS 00095)
Salisbury (Cecil) Manuscripts	*Calendar of the Manuscripts of the Most Honourable the Marquess of Salisbury Preserved at Hatfield House, Hertfordshire*, ed. M. S. Giuseppi and

	D. McN. Lockie (London: HMSO, 1965)
St Martins/Leicester Square Scrapbook	St Martins/Leicester Square Scrapbook, vol. II, pt 2 (Westminster Archive Centre)
Stubbs (1984)	*George Stubbs 1724–1806* (London: Tate Gallery, 1984)
TNA	The National Archives
Theatre Buildings file	Theatre Buildings file, Enthoven Collection, Theatre Museum (Victoria and Albert Museum)
Walpole, *Correspondence*	H. Walpole, *The Yale Edition of Horace Walpole's Correspondence*, ed. W. S. Lewis (New Haven, CT: Yale University Press, 1937–8)

NOTES

CHAPTER 1. THE NORMANS TO THE TUDORS

1. Plot (1677).
2. Rybot (1972: 47). Dobson (1962).
3. Britton and Bayley (1830: 353–4). Anonymous MS notes in the Fillinham Collection, vols 7–8. Pennant (1781: 157). For much more information on the Tower menagerie, see Watson (1978), Parnell (1999), and Hahn (2003).
4. Yalden and Albarella (2009: 107).
5. Yapp (1981: plates 4, 5, 10, 11a, 11B, 38).
6. O'Regan (2002: 15).
7. Power (1964: 61, 306).
8. McEvedy (1961: 2–3).
9. Renault (1959: 197).
10. Gorgas (1997: 216). Power (1964: 151–6). McEvedy (1961: 88). Laufer (1928: 35).
11. McKendrick (1968).
12. George (1980). Asúa and French (2005: 13–14).
13. Fontes da Costa (2009).
14. Bedini (2000). Gorgas (1997).
15. Lai Yu-chih (2013). Ray (1678: 118).
16. Wing (1977). Müller-Haye (1984: 253). Jackson (1994: 36). Clutton-Brock (2012: 129).
17. Pigière et al. (2012). Hamilton-Dyer (2009).
18. 'Earliest Portrait of a Guinea Pig Discovered as Exhibition Reveals Unseen Painting' <http://www.npg.org.uk/about/press/news-release-earliest-portrait-of-a-guinea-pig> (accessed 5 September 2013).
19. Morales (1994; 1995: 3).
20. Bedini (2000: 222). Baratay and Hardouin-Fugier (2002: 22). Carrington (1962: 206).
21. Rye (1865: ii. 19–20).
22. Watson (1978).

23. Caius (1576: 23–4).
24. Gesner (1560: 68–70, 127). Caius (1570a: 26–31). Topsell (1607: 488, 568–9).
25. Ray (1678: 109).
26. *The Family of Henry VIII*, artist unknown, RCIN 405796, in the Royal Collection at Hampton Court.
27. Byrne and Boland (1985: 62).
28. Byrne and Boland (1985: 50).
29. The painting, formerly attributed to Hans Eworth, is now thought to have been painted by the 'Master of the Countess of Warwick' (fl. 1567–9). Van der Stighelen and James (2000).
30. Quoted from Strong (1998: 69).
31. Gorgas (1997). Belozerskaya (2006: 198).
32. 'Cecil Papers, April 1596, 16–30', in *Calendar of the Cecil Papers in Hatfield House*, vi. *1596*, ed. R A. Roberts (London, 1895), 145–64 <http://www.british-history.ac.uk/cal-cecil-papers/vol6/pp145-164> (accessed 19 March 2014).
33. Holinshed (1587: 208).
34. John Nichols (1977: i. 426–523).
35. The garden at Kenilworth has recently been reconstructed by English Heritage, but, as it is now illegal to cage wild birds, all the birds in the aviary are domesticated, having been selected to represent species newly introduced into Europe, such as guinea fowl.
36. Birkhead (2003: 85–6). Barbagli and Violani (1997).
37. 'Letters and Papers: Miscellaneous, 1539', in *Letters and Papers, Foreign and Domestic, Henry VIII*, vol. 14, pt 2, *August–December 1539*, ed. James Gairdner and R. H. Brodie (London, 1895), 303–58 <http://www.british-history.ac.uk/letters-papers-hen8/vol14/no2/pp303-358> (accessed 4 January 2012).
38. Gesner (1555: 234). Translated in MS by Topsell; see also Topsell (1972: 118–19).
39. Festing (1988), after Macray (1894: i. 24–7, 64). *The Shorter Oxford English Dictionary* (1966). Problems of vernacular terminology are discussed in the Glossary.
40. F. M. Nichols (1918: iii. 392).
41. Norrington (1983: 33, 58 (quoting from Erasmus' *Colloquy on Amity*)). Ackroyd (1998).
42. National Trust Inventory Number 960059.
43. Lewis (1998). Foister (2006: 34).
44. MacCarthy (1989: 247). Yorke (1981: 210. Speaight (1966). Many letters in the *Catholic Herald*.
45. See Hoeniger and Hoeniger (1969a) for a brief outline of sixteenth-century biology; also McLean (1972), which is brilliant; and, for exploration, George (1980).
46. Gesner (1554: addendum 18–19; 1602: 847). Caius (1570a: 43). Topsell (1607: 659).
47. Gesner (1560: 28). Caius (1570a: 26). Topsell (1607: 161).
48. Gesner (1560: 15). Caius (1570a: 39–40). Topsell (1607: 599–600).
49. Gesner (1560: 126). Caius (1570: 31–2). Topsell (1607: 757).

50. *Ammotragus lervia*. Gesner (1560: 38–9). Caius (1570a: 34–5). Topsell (1607: 119–21). Pennant (1781: 47).
51. Gesner (1602: 121). Caius (1570a: 36–7). Topsell (1607: 66).
52. Gesner (1560: 55–6). Caius (1570a: 37). Topsell (1607: 326–7).
53. Caius (1570a: 43–4, 55).
54. Caius (1570b: 209–10). Spano and Truffi (1986).
55. Caius (1576: 33).

CHAPTER 2. THE STUARTS, 1603–1688

1. Howes (1631: 835–6).
2. Bayley (1821: 271). Britton and Bayley (1830: 359). Harrison (1946: 207).
3. Britton and Bayley (1830). O'Regan (2002). Watson (1978). Rybot (1972: 58).
4. Harrison (1946: 154, 191).
5. *CSP Domestic* (James I, 1603–10), 16 (October 1605), 227–38.
6. There were several outbreaks of the Plague in James's reign.
7. *Salisbury (Cecil) Manuscripts*, pt 19 (1607), 258.
8. Rye (1865: iii. 132–3).
9. Larwood (1881: 327).
10. Weinstein (1980). Symonds (1912, 1988). *CSP Domestic* (James I, 1603–10), 13 (March–April 1605), 211.
11. David R. Ransome, 'Newport, Christopher (bap. 1561, d.1617)', *Oxford Dictionary of National Biography* (Oxford University Press, 2004); online edn, January 2008 <http://www.oxforddnb.com/view/article/20032> (accessed 3 April 2014).
12. *Issues*, 86.
13. *Salisbury (Cecil) Manuscripts*, pt 20 (1608), 36.
14. Foster (1926: 87).
15. Howes (1615: 904).
16. Larwood (1881: 328). Weinstein (1980).
17. Strong (2000: 56, 157–8).
18. Foster (1926: 87).
19. William Harvey (1653: 25–7).
20. Tradescant (1656: 3).
21. Larwood (1881: 328). Weinstein (1980). *CSP Domestic* (James I, 1603–10), 50 (December 1609), 573.
22. Weinstein (1980), from Rochester's Account preserved in the London Museum, accession no. Z6312. *CSP Domestic* (James I, 1611–18), 65 (July 1611), 51–65.
23. Symonds (1912, 1988).
24. *Issues*, 147–8, 166–7. Foster (1926: 83).
25. Zagorodnaya (2006: 90).
26. Foster (1926: 87–8). *CSP Colonial, East Indies, China and Japan*, 3 (1617–21), 7–81.
27. J. A. Allen (1880: 140).
28. Grant (2002). Hosking (1952: 124). *Issues*, 143.
29. Richardson (2000: 122).
30. John Nichols (1828: ii. 433).
31. *Issues*, 149, 287, 296.

32. Goodman (1839: ii. 237).
33. Hill (1988: 5, 19, 51).
34. Walford (1885: 382).
35. Larwood (1881: 326–7). BL, Harleian MS 6987/37.
36. *CSP Domestic* (James 1, 1623–5), 148 (1–17 July 1623), 1–21; 150 (1–17 August 1623), 40–57.
37. Foster (1926: 83–4). J. Edwards (1974: 27).
38. Haynes (1989: 175).
39. Knox (2002).
40. Browne (1836: 281).
41. John Nichols (1977: ii. 77). Strong (2000: 76–7).
42. Bergengren (2001: fig. 34).
43. Strong (2000: 200, fig. 12).
44. M. A. E. Green (1909: 19–20).
45. M. A. E. Green (1909: 115). Haynes (1989: 76). I have not been able to trace this portrait.
46. Oman (1964: 73).
47. M. A. E. Green (1909: 151).
48. M. A. E. Green (1909: 116).
49. Bacon (1937: 187–95).
50. Bacon (1658: 28–9).
51. *CSP Domestic* (Charles I, 1625–6), 3 (June 1625), 33–52.
52. *CSP Domestic* (Charles I, 1625–6), 1 (April 1625), 4–16.
53. *CSP Domestic* (Charles I, 1635), 304 (14–22 December 1635), 574–5.
54. Browne (1836: 281).
55. *CSP Domestic* (Charles I, 1625–6), appendix, 533–82.
56. Rybot (1972: 70–1).
57. Foster (1926: 88). *CSP Colonial, East Indies and Persia*, viii (1630–4), 7–49.
58. Oman (1951: 72, 83).
59. National Gallery of Art, Washington 1952.5.39. A second version is in Marble Hill House in Twickenham, where the label states that the monkey's name was Pug.
60. Uglow (2004: 77–8, 86).
61. Potter (2006).
62. R. S. Nichols (2003: 147–9).
63. Arnold (2000: 112). Peck (2005: 152).
64. F. P. Verney (1892: 132–3).
65. Browne (1835: 174).
66. Nicholls (2013). Ovenell (1992). It is possible that the surviving bits of dodo may be those of a second bird seen in the Anatomy School in Oxford in 1634 (MacGregor 1983: 68, 340).
67. Murray (1826: 168–70).
68. Larwood (1881: 333–4).
69. Britton and Bayley (1830: 361).
70. Bush (2003).
71. Evelyn (2000: iii. 93, 194–5).
72. Evelyn (2001). Amherst (1896: 203).
73. Foster (1926: 89–95).
74. Lambton (1985: 134). Larwood (1881: 344). Pattacini (1998). Picard (1998: 61).
75. *CTB* i. *1660–7*, 259–70.

76. Pepys (1985: iii. 297).
77. Temple and Ansley (1936: v. 156–8).
78. Monconys (1665: xx. 2–3).
79. *CSP Relating to English Affairs in the Archives of Venice*, xxxiii. *1661–1664*, ed. Allen B. Hinds (London, 1932), 248–50 <http://www.british-history.ac.uk/cal-state-papers/venice/vol33/pp248-250> (accessed 21 March 2014).
80. Evelyn (2000: iii. 398–400).
81. Foster (1926: 89–95).
82. Ray (1678: 59, 66–7, 82, 99–100).
83. Foster (1926: 89–95).
84. Daniel (1801: 408–10). Lever (1977: 357).
85. Evelyn (2000: iv. 265–6). <http://www.saudiaramcoworld.com/issue/201201/an.extreamly.civile.diplomacy.htm> (accessed 20 June 2013).
86. Browne (2006: 131–5). J. Bennett (2010: 45–6). Browne (1964: iv. 206-10).
87. Written over many years from about 1657 and unpublished until the manuscript was transcribed and published in 2001 (Evelyn 2001: 253–69).
88. Worlidge (1677: 68–9).
89. Evelyn (2000: iv. 143–4).
90. <http://en.wikipedia.org/wiki/Musaeum_Clausum> (accessed 26 September 2013).
91. Macaulay (1864).
92. Browne (1880: 337). See Hoeniger and Hoeniger (1969b) for details of natural history in Stuart times.
93. Evelyn (2001: 258). Jackson (1994: 10, 48).
94. Ray (1678: 266, table XLVI).
95. Derham (1718: 23).
96. Derham (1718: 131–2).
97. Ray (1678: 112–13, table XVI).
98. Pepys (1985: v. 131–2). See also footnote to the Wheatley edition of Pepys Diary (Pepys 1893–9: iv. 118).
99. Pepys (1985: iii. 25).
100. Pepys (1985: iii. 95, 128). Ollard (1994: 113–14).
101. Pepys (1985: vi. 111). Pennant (1812: iii. 491–3). Hervey (1968: 58–9).
102. Pepys (1985: ii. 17).
103. Pepys (1985: ii. 23; vi. 8; x. 356).
104. Pepys (1985: v. 352).
105. Bryant (1948: ii. 135).
106. Pepys (1985: ii. 160).
107. Farrington (2002: 64, 119). Rookmaaker (1998: 82–3). Clarke (1986: 37–41). Altick (1978: 37).
108. Evelyn (2000: iv. 389–91). Bedini (2000: 223).
109. Clarke (1986: 39).
110. Foster (1926: 84–5).
111. Verney (1899: 269).
112. Hooke (1935: 174, 184). Anon. (1675).
113. Foster (1926: 84–5).
114. Plot (1677: 136). Wilkin (1836: 255; 1835: 390–1). Hooke (1935: 423).
115. Mullen (1682). Moriarty (2012). Lansdowne (1928: 94–6).
116. BL, 74/C.161.f.1.(40*).
117. Hooke (1935: 160).

118. Evelyn (2000: iii. 255–6).
119. Pepys (1985: iv. 298).
120. Hooke (1935: 5, 321).
121. Strype (1720: bk 1, p. 119).
122. Ray (1678: 59, table I).
123. Pepys (1985: iii. 76).
124. Strype (1720: bk 1, p. 119).
125. Fillinham Collection, vols 7–8.
126. Luttrell (1857: i. 371).
127. As many as fourteen ships might arrive in the Port of London in a single week bringing wine from the Canary Islands (*Severall Proceedings of State Affaires*, 14–21 December 1654).
128. Ray (1678: 262–3, table XLVI). Blagrave (1675).
129. 'Entry Book: April 1686, 21–30', *CTB* viii: *1685–1689* (1923), 701–17.
130. Ormrod (1973).
131. Pieters (1998). George (1985: 185). Loisel (1912: ii. 54).

CHAPTER 3. WILLIAM AND MARY TO GEORGE II, 1688–C.1760

1. Ormrod (1973, 2003).
2. Birkhead (2003: 101–2).
3. Ray (1678: 262).
4. Flageolet: a small high-pitched wind instrument.
5. Hervieux de Chanteloup (1709).
6. Albin (1737: 84–6).
7. Cutting in the Menagerie memorabilia box.
8. Burt (1791).
9. <http://www.geheugenvannederland.nl/hgvn/webroot/files/File/extra/atlanticworld/atlanticworld1EN/tentoon7.html> (accessed 5 January 2014). Pieters (1998). George (1985: 185). Loisel (1912: iv. 54).
10. Albin (1738: 11).
11. Albin (1731: 13; 1734: 41).
12. 'Hackney Petty Sessions Book: 1732 (nos 437–603)', in *Justice in Eighteenth-Century Hackney: The Justicing Notebook of Henry Norris and the Hackney Petty Sessions Book*, ed. Ruth Paley (London, 1991), 83–103 <http://www.british-history.ac.uk/report.aspx?compid=38826&strquery=Bird> (accessed 10 January 2012). A pub called the Birdcage still exists on Stamford Hill, though it is not the original building.
13. According to the *Weekly Miscellany*, 14 April 1739: 'Mr. Michael Bland, famous for his wild Beasts, Variety of Birds, &c.', died on 9 April 1739.
14. George Edwards (1743–51: i. 2).
15. Beilby and Bewick (1790: 244–5). Israel (1990). Donald E. Wilkes, Jr, *Collection: Daniel Defoe* <http://libguides.law.uga.edu/c.php?g=177206&p=1164804> (accessed 17 April 2015).
16. Wortley Montague (1906: 125).
17. Newspaper advertisement in Frost (1874: 159).
18. Fillinham Collection, vol. 6.
19. Lysons, *Collectanea*, ii.
20. Note the callous reference to a 'parcel of slaves'; one wonders what became of them.

21. Boreman (1739: 22–8). Julia Allen (2002: 143–4).
22. Boreman (1739: 27).
23. BM 1914, 0520.691.
24. Caroline Grigson (2015).
25. George Edwards (1758–64: i. 24). Clarke (1986: 42, fig. 22).
26. Glasgow University Library (Hunterian Av.1.17). Parsons also produced an oil painting of the rhino, first recognized by L. C. Rookmaaker, which is in the Natural History Museum in London. Clarke (1986: 41–6).
27. Parsons (1743). Rookmaaker (1973, 1998).
28. Parsons (1743: 535).
29. Frost (1874: 167). *Gentleman's Magazine*, 21 (1751), 571.
30. Oxford Digital Library, Animals on Show 2 (52c), 'Newscutting Announcing a Rhinoceros' (1756).
31. Rookmaaker (1973, 1998). Clarke (1986: 47–68). Caroline Grigson (2015).
32. Blair (1710). Keeling (1989: 20).
33. Hans Sloane (1728).
34. Bradley (1721: 92–3, plate XVII, fig. 1). Stukely (1723).
35. Boreman (1739: 29–32, fig. 5).
36. Walpole letter to Sir Horace Mann, 10 May 1758 (Walpole, *Correspondence*, v. 199).
37. MacGregor (2012: 198–215).
38. Frost (1874: 88).
39. Kim Sloane (2003). MacGregor (1994). MacGregor, 'Sloane, Sir Hans, Baronet (1660–1753)', *Oxford Dictionary of National Biography* (Oxford University Press, 2004) <http://www.oxforddnb.com/view/article/25730> (accessed 9 August 2011). Catesby (1731) <http://www.britishmuseum.org/about_us/the_museums_story/sir_hans_sloane.aspx> (accessed 29 August 2013).
40. George Edwards (1743–51, 1758–64).
41. BM, SL, 5261.1–167.
42. Hans Sloane, 'Catalogue of Fossils. Pisc au ou Quadruped' (Natural History Museum Library, MSS SLO, vol. 5).
43. Clutton-Brock (1994).
44. Frost (1874: 88).
45. Derham (1718: 302–3).
46. The problems over the use of the word 'tiger' are given in the Glossary.
47. Hans Sloane, 'Catalogue' (Birds), no. 235. George Edwards (1743–51: ii. 68). Albin (1738: no. 36).
48. Lever (1977: 259; 1992: 12).
49. Albin (1738: nos 38, 47, 48). Fox (1827: 149).
50. Clutton (1967–71). Ramsbottom (1938).
51. Miles and Grigson (1990: 362).
52. Armstrong (2002: 7).
53. Hans Sloane, 'Catalogue' (Quadrupeds), no. 1284.
54. Hans Sloane, 'Catalogue' (Quadrupeds), no. 850.
55. Hans Sloane, 'Catalogue' (Quadrupeds), no. 1079.
56. Hans Sloane, 'Catalogue' (Quadrupeds), no. 856.
57. Mason (1992). MacGregor (1994).
58. Kent (1896). Festing (1988: 115). McCann (1994).

59. McCann (1994). Lennox (1911).
60. Rosemary Baird (2007). McCann (1994).
61. Lennox (1911).
62. Gascoigne (1994: 79–80).
63. McCann (1994). Lennox (1911).
64. Jacques (1822: 87). Lennox (1839).
65. McCann (1994). Festing (1987: 124).
66. Catesby (1747: 18); the original drawing is in the Royal Collection: RCIN 926087.
67. Rosemary Baird (2007).
68. MacGregor (1994: 31–2).
69. McCann (1994).
70. Buffon (1780: vi. 32 n.).
71. Roberts (1997: 46).
72. Fothergill (1781).
73. See Armstrong (2002) for information on Collinson, Wager, and their colleagues. Phillip Miller (1732). Evelyn (1786: 186). Daniel A. Baugh, 'Wager, Sir Charles (1666–1743)', *Oxford Dictionary of National Biography* (Oxford University Press, 2004); online edn, January 2008 <http://www.oxforddnb.com/view/article/28393> (accessed 9 August 2011).
74. Armstrong (2002). MacGregor (1995).
75. Lever (1977: 289). Perry Gauci, 'Decker, Sir Matthew, First Baronet (1679–1749)', *Oxford Dictionary of National Biography* (Oxford University Press, 2004); online edn, May 2009 <http://www.oxforddnb.com/view/article/7408> (accessed 6 July 2010). Macky (1722: 67).
76. Lever (1977: 289).
77. Burt (1791).
78. Lever (1977: 290).
79. George Edwards (1743–51: iii, no. 209). Pennant (1812: iii. 490).
80. Collins Baker (1949: 63–5).
81. Collins Baker (1949: 65).
82. Collins Baker (1949: 39, 63–5). Festing (1988: 104).
83. Welch (1972).
84. <http://www.baroquemusic.org/chandos.html> (accessed 9 August 2011). Joan Johnson, 'Brydges, James, first duke of Chandos (1674–1744)', *Oxford Dictionary of National Biography* (Oxford University Press, 2004); online edn, September 2010 <http://www.oxforddnb.com/view/article/3806> (accessed 9 August 2011).
85. Jenkins (2007). Collins Baker (1949: p. xvi).
86. Collins Baker (1949: 127–8).
87. Jenkins (2007: 83).
88. Collins Baker (1949: 127–8).
89. Walpole, *Correspondence*, xxxvii. 439–40.
90. Mitchell (2013).
91. Johnson (1984: 148).
92. Letter of 24 September 1728 (Cuthbert 1941: 60–3).
93. Jackson (1994: 34–6).
94. George Edwards (1758–64: i, no. 216).
95. George Edwards (1743–51: iv, no. 191).

96. Jacques and van der Horst (1988).
97. Loisel (1912: ii. 31). <http://api.rijksmuseum.nl/tentoonstellingen/hondecoeter?lang=nl> (accessed 17 July 2013). Mountague (1696: 53).
98. Loisel (1912: ii. 36).
99. Bezemer-Sellers (1990: 110–11).
100. Walter Harris (1699).
101. Switzer (1718: i. 76).
102. <http://archive.org/stream/historyhamptonc003lawe/historyhamptonc003lawe_djvu.txt> (accessed 20 November 2013). Accounts of William Bentinck, first Earl of Portland, 157 DD/5P/8, Account book for 1692 (Nottinghamshire Archives).
103. <www.bdonline.co.uk/kensington-palace-refurbishment-by-john-simpson-and-todd-longstaffe-gowan/5036691.article> (accessed 2 January 2013).
104. Switzer (1718: i. 76).
105. *CSP Colonial, America and West Indies*, 16 (1697–1698), 191–206, 328–44.
106. Switzer (1718: i. 76–7). Jacques and van der Horst (1988: 85). Festing (1988: 112). Longstaffe-Gowan 2005: 59. 'The Household below Stairs: Other Offices', in *Office-Holders in Modern Britain: Volume 11 (revised): Court Officers, 1660–1837*, ed. R. O. Bucholz (London, 2006), 595–601 <http://www.british-history.ac.uk/office-holders/vol11/pp595-601> (accessed 14 April 2013).
107. R. Desmond (1995: 74).
108. Coombs (1997).
109. R. Desmond (1995: 47). <http://www.kew.org/heritage/timeline/1700to1772> (accessed 4 March 2013).
110. George Edwards (1758–64: i. 222, 223). Barnaby (1996: 70–1).
111. <http://www.londonlives.org/browse.jsp?div=LMSLPS15003PS150030020> (accessed 4 May 2012).
112. J. P. Malcolm (1811: 398).
113. Altick (1978: 88). Hahn (2003: 151–3).
114. Strype (1720: 119).
115. Saussure (1902: 85–6).
116. Hone (1832: 1035).
117. Judy Egerton, 'Ellys, John (1701–1757)', *Oxford Dictionary of National Biography* (Oxford University Press, 2004) <http://www.oxforddnb.com/view/article/8728> (accessed 11 June 2011). *European Magazine* (September 1803: 179). *Journals of the House of Commons*, 32 (1803), 473, 481, 489, 497, 504, 513.
118. Albin (1734: 4, plate 4). George Edwards (1743–51: i. 2). Boreman (1736: 12, fig 9). Wilkinson (2011).
119. The egg in the Leverian Museum was probably the one laid when John Ellis was Keeper, which was said to have been presented by his son Charles to James Parkinson, who displayed it in his museum—that is, the erstwhile Leverian Museum, which Parkinson won in a lottery in 1786 (Campbell et al. 1816: 521). It was part of Lot 5429 on the 46th day of the Leverian Sale in 1806.
120. Maitland (1775: i. 172). *Historical Description* (1754).

CHAPTER 4. GEORGE III, C.1760–1811

1. TNA: PROB 11/735.
2. George Edwards (1758–64: ii. 123).

3. *Public Advertiser*, 15 September 1758.
4. Kinch (1986).
5. Northumberland (1775: 40–1). Burt (1791).
6. St James's Piccadilly, Watch Rate Collection Ledgers (Westminster Archive Centre).
7. *St James's Chronicle*, 22 November 1768.
8. This may have been Edward Lambert described by Baker (1755) and George Edwards (1758–64: i) as having a rugged coat or covering, or, more probably, given the date, one of Lambert's sons.
9. TNA: PROB 11/1027. Plumb (2010a).
10. St James's Piccadilly, Watch Rate Collection Ledgers (Westminster Archive Centre).
11. *Not* in Tottenham Court Road, as sometimes stated. Tottenham *Court* itself lay at the northern end of Tottenham Court Road; the New Road built in 1756 ran west from Tottenham Court towards Marylebone; it was later extended east towards Euston and was renamed the Euston Road in 1857.
12. R. Horwood, 'Plan of the Cities of London and Westminster the Borough of Southwark and Parts Adjoining Shewing Every House', 1795–9, BL, Maps 148.e.7.
13. Burt (1791).
14. Bills paid by George, third Earl of Egremont himself, at Petworth House, to Joshua Brookes for animals and birds 1774–98, West Sussex Archives (via TNA).
15. *Gazetteer and New Daily Advertiser*, 2 June 1784.
16. The sequence of events reconstructed from contemporary newspapers and the papers of Joseph Banks held at the State Library of New South Wales (series 30) <http://www2.sl.nsw.gov.au/banks/series_30/30_view.cfm> (accessed 7 July 2012). See also Collins (1802: 534).
17. Philip Castang married Sarah Brookes before 1794.
18. David Collins (1802: 534).
19. TNA: PROB 11/1385.
20. *Morning Post*, 1 August 1804.
21. <www.british-history.ac.uk/report.aspx?compid=45238 & 65194> (accessed 19 January 2012).
22. Banks (2007: ii. 205).
23. Rolfe and Grigson (2006). Caroline Grigson et al. (2008). Egerton (2007: 334, no. 141) follows William Clift's erroneous note that these monkeys had been in the possession of the Earl of Shelburne. John Hunter (1861: ii. 16–17). W. Clift, 16 September 1828, marginal addition to *A Copy of the Oldest Portion of Catalogue in 8vo*, Royal College of Surgeons MS0007/1/1/1/15, fo. 71.
24. For 'Nature's Chain', see Chapter 7 and the Glossary.
25. Sherwin (1958: 454).
26. Rolfe and Grigson (2006).
27. St James's Piccadilly, Watch Rate Collection Ledgers (Westminster Archive Centre).
28. BM, Coins & Medals G 871. Advertisements in Lyson, *Collectanea*, ii.
29. John Hunter (1787).
30. Old Bailey Proceedings Online (<www.oldbaileyonline.org>, version 6.0, 12 January 2012), January 1807, trial of Mazarine Bell (t18070114-100).
31. St James's Piccadilly, Watch Rate Collection Ledgers (Westminster Archive Centre).

32. *The Four Important Trials at Kingston Assizes, April 5, 1816: I. Eliz. Miller, for Poisoning the Children at Kennington. II. J. Brookes, for Shooting E. Thompson, in Lombard Street*... (London: W. Hone, 1816).
33. It is sometimes stated that *John* Brookes was present at the death of Chunee the elephant at the Exeter Change in 1826, but such statements confuse him with his brother, the anatomist Joshua Brookes.
34. Pidcock (1778). Garner (1800a: 42–4). Some of this information is bound in with a catalogue of the animals preserved in the Museum of the Zoological Society, April 1829 (Natural History Museum).
35. Handbill in Exeter Change File.
36. Lysons, *Collectanea*, ii.
37. Letter Thomas Bewick to John Bewick, 9 January 1788 (Robinson 1887: 90). Thomson (1882: 96). Uglow (2006: 162).
38. Beilby and Bewick (1790: 164, 70).
39. Tyne & Wear Archive, Newcastle-upon-Tyne, Acc. 1269/48 (I am indebted to Iain Bain for this information).
40. Now the site of the Strand Palace Hotel.
41. *Morning Chronicle*, 26 April 1788. Frost's frequently quoted statement (Frost 1874: 186) that Pidcock displayed animals from Cross's collection at the Exeter Change at fairs in London during and after 1769 is clearly erroneous, but has become embedded in the literature and in web pages.
42. Lysons, *Collectanea*, ii.
43. Phillip (1789: 277).
44. Phillip (1789: 277–8); rufous rat-kangaroo (*Aepyprymnus rufescens*); Bewick's figure of the kangaroo rat in his *Quadrupeds* (Beilby and Bewick 1790: 379) is copied from Phillip. Plumb (2010b) seems to have confused them with kangaroos, first imported in 1791, not 1789. Handbill in Lysons, *Collectanea*, ii.
45. Handbill in the Menagerie memorabilia box. Burt (1791: 10 and figure). Clarke (1986: 70–1, quoting Bingley 1805). The painting is reproduced by Egerton (2007: 512, no. 278).
46. Burt (1791: 43 and figure). Garner (1800a: 29–31 and figure).
47. Newspaper cutting in Theatre Buildings file.
48. *London Chronicle*, 29 January 1791.
49. Burt (1791). Garner (1800a: 41–2). The skull of Clark's paca was no. 2042 in the Osteological Series in the Hunterian Museum in the Royal College of Surgeons, but no longer exists (Hunter 1861: ii. 216).
50. Not the 'Irish Giant', whose skeleton occupies pride of place in the Hunterian Museum in the Royal College of Surgeons, but another 'giant' of the same name.
51. *Bath Chronicle*, 11 August 1791.
52. Pidcock's vehicles are variously described in his advertisements as caravans, wagons, or carriages.
53. Garner (1800a, b).
54. Handbill in Lysons, *Collectanea*. Pennant (1793: ii. 31). Roberts (1997: 366). Blunt (1976: 67).
55. Granger and Caulfield (1804: 711–15).
56. Menagerie memorabilia box.
57. OED: *Carbonado* 1. To score across, and boil or grill. 2. To cut, slash, hack. Basil Taylor (1971: 215–16). Egerton (2007: 566) mistakenly suggests that the tiger may have been obtained for Stubbs by John Hunter (d.5 May 1793).

58. London Metropolitan Archives, MS 11936/395/614342.
59. NB Cosham, not Corsham as frequently stated. Bingley (1805: 487–9).
60. Rookmaaker (1998: 84).
61. Garner (1800a: 3).
62. London Metropolitan Archives, MS 11936/395/614342.
63. Altick (1978: 39).
64. Garner (1800a: 9).
65. *The True Briton*, 23 December 1795.
66. Garner (1800a: 3).
67. Pie Powder Court Book. Moiser (2005). Mark Sorrell, 'Pidcock, Gilbert (d. 1810)', *Oxford Dictionary of National Biography* (Oxford University Press, 2004); online edn, Jan 2008 <http://www.oxforddnb.com/view/article/73321> (accessed 28 May 2014).
68. Much of the information about Pidcock's second tour of Scotland, including the composition of the caravan, is taken from R. Edwards (2006).
69. Garner (1800a: 29–31). Sale catalogue, Messrs King & Lochee, 19 March 1810.
70. Garner (1800a: 19–21).
71. Transcripts of Bewick's correspondence kindly supplied by Iain Bain. See also Tattersfield (2011: ii. 121–2; iii. 31). Hugo (1866: 450).
72. Uglow (2006: 268). Iain Bain, 'Bewick, Thomas (1753–1828)', *Oxford Dictionary of National Biography* (Oxford University Press, 2004); online edn, May 2005 <http://www.oxforddnb.com/view/article/2334> (accessed 28 May 2014).
73. Garner (1800a).
74. Garner (1800b) (it seems that the only publically owned copy in Britain is in the Enthoven Collection in the Victoria and Albert Museum).
75. In a copy of the poster, wrongly dated by hand 1795, is in the Victoria and Albert Museum (S.516-1996).
76. Barnard (1994). Garner (1800a, b). The whiptail wallaby was described and illustrated by Lambert (1807); see also Plumb (2010b). The camel, a two-humped Bactrian, wrongly described in the catalogue of the Royal Collection as 'The Dromedary' (RCIN 913312), was drawn at the Exeter Change by Johannes Eckstein in 1798.
77. Also known as Alpy or Alpey.
78. Knight (1829–40: iii. 8–9). Leigh Thomas (1801).
79. William Wood (1807: i. 225–6).
80. Rieke-Müller and Dittrich (1999: 27–33). Garner (1800a: 15–17, 19–21). Bingley (1813).
81. Known before the revolution as Jean-Claude-Michel Mordant de Launay. Anon. (1804). Anon. (1823: xiv. 578).
82. Pie Powder Court Book. Cutting in the Osborne Scrapbook, 26.
83. *Oracle and Daily Advertiser*, 3 October 1800.
84. Altick (1978: 307).
85. Lysons, *Collectanea*, ii. Charles Blagden letter from Paris to Joseph Banks (Natural History Museum, Dawson Turner Copies, iii. 170–4).
86. In 1792 Banks presented a number of stuffed animals to John Hunter for his museum, among them a black swan, which he had probably acquired in a dried or pickled state.

87. Lysons, *Collectanea*, ii.
88. Smith (1806: i. 235–6).
89. A handbill dated 1810 confirms that the skeleton was that of the elephant that died in 1806 (Exeter Change File).
90. Westminster Archives: WCCDEP358270260. Frost (1874: 209).
91. Home (1808).
92. Bell's paintings and his surviving anatomical and pathological preparation are in the Royal College of Surgeons of Edinburgh. Bell (1806). L. S. Jacyna, 'Bell, Sir Charles (1774–1842)', *Oxford Dictionary of National Biography* (Oxford University Press, 2004); online edn, January 2008 <http://www.oxforddnb.com/view/article/1999> (accessed 29 July 2012). Gibson (1841: 139–40).
93. Elwin (1950). Altick (1978: 310).
94. TNA: PROB. 11 /1509.
95. Wirksworth Parish Records 1600–1900, from *Pedigrees* by Thomas Norris Ince (1799–1860) <http://www.wirksworth.org.uk/> (accessed 7 March 2011).
96. 'Sale Catalogue of the Valuable Menagerie and Museum of Natural Curiosities, Late the Property of Mr Gilbert Pidcock', 19 March 1810, Messrs King and Lochee, with sums paid on days two and three added in manuscript; also a handbill issued by the executors in February 1810 (both in the Exeter Change File).
97. Pie Powder Court Book.
98. Pie Powder Court Book. Mark Sorrell, 'Polito, Stephen (1763/4–1814)', *Oxford Dictionary of National Biography* (Oxford University Press, 2004); online edn, January 2008 <http://www.oxforddnb.com/view/article/73320> (accessed 7 June 2012). Polito's Will (TNA: PROB 11/1556/131).
99. Pie Powder Court Book.
100. Pie Powder Court Book.
101. Polito (1803).
102. Tattersfield (2011: ii. 822).
103. Tattersfield (2011: ii. 821–3; iii. 136–7). I am indebted to Iain Bain for transcripts of letters sent to Bewick.
104. Letters Polito to Bewick, 2, 6, 17 March 1808.
105. Letter Polito to Bewick, 20 March 1808. Tattersfield (2011: ii. 822–3). The poster is in Nigel Tattersfield's private collection.
106. Beilby and Bewick (1790: 164).
107. Letter Elizabeth Robinson, 24 June 1740 (Climenson 1899: i. 49–50).
108. His widow sold the library in 1744 and the manuscripts to the British Museum in 1753.
109. Festing (1986a, b). Gascoigne (1994: 94–5). Pat Rogers, 'Bentinck, Margaret Cavendish [Lady Margaret Cavendish Harley], duchess of Portland (1715–1785)', *Oxford Dictionary of National Biography*, (Oxford University Press, 2004); online edn, October 2006 <http://www.oxforddnb.com/view/article/40752> (accessed 27 May 2014). Pennant (1777).
110. The drawing is reproduced by Laird and Weisberg-Roberts (2009: 185). Llanover (1861: ii. 527). Catesby (1747).
111. Llanover (1861: iii. 241–2, 293). The drawing is in a private collection, but reproduced by Festing (1988: 105), by Laird and Weisberg-Roberts (2009: 185), and at <http://venetianred.net/tag/mary-delany> (accessed 12 June 2013).

112. Gilpin (1789: 188–9).
113. Festing (1986a). Laird (1999: 22–4, fig. 133).
114. Festing (1986a; 1988: 107, 111). Hayes (1775: 14). George Edwards (1758–64: ii, no. 295). Llanover (1862: i. 161; iii. 459–60).
115. Climenson (1899: 121–2).
116. Latham (1823: viii. 118–19, 249). Festing (1986a, b; 1988: 114). Llanover (1862: iii. 158–60).
117. Llanover (1862: ii. 433–4).
118. Latham (1823: viii. 190–1, 293).
119. Letters from Collinson to Duchess of Portland, September 1758, 23 January 1768 (Longleat, TNA PO/Vol. XIV, fos 99, 112). Festing (1988: 111). Pontoppidan (1755: ii. 10).
120. Llanover (1862: ii. 433–8). Pennant (1793: i. 45). Boswell (1768: 41).
121. *Morning Herald*, 13 May 1786.
122. Nelson (2000: 19). Tillyard (1995: 86–7).
123. Buffon (1785: vi. 238).
124. Buffon (1785: vi. 201–2).
125. Buffon and Daubenton (1771: xv, plate III; 1784 xv, 15: plate XIII). Rookmaaker (1992). A translation of Allamand's description, but not the plate, is included in the third edition of William Smellie's translation of Buffon (Buffon 1791: vi. 352).
126. Nelson (2000: 31).
127. Most of the information regarding the Duke's moose is taken from Ian Rolfe's scholarly interpretation of William Hunter's unpublished notes (Rolfe 1983a).
128. Buffon and Daubenton (1771: xv, plate II; 1784: xv, plate XII). Rookmaaker (1992).
129. Buffon (1791: vi. 351).
130. The painting is in the Hunterian Museum in Glasgow. Hunter's detailed description of this animal and the next lay unpublished among his papers in Glasgow (Rolfe 1983a). *Stubbs*, 118–19. Egerton (2007: 283–4).
131. Letter from third Duke of Richmond to the Marquis of Rockingham (Sheffield Archives WWM/R/1/1316). Egerton (1978: 60). Pennant (1771: 40–1; 1781: 94; 1793: i. 106).
132. Brock (1996: 298). Raat (2010).
133. For the drawing of the second bull moose, William Hunter's own notes on the moose, and much additional information, see Rolfe (1983b). Stubbs's drawing of the second bull moose is in Glasgow University Library (HF. 234). See also Hunter (2008: ii. 7–9, 131).
134. Peter Mazell (fl. 1761–1802); John Frederic Miller (fl. 1772–96), son of John Sebastian Miller. BM 1914,0520.204, 205. Shaw (1796).
135. Possibly a musk deer; no. 2818 in the MS 'Catalogue of Anatomy and Physiology forming the Original Catalogue of the Hunterian Museum' (RCSEng MS0189/2/8—as *Moschus*).
136. Egerton (1978: 76–7 (quoting Ozias Humphrey)). Humphrey and Meyer (2005: 75). Both paintings are in the Yale Center for British Art; the second painting should not be confused with *Lion Attacking a Horse*, also commissioned by Rockingham, but painted in 1762.
137. Fitzmaurice (1912). Festing (1988: 111). Kerry (1922).
138. John Hunter (1861: ii. 8).

139. Festing (1988). Kerry (1922).
140. Fitzmaurice (1912). Kerry (1922).
141. Bence-Jones (1974). Robert Harvey (1998). Edwardes (1991). Farrington (2002: 100).
142. The house still exists.
143. Bence-Jones (1974: 177). BL, MS Eur F128/27.
144. Walpole, *Correspondence*, xxi. 378. J. Malcolm (1836: ii. 119–25). Chaudhuri (1975: 266). Spear (1975: 211). Parsons (1760). *London Magazine*, 28 (1759), 664. *Gentleman's Magazine*, 31 (1761), 272–3; 37 (1767), 489. J. Hunter (1861: ii, pp. iv, 50). Dobson (1962). Beilby and Bewick (1790: 199–200).
145. Heron (1850: appendix).
146. Egerton (2007: 240–1, no. 77). It is odd that the *Evening Post* referred to Clive as his Lordship, since his Irish peerage 'Lord Clive of Plassey' was not awarded until March. Lennox-Boyd et al. (1989: 121). David Green (1951: 282).
147. J. Malcolm (1836: ii. 256–8). Foster (1926: 86). *Gentleman's Magazine*, 33 (1763), 415, 506–9. Picard (2000: 32).
148. Bence-Jones (1974: 232). Clive's Household Management Papers (British Online Archives H5/1, image 89) <http://www.britishonlinearchives.co.uk> (accessed 16 June 2011).
149. Chaudhuri (1975: 324). Bence-Jones (1974: 220, 230, 232).
150. Bence-Jones (1974: 230, 243).
151. Inventory of Claremont (British Online Archives, CR4/1, image 9). Pennant letter to Banks, 30 May 1768 (Banks 2007: i. 37–8). Pennant (1771: 52).
152. Parsons (1745: 465–7, plate III, fig. 9). Parish (1991).
153. The best account of the history of the garden is *Claremont Landscape Garden* (Anon. 1995). Walker (1968).
154. Bence-Jones (1974: 257–8). Latham (1785: 449, 453–4).
155. Bence-Jones (1974: 257–8). BL, MS Eur F128/93, fo. 80. This was probably the animal dissected by John Hunter, whose sectioned eyeball is still in William Hunter's collection in Glasgow.
156. Hunter (2008: 45). William Hunter (1771). 'Eye of Neel-Gaw' Anatomical Preparations no. 23.66 (Marshall 1970: 357).
157. BL, MS Eur F128/93, fo. 97.
158. Egerton (2007: 334, no. 142).
159. John Hunter (1861: ii. 145). Rolfe quoted by Egerton (2007: 334).
160. John Hunter (1861: i. 194).
161. Banks (1799: 267–8).
162. Festing (1988: 112). Walker (1968: 93). Claremont Account for the year 1777 (British Online Archives, EC1/5).
163. Pennant (1781: ii. 564).
164. *Claremont Landscape Garden* (1995). Robert Harvey (1998: 376). Bence-Jones (1974: 300–1). Parish (1991: 13–28).
165. BL, MS Eur 226/5, E226/77(h).
166. Turner (1800: 200, 204–5). Grier (1905: 67, 69, 93–4).
167. Impey (1963). Moorhouse (1971: 40).
168. Archer and Archer (1955). Archer and Parlett (1992). Pennant (1798). Some images and additional information from various sale catalogues and collections (Gorringes, 7 September 2011; Christie's, May 2008; Sotheby's June 1963, April 1972; Victoria and Albert Museum; British Library; Radcliffe

Science Library, Oxford; San Diego Museum; Metropolitan Museum of Art, New York).

169. Latham (1787: esp. 208–9; 1790: ii. 632; 1801: 227–8).
170. Pennant (1798: ii. 156, 158, 159, 160, 247, 252, 344). I have not been able to trace any of these paintings.
171. Williams (1933). Pennant (1790: 146; 1793: i. 283, plate LV).
172. Trotter (1878: 299).
173. Letter to Thompson, 20 May 1786, letter to Anderson, 9 October 1786 (Gleig 1841: 243, 319–20).
174. Lawson (1895: 48, 114).
175. Egerton (2007: 520: no. 284); no. 285 in the Royal College of Surgeons of England is a simplified version, probably commissioned by John Hunter.
176. A second, smaller version of the yak, probably commissioned by John Hunter and now in the Royal College of Surgeons in London, shows much less of the landscape, and the palace is scarcely discernible.
177. Some of these views survive in the Victoria Memorial in Calcutta.
178. Michael Aris, 'Turner, Samuel (1759–1802)', *Oxford Dictionary of National Biography* (Oxford University Press, 2004) <http://www.oxforddnb.com/view/article/27864> (accessed 27 May 2014). Hussey (1970). Verey (1979: 207–9). Egerton (2007: 520–1). Lennox-Boyd et al. (1989: 282). Turner (1800). Aris (1982: 28).
179. Macaulay (1841). Clutton-Brock (1981: 140–1). Roberts (1997: 366).
180. Festing (1987: 125; 1988: 115). Hayes (1775).
181. Llanover (1861: iii. 440).
182. Burt (1791).
183. Delany to Mrs Dewes, 8 November 1760 (Llanover 1861: iii. 611).
184. Mrs Boscowen to Mrs Delany, 25 September 1783 (Llanover 1862: iii. 141–2).
185. Walpole, *Correspondence*, ix. 142–3; xxxvii. 344–5; x. 94.
186. *Foll-e*, 5 (July 2008), 1–2 <www.follies.org.uk> (accessed 14 July 2013). Cherry and Pevsner (1973: 140).
187. Lambton (1985: 152). Toynbee (1927–8: 52–3). Festing (1988: 110, 112). Keeling (1989: 23).
188. Walpole, *Correspondence*, x. 334.
189. Festing (1988: 107, 112). Rodenhurst (1802). <http://www.hawkstoneparkfollies.co.uk/history> (accessed 4 September 2012).
190. Festing (1988: 112). Walpole, *Correspondence*, x. 80.
191. Forster (1778).
192. Sold at Bonham's London for £43,020 in 2004. Jackson (1985: 127). Earl Spencer's menagerie was probably at Wimbledon Park.
193. Sold at Christies, 2004.
194. Owned by the National Trust.
195. Hayes (1794–9).
196. Festing (1987: 125. 1988: 115).
197. Loisel (1912: ii. 16). The Aviary farm and a large section of the park are said to be owned by the Sultan of Brunei and are very, very private; however, aerial photographs show that the building was still much the same in 1951 as when depicted by Hayes in the 1780s; see 'Britain from Above': The Aviary, Norwood Green, 1951 <http://www.britainfromabove.org.uk/image/EAW041321> (accessed 20 June 2013), but in pictures taken in

2012 the building is unrecognizable <https://www.flickr. com/photos/ 77076455@N07/7240204580/> (accessed 22 June 2013).

198. Cherry and Pevsner (1991: 438–42).
199. <http://osterleynationaltrust.wordpress.com/2014/05/08/curious-birds-in-the-trappings-of-trade> (accessed 20 June 2013).
200. Lister (2001). <http://osterleynationaltrust.wordpress.com/2014/05/08/curi ous-birds-in-the-trappings-of-trade> (accessed 20 June 2013).
201. Climenson (1899: 230).
202. Frank Buckland (1881). The best accounts of John Hunter's menagerie are by Thomas Baird (1793: 38–42), Pasmore (1976–8), and Schupbach (1986). A scurrilous biography by Foote (1794) contains many inaccuracies. Hunter's most recent and best biography is by Wendy Moore (2005). One of the most illuminating of the many analyses of Hunter's work is Chaplin (2009). Many details of his life and possessions are contained in manuscript notes made by William Clift stored by the Royal College of Surgeons of England.
203. Cave (1941). Wood Jones (1949, 1951).
204. Allen et al. (1993).
205. Clift, MS catalogue of Hunterian Pictures, 1816 (Royal College of Surgeons MS0007/1/4). Le Fanu (1931).
206. Letter Dr W. Irvine to Prof Thomas Hamilton, 17 June 1771 (*Lancet*, 18 February 1928, 354–60).
207. John Hunter (1861: ii. 260).
208. John Hunter (1787: 257).
209. Not the miniature breed now known as Pomeranian, but a large dog equated by Pennant to Buffon's *chien-loup*, perhaps the progenitor of the German Shepherd.
210. Pennant (1781: 222).
211. John Hunter (1787).
212. See Moore (2005: 213–17, 292–3). Paterson (1789: 126–7). Hickey (n.d.: ii. 108, 223–7, 267, 290).
213. Clift, 'Account Book' (Royal College of Surgeons MS0007/1). Clift, 'Notes on Natural History' (Royal College of Surgeons MS0007/1/7/2/1, p. 127 rev., marginal note).
214. 'Diary of a London Resurrectionist' (Royal College of Surgeons MS0024).
215. For more on Brookes's life, see Anon. (1834). Feltoe (1884: 104).
216. One of these quaggas was painted by Agasse (Royal College of Surgeons RCSSC/P 270). The skulls were in the college's museum, but were destroyed when the college was bombed in 1941. See also Griffith et al. (1827: iii. 465).
217. Brookes's collection is itemized in three sale catalogues: Brookes (1828a, 1828b, 1830).
218. From: 'Great Marlborough Street Area', Survey of London, vols 31 and 32: St James Westminster, pt 2 (1963), 250–67 <http://www.british-history.ac.uk/report.aspx?compid=41476> (accessed 30 December 2011).
219. *The Field*, 29 May 1869.
220. Sharpe (1832: 266). Vevers (1976: 16–17).
221. Keeling (1984: 19).
222. Unless otherwise attributed, most of the information relating to Windsor is taken from Roberts (1997).
223. Walpole, *Correspondence*, xxxvii. 383.
224. Royal Collection: RCIN 451431, RCIN 914626.

225. Llanover (1861: iii. 461).
226. *London Magazine*, 34 (1764), 346.
227. The painting, now entitled *Cheeth with two Indian Attendants*, is in Manchester City Art Gallery <http://www.manchestergalleries.org> (accessed 20 September 2013). *Stubbs*, 114–15, no. 79. Egerton (2007: 236–9, no. 76). Basil Taylor (1970; 1971: 208). Geoffrey Grigson (1972: 36–9). Clutton (1970). *La Gazette de France*, 16 July 1764: 230; and many other newspapers.
228. Information from Adrian Lister.
229. Fisher (2004: 64–5). Ullah and Eversley (2010). *Old Bailey Proceedings Online* (<www.oldbaileyonline.org>, version 6.0, 02 December 2012), February 1765, trial of John Ryan Jeremiah Ryan Mary Ryan (t17650227–5).
230. *Historical Description* (1768).
231. *London Magazine*, 31 (1762), 347 and plate.
232. Foster (1926: 86). *Gentleman's Magazine*, 30 (1763), 506–9. Picard (2000: 32).
233. Picard (2000: 38).
234. Cutting in the Menagerie memorabilia box. Picard (2000: 306).
235. Smollett (1771).
236. Goldsmith (1774: ii. 393, 398).
237. Boreman (1768: 27).
238. Laskey (1813).
239. Walpole, *Correspondence*, xxiv. 90–1. Plumb (2010a, 2011).
240. Foster (1926: 86–7).
241. *Gentleman's Magazine*, 43 (1773), 409.
242. Home (1794: 250; 1800; 1823: 250). Robert Chambers (1835: iii. 161). Glasgow University Library, MS Hunter H140.
243. Laskey (1813). Jackson (1998: 29 (figure), 110).
244. Egerton (2007: 283, no. 105).
245. Brock (1996: 45). William Hunter (1771). The painting was engraved and used to illustrate Hunter's description in the *Philosophical Transactions of the Royal Society* as well as being reproduced in the *Gentleman's Magazine*, 42 (1772), 409.
246. Mr Chambers (1763: 2). Ray Desmond (1995: 44–63).
247. Anon. (1792).
248. Two drawings of an obviously stuffed springbok by Frederick Birnie dated 4 December 1778 in Hunter's collection at his anatomy school in Great Windmill Street may represent this animal (Rookmaaker 1989: 59).
249. Solander letter to Banks, 14 August 1775 <http://www2.sl.nsw.gov.au/banks/series_72/72_181.cfm> (accessed 7 July 2012). Rookmaaker (1989: 59). Latham (1781: 31).
250. Chisholm (2001: 301, n. 23).
251. Home (1794).
252. Ray Desmond (1995: 75). Loisel (1912: ii. 16). See also Festing (1988: 115).
253. Recently acquired by the National Maritime Museum, London. Egerton (2007: 336–7, no. 143).
254. Home (1795: p. xxv).
255. I thank Ray Desmond for this information.
256. *Gentleman's Monthly Intelligencer*, 44 (1775), 601. *British Magazine and Review* (1783).

257. *Historical Description* (1784).
258. *Historical Description* (numerous editions). It is thought that the anatomist John Hunter had first refusal on the corpses of the animals that died in the Tower. Dobson (1962). Beilby and Bewick (1790: 169).
259. *Historical Description* (1759). *Gentleman's Magazine*, 45 (1775), 202, 326.
260. *Historical Description* (1784, 1800).
261. Pennant (1790: 259).
262. *Historical Description* (1787).
263. *Historical Description* (1787). Bingley (1814: 36–7). Pennant (1790: 259).
264. Burt (1791). *Historical Description* (1800, 1809).
265. Pennant (1790: 146; 1793, i. 283, plate LV). *Historical Description* (1800).
266. Catton (1778: no. 9).
267. *Historical Description* (various editions). Beilby and Bewick (1790: 248). Burt (1791).
268. *Historical Description* (1896, 1800, 1804, 1806).
269. *Historical Description* (various editions). I am indebted to Iain Bain for a transcript of the letter from John to Thomas Bewick. Polito (1803).
270. *Salisbury & Winchester Journal*, 10 July 1786.
271. John Hunter (1861: ii. 59). Dobson (1962). Pennant (1793: i. 272). Beilby and Bewick (1800: 301–2). Smith (1806: i. 51–2).
272. Beilby and Bewick (1790: 241). *Historical Description* (1788). There is no reason to suppose that Bewick drew the animal himself, nor that its corpse was acquired by John Hunter, as conjectured by Dobson (1962).
273. Smith (1806). *Historical Description* (1788, 1796, 1800).
274. John Hunter (1861: ii. 273).
275. *Historical Description* (1782, 1787).
276. Boreman (1741). Anon (*c.*1743). *Historical Description* (various editions).
277. *British Magazine and Review* (1783). *Town and Country Magazine* (January 1785: 329). Keating (1816). Austen (2011: 560).
278. Bartholomew Fair handbill, 1794, from Morley (1880: 362).
279. *Historical Description* (1796, 1800).
280. Hyde (1977: 150).
281. Uglow (2006: 98–108). I am indebted to Iain Bain for a transcript of the letter from John to Thomas Bewick.
282. Williams (1933: 126–7).
283. Payne was a family friend of Jane Austen (Austen 2011: 560).
284. This zebra is sometimes confused with a tame zebra that arrived several years later. Goodrich (1845: 227–8). Bingley (1820: 140). *Historical Description* (1806). Agasse's zebra cannot have been painted at the Exeter Change, as usually stated, because the menagerie there did not have a zebra at that time; see *Agasse*, no. 14.
285. Hughson (1805). *Historical Description* (1806). Tipu's cheetah: BL, NHD32/3.
286. Hughson (1805).
287. Simond (1815: 155–9).

CHAPTER 5. GEORGE IV AS REGENT AND KING, C.1811–1830

1. *La Belle Assemblée*, 1 May 1810: 35.
2. I am indebted to Iain Bain for a transcript of this letter.

3. Lysons, *Collectanea*, ii.
4. *New Monthly Magazine*, 2/3 (1827). *Historical Register* (1827: 478). *Literary Gazette* (1827), 623.
5. *Agasse*, 96, no. 14. Nygren and Pressly (1977: no. 100). There is no evidence for Egerton and Snelgrove's statement (1978) that the zebra was painted at Herring's menagerie in the New Road; what is known is that Herring *owned* the portrait of the zebra painted in 1803, almost certainly at the Tower, as there were no other zebras in England at the time, and that Agasse painted Herring's portrait at the New Road in about 1838 (Baud-Bovy 1904: 125).
6. John Taylor (1826); Altick (1978: 312) and many others, varying in detail.
7. Keeling (1991a: 16–17).
8. Keeling (1991b: 7–8; 2000: 75–6).
9. Old Bailey Proceedings Online (<www.oldbaileyonline.org>, version 7.0, 23 July 2014), April 1838 (t18380402).
10. Castang (1819).
11. <www.richardfordmanuscripts.co.uk/catalogue/8899> and <www.darwinproject.ac.uk/entry-1975> (both accessed 24 January 2012).
12. Thompson (1934). <www.circushistory.org/History/Bios.htm> (accessed 16 April 2011). Cached
13. Egerton and Snelgrove (1978: 187). *Agasse*, 96. The present whereabouts of the painting are unknown.
14. Print in <www.rhinoesourcecenter.com> (accessed 20 May 2011). Newspaper cuttings in the Menagerie memorabilia box.
15. 'Mr Polito's private apartment' suggests that he was in Edinburgh himself, but, given his activities in London at the time, this seem unlikely. One possibility is that this refers to *John* Polito, whose name appears for the first time in 1814.
16. Lysons, *Collectanea*, ii. 30. Clift, *Diary* (1811).
17. *Trewman's Exeter Flying Post*, 17 October 1811.
18. Miles handbill, 1811, Lysons, *Collectanea*, ii.
19. Clift, *Diary* (1812).
20. *Bentley's Miscellany* (1840: vii. 376–8).
21. Engelbach (1915: 11).
22. The *Astel* arrived in July 1811, not 1810; the present account is based on contemporary newspaper reports, not on the much less reliable memoir of Chunee written by one of his keepers, fifteen years later (John Taylor 1826). Handbills advertising *Harlequin & Padmanaba*, 26 December 1811–20 February 1812, Enthoven Collection (Victoria and Albert Museum).
23. Ackermann (1812: 27–30 and plate 2).
24. Clift, *Diary* (1812).
25. *Universal Magazine*, 14 (1810), 44. The much repeated statement (e.g. Egerton and Snelgrove 1978: 114, 180) that Cross married Stephen Polito's sister is erroneous. It stems from Baud-Bovy's essay (1904) on Agasse.
26. Lysons, *Collectanea*, ii. London Metropolitan Archive MS 11936/459/873095. Clift, *Diary* (1812).
27. Altick (1978: 309).
28. Louis XVIII, briefly King of France. St Martins/Leicester Square Scrapbook. Cutting in the Exeter Change File.
29. St Martins/Leicester Square Scrapbook. Cutting in the Exeter Change File.
30. TNA: PROB 11/1556/131.

31. London Metropolitan Archive: X 020357, p. 768.
32. Most of this information comes from Rookmaaker (1973) and the text accompanying a print depicting Alpy's (= Alpi's) rhinoceros <http://www.rhinoresourcecenter.com/images/Amsterdam-1814_i1182594300.php?type=all_images&sort_order=desc&sort_key=year> (accessed 16 March 2009). However, Rieke-Müller and Dittrich (1999) suggest that it may have been a Javan, rather than an Indian rhino, and that it may have been in possession of Jacques and Philippine Tourniaire as early as 1811.
33. *Morning Chronicle*, 24 May 1816.
34. *Agasse*, 96. Egerton and Snelgrove (1978: 180). Baud-Bovy (1904: 114).
35. Poster in the Menagerie memorabilia box.
36. Abel (1818: 318–30; 1825). Bingley (1820: 62–7). Jardine (1848: 110–24).
37. *Morning Chronicle*, 30 August 1817.
38. Donovan (1834: ii, unnumbered pages relating to plate LVII). Donald (2007: 106, 170). Handbill, Lysons, *Collectanea*, iv. St Martins/Leicester Square Scrapbook. Baud-Bovy (1904: 118–20 and fig. 82). Musée d'art et d'histoire in Geneva 1920–0016.
39. Anon. (1831: 33, no 155). None of her remains survive.
40. Bingley (1820: 67). Anon. (1831: 32, no. 152). The college paid Cross £21 for the corpse (Clift, *Diary* (1819)).
41. William Smart Herring married Ann Cross (Edward's sister), St Andrews Holborn, 5 October 1817, E. Cross and Mary Wolf witnesses. Confirmed by Cross referring to Herring as his brother-in-law (Griffith et al. 1827: iii. 353).
42. Animals listed in Cross (1820): elephant, lion, panther, leopard, cheetah, lynx, striped hyaena, jackal, civet, coati-mundi, polar bear, racoon, agouti, dog-faced baboon, porcupine, beaver, nylghai, gnu, capybara, quagga, kangaroo, emu, adjutant bird, spoonbill, crowned crane, rose-coloured pelican, black swan, vulture, parrot, llama, and zebu. See also the *Morning Chronicle*, 28 December 1820.
43. Compare Exeter Change poster of 1823 with several issued in and after 1824 in the Exeter Change File.
44. Lysons, *Collectanea*, iv. Hone (1826: 1191–2). Cuttings in Osborne Scrapbook and Exeter Change File. *A Collection of Playbills for Astley's Royal Amphitheatre for the Years 1821–1845* (BL, Playbills 171) <http://www.circopedia.org/index.php/The_Samwell_Family> (accessed 17 September 2012).
45. Clift, *Diary* (1820).
46. Griffith et al. (1827: ii. 447).
47. Reichenbach (2002). Keeling (1984: 43; 2001: 65).
48. *Pallot's Marriage Index* (available through <www.ancestry.com>). John Taylor (1826). Parish Records, St Andrew Holborn and St Clement Dane. Griffith et al. (1827: iii. 353). Edward Cross will, TNA: PROB 11/2198. Much confusion has been added to the relationships of Edward Cross by too literal translations from the text of Baud-Bovy (1904: 125).
49. Handbill, 1823/4, Lysons, *Collectanea*, iv.
50. Cutting in the St Martins/Leicester Square Scrapbook.
51. Griffith (1821: ii. 58).
52. Alexander (1986: fig. 2). Brookes (1828a: 10; 1828b: 16, 34). Griffith et al. (1827: v. 151). Hamilton Smith (1846: 261–5).
53. Ormond (1981). J. C. Wood (1973). The lions above Lambeth Bridge and in the halls of Tate Britain are also attributed to Edwin Landseer (Lennie 1976).

54. Royal Academy of Arts Collection.
55. e.g. *The Age*, 4 December 1825.
56. Griffith et al. (1827: iii. 170–1). Brookes (1829: 95–104). Burmeister (1879: 247).
57. Griffith (1821: i. 37; 1827: ii. 450–1). Christiansen and Kitchener (2010).
58. Griffith et al. (1827: iv. 66, 112).
59. Clift, *Diaries* (various dates). Anon. (1831).
60. Home (1823: 283–4). Letter from Sir Stamford Raffles to Sir Everard Home, 6 July 1822 <http://static.zsl.org/files/raffles-letter-1172.PDF> (accessed 29 July 2014).
61. Frank Buckland (1860: 48–53). It is clear from William Buckland's *Reliquiae Diluvianae* that his experiments on bone-cracking were carried out using a hyaena in Wombwell's menagerie, not in Cross's, as stated by Frank Buckland.
62. Cutting in the Menagerie memorabilia box. Heron (1850: 157).
63. Cutting in the Menagerie memorabilia box. Roberts (1997: 364–5). Handbill, 1828, Exeter Change File.
64. John Taylor (1826).
65. *Gentleman's Magazine* (1825: 475). Cuttings in the Menagerie memorabilia box.
66. John Taylor (1826).
67. Hone (1826: 321–36). John Taylor (1826). Bondesen (1999: 64–95). Hancocks (2003: 123). Altick (1978: 307–12). Keeling (1991b: 4–9, 41–3). Cuttings in the Menagerie memorabilia box. Anon. (1853: 471).
68. Scherren (1905). Blunt (1976: 25–31). Adrian Desmond (1985: 224–6).
69. St Martins/Leicester Square Scrapbook.
70. Audubon (1897: 279–80).
71. Roberts (1997: 364–7) <www.worldnavalships.com> (accessed 25 September 2012). Keeling (1991b: 9). Exeter Change handbills, 1828, Exeter Change File.
72. The first and third 1828 handbills are in the Exeter Change File; the second is reproduced by Altick (1978: 308) and is in the Westminster Archive Centre. A collection of playbills for Astley's Royal Amphitheatre for the years 1821–45 (BL, Playbills 170).
73. *Pallot's Marriage Index*.
74. Keeling (1995: 5–7).
75. Horsfield (1825).
76. Plumb (2010a: 123–4). Cowie (2013).
77. *Liverpool Mercury*, 7 March 1823.
78. Fillinham Collection, vols 7–8. Mark Sorrell, 'Polito, Stephen (1763/4–1814)', *Oxford Dictionary of National Biography* (Oxford University Press, 2004); online edn, January 2008 <http://www.oxforddnb.com/view/article/73320> (accessed 12 August 2014).
79. Bostock (1927) and Frost (1874) are not always to be relied on; much new information is being made available online at <http://www.georgewombwell.com> (accessed 8 November 2013). For more up-to-date thinking on nineteenth-century menageries, see especially Ritvo (1990), Simons (2012), and Cowie (2013, 2014).
80. Robert Chambers (1832: 586).
81. Osborne Scrapbook. Lysons, *Collectanea*, iv.

82. See especially Hone (1837: i. 978–1000).
83. This much-repeated story originates with Frost (1874: 257–9), where it is undated.
84. 'Gillman and Atkins's Grand Menagerie' handbill, 1815, Lysons, *Collectanea*, iv. Many other newspaper advertisements and handbills.
85. Newspaper cutting (1829), Osborne Scrapbook.
86. The house is now a hotel <http://www.georgianindex.net/Oatlands/Oatlands.html> (accessed 12 June 2012).
87. Plumb (2010a: 231).
88. Festing (1988: 109, 115).
89. Raikes (1858: i. 89–90).
90. Clift, *Diary* (1822).
91. Royal Collection: RCIN 420214.
92. Sometimes spelled Bowditch; later Mrs Lee. Lee (1833).
93. Bowdich (1828). Knight (1829–40: i. 193). Jardine (1834: 163–71).
94. Exeter Change handbill, 1828(?), Exeter Change File.
95. The house is now a home for people with learning disabilities, and is not open to the public.
96. Hinde (1981: 173).
97. Ione Leigh (1951: 223–4).
98. Bew (2011: 380–2).
99. Rush (1833: 369).
100. Rush (1845: 185–6).
101. Stewart (1958: 66).
102. Arbuthnot (1950: 36)
103. MS note in the Fillinham Collection, vols 7–8.
104. *Morning Post*, 24 December 1822.
105. Edward Bennett (1829: 48).
106. Bew (2011: 546). Weinreb and Hibbert (1995: 742).
107. *Morning Chronicle*, 21 January 1824.
108. Anon. (1833: 35). Vigors and Broderip (1829: 10). Sharpe (1832: 223). Keeling (1991a: 11).
109. Vigors and Broderip (1829: 31). Edward Bennett (1831: ii. 57–8).
110. Croke (1997). Waterhouse (1838: 34, 49). Sharpe (1832: 325).
111. Lysons (1792: ii. 194). Clegg (2011). The gardens are currently being restored.
112. Knight (1829–40: ii. 7–9).
113. *Morning Chronicle*, 3 October 1818.
114. Knight (1829–40: ii. 9). Clegg (2008).
115. Surtees (1990: 145). The 'elks' referred to may have been American red deer or moose.
116. Surtees (1990: 149).
117. Pückler-Muskau (1832: iii. 57).
118. Scott (1998: 534).
119. Holder (1886: 190–1). Sharpe (1832: 246–8).
120. Clegg (2008).
121. Loudon (1846: 975).
122. Knight (1829–40: i. 47).
123. Sharpe (1832: 302).
124. Pearson (1983: 127).

125. Clegg (2008).
126. Keeling (1984: 18). Clegg (2008).
127. Scherren (1905).
128. This account published in 1846, three years after Loudon's death, probably relates to the menagerie much earlier in the century (Loudon 1846: 975).
129. Brookes (1828b: 60).
130. Huish (1830a).
131. Flinders (1814).
132. Anon. (1836: 105).
133. Scherren (1901).
134. Sharpe (1832: 361–2).
135. Warne (1802: 82).
136. Dutton (1949: 112).
137. Much of this information in this chapter is taken from Robert (1997). See also Knight (1829–40: i. 25–6).
138. John Clark(e) with the animals at Sandpit Gate. Royal Collection: RCIN 403397; sometimes also referred to as Thomas Clarke.
139. Most of the information about Cross's charges is taken from transcriptions of the original records by Mark Sorrell, which are included in Keeling (1984, 1993). See also Scherren (1901: 9) and Roberts (1997: 367, 578–9).
140. Knight (1829–40: i. 323, 358–60). Sharpe (1832: 4).
141. Edward Bennett (1829: 190). *Agasse*, 158, no. 60. Anon. (1966–7: 21–2). Royal Collection: RCIN 404395.
142. Edward Bennett (1831: ii. 324).
143. Atkins (1841). Huish (1830a). Hone (1826: 1175–82). Several publications, including Griffith et al. (1827: ii), include pictures of the cubs.
144. *Agasse* (1988: 150, no. 57).
145. Huish (1830b).
146. Laufer (1928: 88). Lambourne (1965). Blunt (1976: 74–9). Most of the information about the French giraffe is taken from Allin's charming and informative biography *Zarafa* (1998).
147. Fillinham Collection, vol. 6.
148. Roberts (1997: 364–5).
149. Lambourne (1965: 1498–1502). Roberts (1997: 365).
150. I have not been able to trace Van Worrell's painting, although it is known that it was sold at Sotheby's in 1983. It was reproduced by Dardaud (2007) and was also engraved.
151. Royal Collection: RCIN 404394. Roberts (1997: 578 and plate 375). *Agasse*, 154–7, no. 59.
152. Knight (1829–40: i. 25–6).
153. Roberts (1997: 578).
154. Knight (1829–40: i. 342). Home (1829: 244–9). *London Magazine*, 10 (1828), 353.
155. BM 1868,0808.8815, attributed to William Heath.
156. BM 1868,0808.8917.
157. *Morning Chronicle*, 15 July 1828.
158. It had previously been assumed that it was intended for the Zoological Society (*Morning Post*, 17 October 1829 and many other newspapers).
159. Home (1830: 85–6 and plates). Clifford (1992: 52–3).

160. Roberts (1997: 367). Keeling (1984: 23–5; 1993: 3–22). Scherren (1901: 9; 1905: 44).
161. *An Improved History and Description of the Tower of London* (1815, 1817). Charlton (1978: 103).
162. *Hereford Journal*, 8 November 1820.
163. *Morning Chronicle*, 19 November 1821.
164. Fillinham Collection, vols 7–8.
165. *Bell's Life in London and Sporting Chronicle*, 10 July 1825.
166. Edward Bennett (1829). Huish (1830a).
167. This must have been Benjamin Wyatt, who was employed by the Duke of Wellington to work on Apsley House.
168. Anon. (1825/6). Bayley (1830: 266).
169. Britton and Bayley (1830: 266). Anon. (1827). Broderip (1826).
170. Griffth et al. (1827: ii. 232 and figure (printed in 1825)). Edward Bennett (1829: 120). Pennant (1793: ii. 243–5). Catton (1788).
171. Handbill 1826 (Parnell 1999: back cover); handbills 1827–9 (Fillinham Collection, vols 7–8). Edward Bennett (1829: 7–10, 13–16, 22–23). Huish (1830a: 31, 107 (with plates)).
172. Handbills, 1828, 1829. Edward Bennett (1829: 33–4). Newspaper cutting in the Fillinham Collection, vols 7–8.
173. Edward Stamford Raffles (1822). Griffith et al. (1827: ii. 237–8). Horsfield (1826). Edward Bennett (1829: 133–6). Handbills, 1826–9. Anon. (1827). Waterhouse (1838).
174. Handbills, 1826–9. Sharpe (1832: 4). Huish (1830a: 11–12). Edward Bennett (1829: 184, 190, 196).
175. Edward Bennett (1829: 20). Goodrich (1845: 227–8.)
176. Handbills, 1826–9.
177. Edward Bennett (1829: 137–50). Sharpe (1832). Huish (1830a:109–10). Tower Handbill, 1826.
178. For Banks's emus, see Section 4.1. Handbills, 1826–9. Huish (1830a: 35–6).
179. Handbills, 1826–9. Edward Bennett (1829: 229–30). Huish (1830a: 53–4).
180. Handbills, 1826–9. Edward Bennett (1829: 199–214). Huish (1830a: 55–6).
181. Broderip (1826). Edward Bennett (1829: 238).
182. Edward Bennett (1829: 231–41). Tower Handbills, 1826–9.
183. Edward Bennett (1829: pp. xvi, and *passim*).

CHAPTER 6. WILLIAM IV, C.1830–1837

1. Sophia Raffles (1835: 289–90).
2. G. Cuvier (1821). Anon. (1824). T. S. Raffles (1822: 239–40).
3. C. M. Turnbull, 'Raffles, Sir (Thomas) Stamford Bingley (1781–1826)', *Oxford Dictionary of National Biography* (Oxford University Press, 2004) <http://www.oxforddnb.com/view/article/23010> (accessed 10 September 2013). Glendinning (2013: 272).
4. Horsfield (1825).
5. T. S. Raffles (1822: 254–5).
6. Sophia Raffles (1835: 289–90, 592–3, 699–701). Bastin (1970). D. P. Miller (1983). Adrian Desmond (1985: 223–50; 1989: 134–44). Ritvo (1990: 205–18). Glendinning (2013: 286). Ito (2014: 14–18).
7. Scherren (1905). Blunt (1976: 21–8). Adrian Desmond (1985: 224–6).

8. Anon. (1829). Scherren (1905).
9. Vevers (1976).
10. Blunt (1976: 32).
11. Scherren (1905). Keeling (1984: 18). Newspaper cutting and handbill of 1829 in the Exeter Change File.
12. Egan (1832: 290–4).
13. <http://www.emich.edu/english/adelphi_calendar/hst1829.htm> (accessed 4 June 2011). Knight (1829–40: ii. 16–20, with plates on pp. 14 and 21).
14. BL, Playbills, no. 1750.
15. Newspaper cutting, 9 July 1837, in the Fillinham Collection, vol. 6.
16. Knight (1829–40: i. 367). Kelly (1829). Huish (1830a).
17. <http://www.british-history.ac.uk/report.aspx?compid=65448> (accessed 28 September 2012).
18. Handbills, January and April 1831, in the Exeter Change File.
19. <www.preservedproject.co.uk/happy-jerry> (accessed 7 November 2013). Anon. (*c.*1880: i. 64). Egan (1832: 290–4). Jerry's skin has the accession no. ZD.25a (Everest 2011). For Billy, see Section 5.3 and Frank Buckland (1860: 86–93); Billy's skeleton was included in the catalogue of the Royal College of Surgeons (Anon. 1853: 699–700, n. 4446), but has not survived. See also Bartlett (1899).
20. *Kidd's New Guide to the 'Lions' of London; or, The Stranger's Directory* (1832).
21. Loudon (1831, 1832).
22. Ito (2014: 65–9). Anon. (1836: xxvi. 386–7; xxvii. 16). Blunt (1976: 79–83). Warwick (1836). *Standard*, 9 June 1836. *The Times*, 7 July 1836.
23. BL, Evanion Catalogue: Evan 1688 <http://www.bl.uk/catalogues/evanion/Record.aspx?EvanID=024-000000596&ImageIndex=0> (accessed 10 July 2011); Evan 1718 <http://www.bl.uk/catalogues/evanion/Record.aspx?EvanID=024-000001489&ImageIndex=0> (accessed 10 July 2011).
24. Frank Buckland (1860: 86–93).
25. Tower handbill 1832b. Fillinham Collection, vols 7–8.
26. Jardine (1833: 51–3, 92–3, and plate 2).
27. Kotar and Gessler (2011: 117, 129–38). Thayer (2005). *New York Evening Post*, 13 April 1833. Parnell (1999: 30).
28. Watson (1978). Parnell (1999: 30).
29. Handbill for sale of fittings of Menagerie at the Tower of London, 22 October 1835 (Fillinham Collection, vols 7–8). *The Times*, 28 October 1835, quoted by Parnell (1999: 30).
30. Letter from Cops's son-in-law, (illeg.) Bennett, to Edward Cross, 8 July 1853 (Fillinham Collection, vols 7–8).

REFERENCES

Abel, C. (1818). *Narrative of a Journey in the Interior of China: And of a Voyage to and from that Country, in the Years 1816 and 1817, Containing an Account of the Most Interesting Transactions of Lord Amherst's Embassy to the Court of Pekin and Observations on the Countries which it Visited*. London: Longmans Green.

Abel, C. (1825). 'Some Account of an Orang Outang of Remarkable Height Found on the Island of Sumatra', *Asiatic Researches*, 15: 389–497; 5 plates.

Ackermann, R. (1812). 'Polito's Royal Menagerie, Exeter Change', *Repository of Arts, Literature, Commerce, for July 1812*, 8: 27–30.

Ackroyd, P. (1998). *The Life of Thomas More*. New York and London: Doubleday.

Albin, Eleazar (1731). *A Natural History of Birds. Illustrated with a Hundred and One Copper Plates, Curiously Engraven from the Life*. London.

Albin, Eleazar (1734). *A Natural History of Birds. Illustrated with a Hundred and Four Copper Plates, Engraven from the Life*, ii. London.

Albin, Eleazar (1737). *A Natural History of English Song-Birds, and Such of the Foreign as are Usually Brought over and Esteemed for their Singing*. London.

Albin, Eleazar (1738). *A Natural History of Birds. Illustrated with a Hundred and One Copper Plates, Engraven from the Life*, iii. London.

Alexander, R. M. (1986). 'Drawings of Vertebrate Animals from the Collection of Charles Hamilton Smith', *Archives of Natural History*, 13/1: 39–70.

Allen, E., Turk, J. L., and Murley, R. (1993). *The Case Books of John Hunter FRS*. London: Royal Society of Medicine.

Allen, J. A. (1880). *History of North American Pinnipeds: A Monograph of the Walruses, Sea-Lions, Sea-Bears*. US Department of the Interior. Miscellaneous. Publication 12.

Allen, Julia (2002). *Samuel Johnson's Menagerie: The Beastly Lives of Exotic Quadrupeds in the Eighteenth Century*. Banham: Erskine Press.

Allin, Michael (1998). *Zarafa*. London: Headline.

Altick, R. D. (1978). *Shows of London*. Cambridge, MA: Belknap Press of Harvard University Press.

Amherst, A. M. T. (1896). *A History of Gardening in England*. London: Bernard Quaritch.

Anon. (1675). *A True and Perfect Description of the Strange and Wonderful Elephant*, BL B.424.(2).

Anon. (*c*.1743). *The Famous Curiosities Lodged in the Tower of London*. London: Printed and sold at the Printing-Office in Bow-Church-Yard.

Anon. (1762–1809). *Historical Description of the Tower of London* (numerous editions). London.

Anon. (1792). *Ambulator: Or, a Pocket Companion in a Tour round London, within the Circuit of Twenty-Five Miles*. London: J. Bew.

Anon. (1804). *Notice des animaux vivans de la ménagerie: Leur origine et leur histoire dans cet établissement*. Paris: Levrault, Schoell et Cie.

Anon. (1815, 1817). *An Improved History and Description of the Tower of London*. London: P. & F. Hack.
Anon. (1823). 'History of the Garden of Plants. Part II', *Edinburgh Monthly*, 14: 578.
Anon. (1824). 'Notice sur le voyage de M. A. Duvaucel, dans l'Inde', *Journal Asiatique*, 4 (March), 137–45.
Anon. (?1825/6). *A New and Improved History and Description of the Tower of London*. London: Brook & King.
Anon. (1827). *A New and Improved History and Description of the Tower of London*... London: H. Steel.
Anon. (1829). *A Catalogue of the Animals Preserved in the Museum of the Zoological Society April 1829*. London.
Anon. (1831). *Catalogue of the Contents of the Museum of the Royal College of Surgeons in London. Part III. Comprehending the Human and Comparative Osteology*. London.
Anon. (1833). *List of the Animals in the Gardens of the Zoological Society June 1833*. London: Richard Taylor.
Anon. (1834). 'Joshua Brookes', *The Annual Biography and Obituary for the Year 1834*, 18: 282–95.
Anon. (1836). *The Mirror of Literature, Amusement and Instruction*. London: J. Limberg.
Anon. (1853). *Descriptive Catalogue of the Osteological Series Contained in the Museum of the Royal College of Surgeons of England. Volume II. Mammalia Placentalia*.
Anon. (*c.*1880). *Cassell's Popular Natural History*. London: Cassell, Petter & Galpin.
Anon. (1966–7). *Animal Painting. Van Dyk to Nolan: The Queen's Gallery, Buckingham Palace*. London: The Queen's Gallery.
Anon. (1995). *Claremont Landscape Garden*. London: National Trust.
Arbuthnot, H. (1950). *The Journal of Mrs. Arbuthot, 1820–1832*, i. London: Macmillan.
Archer, M., and Archer, W. G. (1955). *Indian Painting for the British*. London: Oxford University Press.
Archer, M., and Parlett, G. (1992). *Company Paintings: Indian Paintings of the British Period*. London: Victoria and Albert Museum.
Aris, M. (1982). *Views of Medieval Bhutan: The Diary and Drawings of Samuel Davis, 1783*. London: Serindia; Washington: Smithsonian Institution Press; Chicago: Serindia Publications.
Armstrong, A. W. (2002). *Forget not Mee & My Garden: Selected Letters 1725–1768 of Peter Collinson, F.R.S.* Philadelphia: American Philosophical Society.
Arnold, K. (2000). *Cabinets for the Curious*. Aldershot: Ashgate.
Asúa, Miguel de, and French, Roger (2005). *A New World of Animals: Early Modern Europeans on the Creatures of Iberian America*. Aldershot: Ashgate.
Atkins, T. (1841). *The Visitors' Handbook to the Liverpool Zoological Gardens*. Liverpool: John R. Isaac.
Audubon, Maria R. (1897). *Audubon and his Journals*, i. New York: Charles Schribner's Sons.
Austen, Jane (2011). *Jane Austen's Letters*, ed. Deirdre Le Faye, 4th edn. Oxford: Oxford University Press.
Bacon, F. (1658). *New Atlantis A Work Unfinished*. London.

Bacon, F. (1937). *Essays by Francis Bacon*, edited by Geoffrey Grigson. Oxford University Press.

Baird, Rosemary (2007). *Goodwood: Art and Architecture, Sport and Family*. London: Frances Lincoln.

Baird, Thomas (1793). *General View of the Agriculture of the County of Middlesex*. London.

Baker, Henry (1755). 'Supplement to the Account of a Distempered Skin', *Philosophical Transactions of the Royal Society*, 40: 21–4.

Banks, Joseph, Sir (1799). 'On the Propogation of the Zebra with the Ass', *Nicholson's Journal of Natural Philosophy*, 2: 267–8.

Banks, Sir Joseph (2007) (ed.). *The Scientific Correspondence of Sir Joseph Banks, 1765–1820*, ed. Neil Chambers. London: Pickering & Chatto.

Baratay, E., and Hardouin-Fugier, E. (2002). *Zoo*. London: Reaktion Books.

Barbagli, F., and Violani, C. (1997). 'Canaries in Tuscany', *Bolletino del Museo Regionale di Scienze Naturali di Torino*, 15: 25–33.

Barnaby, D. (1996). *Quaggas and other Zebras*. Plymouth: Bassett Publications.

Barnard, Anne (1994). *The Cape Journals of Lady Anne Barnard 1797–1798*, ed. A. M. L. Robinson, M. Lenta, and B. Le Cordeur. Cape Town: Van Riebeeck Society.

Bartlett, A. D. (1899). *Wild Animals in Captivity: Being an Account of the Habits, Food, Management and Treatment of the Beasts and Birds at the 'Zoo' with Reminiscences and Anecdotes*. London: Chapman & Hall.

Bastin, John (1970). 'The first Prospectus of the Zoological Society of London: New Light on the Society's Origins', *Journal of the Society for the Bibliography of Natural History*, 5/5: 369–88.

Baud-Bovy, D. (1904). *Peintres Genevois du XVIIIe et du XIXe siècle, 1766–1849, IIe série*, ii. Geneva.

Bayley, J. W. (1821). *The History and Antiquities of the Tower of London, with Biographical Anecdotes of Royal and Distinguished Persons, Deduced from Records, State-Papers, and Manuscripts, and from Other Original and Authentic Sources*. London: T. Cadell.

Bayley, J. W. (1830). *The History and Antiquities of the Tower of London, with Biographical Anecdotes of Royal and Distinguished Persons, Deduced from Records, State-Papers, and Manuscripts, and from Other Original and Authentic Sources*. 2nd edn. London: Jennings & Chaplin.

Bedini, S. A. (2000). *The Pope's Elephant: An Elephant's Journey from Deep in India to the Heart of Rome*. London: Penguin.

Beilby, R., and Bewick, T. (1790). *A General History of Quadrupeds*. Newcastle-upon-Tyne: Beilby & Bewick.

Beilby, R., and Bewick, T. (1800). *A General History of Quadrupeds*. 4th edn. Newcastle-upon-Tyne: Hodgson, Beilby & Bewick.

Bell, Charles (1806). *Essays on the Anatomy of Expression in Painting*. London: Longmans.

Belozerskaya, Marina (2006). *The Medici Giraffe and Other Tales of Exotic Animals and Power*. New York: Little, Brown & Co.

Bence-Jones, M. (1974). *Clive of India*. London: Book Club Associates.

Bennett, Edward Turner (1829). *The Tower Menagerie: Comprising the Natural History of the Animals Contained in that Establishment; with Anecdotes of their Characters and History*. London: Robert Jennings.

Bennett, Edward Turner (1831). *The Gardens and Menagerie of the Zoological Society Delineated (Being Descriptions and Figures in Illustration of the Natural History of the Living Animals in the Society's Collection)*. London: Charles Tilt.
Bennett, Joan (2010). *Sir Thomas Browne: 'A Man of Achievement in Literature'*. Cambridge: Cambridge University Press.
Bergengren, Charles (2001). *Review Sheet: Mannerism, Art and Gardens* <http://gate.cia.edu/cbergengren/arthistory/mannerism/> (accessed 16 July 2013).
Bew, J. (2011). *Castlereagh: Enlightenment, War and Tyranny*. London: Quercus.
Bezemer-Sellers, Vanessa (1990). 'The Bentinck Garden at Sorgvliet', in John Dixon Hunt (ed.), *The Dutch Garden in the Seventeenth Century*. Dumbarton Oaks Colloquium on the History of Landscape Architecture, vol. 12. Cambridge, MA, and London: Harvard University Press, 110–30.
Bingley, William (1805). *Animal Biography; or, Authentic Anecdotes of the Lives, Manners and Economy, of the Animal Creation, Arranged According to the System of Linnaeus*. 3rd edn. London: Richard Phillips.
Bingley, William (1813). *Animal Biography; or, Authentic Anecdotes of the Lives, Manners and Economy, of the Animal Creation, Arranged According to the System of Linnaeus*. 3rd edn. London: F. C. & J. Rivington.
Bingley, William (1814). *Animated Nature; or, Elements of the Natural History of Animals, etc*. London: Darton, Harvey & Darton.
Bingley, William (1820). *Animal Biography, or, Popular Zoology. 5th Edition. Volume I. Mammiferous Animals*. London: Rivington.
Birkhead, T. (2003). *The Red Canary*. London: Phoenix.
Blagrave, J. (1675). *The Epitome of the Art of Husbandry: Comprising All Necessary Directions for the Improvement of it*. London: Benjamin Billingsley.
Blair, P. (1710). 'Osteographia Elephantina: Or, a Full and Exact Description of all the Bones of an Elephant, which Died near Dundee, April the 27th 1706, with their Several Dimensions. Pts 1 and 2', *Philosophical Transactions of the Royal Society*, 27/326–7: 51–168.
Blunt, W. (1976). *The Ark in the Park: The Zoo in the Nineteenth Century*. London: Hamish Hamilton.
Bondesen, J. (1999). *The Feejee Mermaid*. London and Ithaca, NY: Cornell University Press.
Boreman, Thomas (1736). *A Description of Three Hundred Animals*. London: R. Ware.
Boreman, Thomas (1739). *A Description of Some Curious and Uncommon Creatures, Omitted in the Description of Three Hundred Animals*. London: Richard Ware and Thomas Boreman.
Boreman, Thomas (1741). *Curiosities in the Tower of London*, i. 2nd edn. London: printed for Tho. Boreman.
Boreman, Thomas (1768). *A Description of Three Hundred Animals: Viz. Beasts, Birds, Fishes*. London.
Bostock, Edward Henry (1927). *Menageries, Circuses and Theatres, etc*. London: Chapman & Hall.
Boswell, James (1768). *An Account of Corsica: The Journal of a Tour to that Island; and Memoirs of Pascal Paoli*. London: Edward & Charles Dilly.
Bowdich, Mrs [Sarah] (1828). 'Anecdotes of a Tamed Panther', *Mirror of Literature*, 12: 36–9.
Bradley, R. (1721). *A Philosophical Account of the Works of Nature. Endeavouring to Set Forth the Several Gradations Remarkable in the Mineral, Vegetable,*

and Animal Parts of the Creation . . . to which Is Added, an Account of the State of Gardening, etc. London: W. Mears.

Britton, J., and Bayley, E. W. (1830). *Memoirs of the Tower of London.* London: Hurst Chance & Co.

Brock, C. Helen (1996). *Calendar of the Correspondence of Dr William Hunter 1740–1783.* Cambridge: Cambridge Wellcome Unit for the History of Medicine.

Broderip, W. J. (1826). 'Some Account of the Mode in which the Boa Constrictor Takes its Prey: And of the Adaption of its Organization to its Habits', *Zoological Journal*, 2: 215–21.

Brookes, Joshua (1828a). *Brookesian Museum: The Museum of Joshua Brooke.* London.

Brookes, Joshua (1828b). *A Catalogue of the Anatomical and Zoological Museum of Joshua Brookes, Esq. F.R.S., F.L.S. etc . . . .* London.

Brookes, Joshua (1829). 'On a New Genus of the Order Rodentia', *Transactions of the Linnean Society of London*, 16 (1833), 95–104.

Brookes, Joshua (1830). *Museum Brookesianum: A Descriptive and Historical Catalogue of the Remainder of the Anatomical & Zootomical Museum, of Joshua Brookes, Esq.* London.

Browne, Thomas (1643). *Religio Medici.* London.

Browne, Thomas (1835). *Sir Thomas Browne's Works Including his Life and Correspondence*, ii, ed. S. Wilkin. London: William Pickering.

Browne, Thomas (1836). *Sir Thomas Browne's Works Including his Life and Correspondence*, i, ed. S. Wilkin. London: William Pickering.

Browne, Thomas (1880). *The Works of Sir Thomas Browne*, ed. S. Wilkin. London: George Bell & Son.

Browne, Thomas (1964). *The Works of Sir Thomas Browne*, ed. Geoffrey Keynes. London: Faber and Faber.

Browne, Thomas (2006). *Sir Thomas Browne Selected Writings*, ed. Claire Preston. Manchester: Carcanet Press.

Bryant, A. W. M. (1948). *Samuel Pepys: The Years of Peril.* London: Collins.

Buckland, Frank (1860). *Curiosities of Natural History. Second Series.* London: Macmillan; New York: Rudd & Carleton; London: Richard Bentley.

Buckland, Frank (1881). *John Hunter at Earl's Court, Kensington, 1764–1793: John Hunter's Residence at Earl's Court, Kensington*, ed. J.J.M. [J. J. Merriman]. London: Wakeham and Sons.

Buckland, William (1824). *Reliquiæ Diluvianæ.* London: John Murray.

Buffon, Georges, Comte de (1780). *Natural History, General and Particular . . . Translated into English . . . by William Smellie.* 1st edn. Edinburgh: William Creech.

Buffon, Georges, Comte de (1785). *Natural History, General and Particular . . . Translated into English . . . by William Smellie.* 2nd edn. London: W. Strahan & T. Cadell.

Buffon, Georges, Comte de (1791). *Natural History, General and Particular . . . Translated into English . . . by William Smellie.* 3rd edn. London: W. Strahan & T. Cadell.

Buffon, Georges, Comte de, and Daubenton, L.-J.-M. (1771). *Histoire naturelle, generale et particulière . . . (Additions de l'éditeur de Hollande . . . Mr Allemand. Observations sur la Renne, par P. Camper)*, xv. Amsterdam: J. H. Schneider.

Buffon, Georges, Comte de, and Daubenton, L.-J.-M. (1784). *De Allgemeene en byzondere Natuurlyke historie, met de Beschryving des Konigs Kabinet*, xv. Amsterdam: J. H. Schneider.

Burmeister, H. (1879). *Description physique de la République Argentine d'après des observations personnelles et étrangères. Tome. III Animaux Vertébrés: Pt 1. Mammiféres vivants et éteints*. Paris.

Burt, N. (1791). *Delineation of Curious Foreign Beasts and Birds in their Natural Colours: Which are to be Seen Alive at the Great Room over Exeter Change and at the Lyceum, in the Strand*. London: N. Burt.

Bush, S. (2003). 'Satisfying Cromwell's Curiosity', *New England Quarterly*, 76/1: 108–15.

Byrne, M. St. C., and Boland, B. (1985). *The Lisle Letters: An Abridgement*. Harmondsworth: Penguin.

Caius, J. (1570a). *De Rariorum Animalium Atque Stirpum Historia, Libellos*, in *The Works of John Caius M.D.*, ed. E. S. Roberts. Cambridge: Cambridge University Press, 1912.

Caius, J. (1570b). Extracts from the work of John Caius, *De Rariorum Animalium atque stirpum historia*, in A. H. Evans, *Turner on Birds* (Cambridge: Cambridge University Press, 1903), 192–211.

Caius, J. (1576). *A Treatise on English Dogges*, in *The Works of John Caius M.D*, ed. E. S. Roberts. Cambridge: Cambridge University Press, 1912.

Campbell, T., et al. (1816) (eds). *The New Monthly Magazine and Universal Register*, v. London: H. Colburn.

Carrington, R. (1962). *Elephants*. Harmondsworth: Penguin.

Castang, Philip (1819). 'Instructions for Breeding Pheasants', in Bonington Moubray (ed.), *Practical Treatise on Breeding, Rearing, and Fattening All Kinds of Domestic Poultry, Pheasants, Pigeons, and Rabbits*. London: Sherwood, Neely & Jones, 121–8.

Catesby, Mark (1731). *The Natural History of Carolina, Florida and the Bahama Islands*, i. Printed for the author.

Catesby, Mark (1747). Appendix included in Catesby's *The Natural History of Carolina, Florida and the Bahama Islands*, ii (London, 1743).

Catton, Charles (1788). *Animals Drawn from Nature and Engraved in Acquatinta*. London: I. & J. Taylor.

Cave, A. J. E. (1941). 'Museum', in *The Royal College of Surgeons of England, Scientific Reports 1940–1941*. London.

Chambers, Mr (1763). 'A Description of the Palace and Gardens at Kew: The Seat of the Princess Dowager of Wales', *London Chronicle*, 30 August–1 September, 1–3.

Chambers, Robert (1832). *The Book of Days: A Miscellany of Popular Antiquities*, ii. London & Edinburgh: Chambers.

Chambers, Robert (1835). *A Biographical Dictionary of Eminent Scotsmen*. Glasgow: Blackie & Son.

Chaplin, S. D. J. (2009). 'John Hunter and the "Museum Oeconomy", 1750–1800'. Ph.D. thesis. King's College, London.

Charlton, John (1978). *The Tower of London: Its Buildings and Institutions*. London: Dept. of the Environment, HMSO.

Chaudhuri, Nirad C. (1975). *Clive of India: A Political and Psychological Essay*. London: Barrie & Jenkins.

Chaudhuri, Kirti Narayan (2006). *The Trading World of Asia and the East India Company: 1660–1760*. Cambridge: Cambridge University Press.

Cherry, B., and Pevsner, N. (1973). *The Buildings of England: Northamptonshire*. Harmondsworth: Penguin.

Cherry, B., and Pevsner, N. (1991). *The Buildings of England: London 3: North-West*. New Haven, CT: Yale University Press.

Chisholm, K. (2001). *Fanny Burney: Her Life*. London: Vintage Books.

Christiansen, P., and Kitchener, A. C. (2010). 'A Neotype of the Clouded Leopard (*Neofelis nebulosa* Griffith 1821)', *Mammalian Biology*, 76/3: 325–31.

Clarke, T. H. (1986). *The Rhinoceros from Dürer to Stubbs, 1515–1799*. London: Sotheby's Publications.

Clegg, Gillian (2008). 'The Duke of Devonshire's Menagerie at Chiswick House', *English Heritage Historical Review*, 3: 123–7.

Clegg, Gillian (2011). *Chiswick House and Gardens: A History*. London: McHugh.

Clifford, Timothy (1992). 'The Plaster Shops of the Rococo and Neo-Classical Era in Britain', *Journal of the History of Collections*, 4: 39–65.

Climenson, Emily (1899) (ed.). *Passages from the Diaries of Mrs Philip Lybbe Powys of Hardwick House, Oxon: AD 1756 to 1808*. London: Longmans, Green.

Clutton, Sir G. (1967–71). 'The Gardeners of the Eighth Lord Petre', *Essex Naturalist*, 32: 201–6.

Clutton, Sir G. (1970). 'The Cheetah and the Stag', *Burlington Magazine* (August), 536–9.

Clutton-Brock, Juliet (1981). *Domesticated Animals from Early Times*. London: Heinemann and British Museum (Natural History).

Clutton-Brock, Juliet (1994). 'Vertebrate Collections', in Arthur MacGregor (ed.), *Sir Hans Sloane, Collector, Scientist, Antiquary*. London: British Museum Press, 77–92.

Clutton-Brock, Juliet (2012). *Animals as Domesticates*. East Lansing, MI: Michigan State University Press.

Collins, David (1802). *An Account of the English Colony in New South Wales: From its First Settlement in 1788, to August 1801*. London: Cadell & Davies.

Collins Baker, C. H. (1949). *James Brydges: First Duke of Chandos*. Oxford: Clarendon Press.

Coombs, D. (1997). 'The Garden at Carlton House of Frederick Prince of Wales and Augusta Princess of Wales: Bills in their Household Accounts 1728–1772', *Garden History*, 25/2: 153–77.

Cowie, Helen (2013). 'Elephants, Education and Entertainment: Travelling Menageries in Nineteenth-Century Britain', *Journal of the History of Collections*, 25/1: 103–17.

Cowie, Helen (2014). *Exhibiting Animals in Nineteenth-Century Britain: Empathy, Education*. New York: Palgrave Macmillan.

Croke, Vicki (1997). *The Modern Ark*. New York: Schribner.

Cross, Edward (1820). *Companion to the Royal Menagerie, Exeter 'Change, Containing Concise Descriptions, Scientific & Interesting, of the Curious Foreign Animals.... Derived from Actual Observation*. London: Tyler & Honeyman.

Cuthbert, N. B. (1941). *American Manuscript Collections in the Huntington Library for the History of the Seventeenth and Eighteenth Centuries*. San Marino: Huntington Library.

Cuvier, G. (1917). *Le Règne animal distributé d'après son organisation*. Paris: Deterville.

Cuvier, G. (1821). 'Notice sur les voyages de MM Diard et Duvaucel, naturalistes français, dans les Indes orientales et dans les îles de la Sonde', *Revue encyclopédique* (June), 10: 472–82.

Daniel, William Barker (1801). *Rural Sports*. London: Bunny & Gold.
Dardaud, Gabriel (2007). *Une Girafe pour le roi*. Bordeaux: Elytis.
Derham, W. (1718). *Philosophical Letters between the Late Learned Mr Ray and Several of his Ingenious Correspondents, Natives and Foreigners: To which Are Added those of Francis Willughby Esq; the Whole Consisting of Many Curious Discoveries and Improvements in the History of Quadrupeds, Birds, Fishes, Insects, Plants, Fossiles, Fountains, &c.* London: William & John Innys.
Desmond, Adrian (1985). 'The Making of Institutional Zoology in London, 1822–1836', *History of Science*, 23: 153–85, 223–50.
Desmond, Adrian (1989). *The Politics of Evolution: Morphology, Medicine and Reform in Radical London*. London: University of Chicago Press.
Desmond, Ray (1995). *Kew: The History of the Royal Botanic Gardens*. London: Harvill Press.
Dobson, Jessie (1962). 'John Hunter's Animals', *Journal of the History of Medicine and Allied Sciences*, 17/4: 479–86.
Donald, Diana (2007). *Picturing Animals in Britain 1750–1850*. New Haven, CT: Yale University Press.
Donovan, Edward (1834). *The Naturalist's Repository, or Monthly Miscellany of Exotic Natural History*. London.
Dutton, R. (1949). *The English Country House*. London: Batsford.
Edwardes, Michael (1991). *The Nabobs at Home*. London: Constable.
Edwards, George (1743–51). *A Natural History of Birds. Most of which have not been Figur'd or Describ'd, and Others Very Little Known from Obscure or too Brief Descriptions*, 4 vols. London.
Edwards, George (1758–64). *Gleanings of Natural History Exhibiting Figures of Quadrupeds, Birds, Insects, Plants, &c.* 3 vols. London.
Edwards, J. (1974). *Cheshunt in Hertfordshire*. Cheshunt: Cheshunt Urban District Council.
Edwards, R. (2006). 'The Exeter Change Tour of 1798', in R. Edwards and C. H. Keeling (eds), *Menagerie Miscellany: Six Essays*. Guildford: Bartlett Society, 2–9.
Egan, P. (1832). 'Happy Jerry! Late of the Surrey Zoological Gardens', in *Pierce Egan's Book of Sports, and Mirror of Life*. London T. T. & J. Tegg, 290–4.
Egerton, Judy (1978). *British Sporting and Animal Paintings 1655–1867: The Paul Mellon Collection*. London: Tate Gallery.
Egerton, Judy (1984). *George Stubbs 1724–1806*. London: Tate Gallery.
Egerton, Judy (2007). *George Stubbs Painter: Catalogue Raisonné*. London: Yale University Press.
Egerton, Judy, and Snelgrove, Dudley (1978). *British Sporting and Animal Drawings, c.1500–1850: A Catalogue*. London: Tate Gallery for the Yale Center for British Art.
Engelbach, Arthur H. (1915). *More Anecdotes of Bench and Bar*. London: G. Richards.
Evelyn, John (1786). *Silva, Or a Discourse of Forest-Trees*. London.
Evelyn, John. (2000). *The Diary of John Evelyn: Now First Printed in Full from the Manuscripts belonging to Mr. John Evelyn*, ed. E. S. de Beer. Oxford: Clarendon Press.
Evelyn, John (2001). *Elysium Britannicum, or The Royal Gardens*, ed. John E. Ingram. Philadelphia: University of Pennsylvania Press.
Everest, Sophie (2011). '"Under the Skin": The Biography of a Manchester Mandrill', in Samuel Alberti (ed.), *The Afterlife of Animals: A Museum Menagerie*. Charlottesville, VA: University of Virginia Press, 74–91.

Farrington, Anthony (2002). *Trading Places: The East India Company and Asia 1600–1834*. London: British Library.
Feltoe, C. L. (1884). *Memorials of John Flint South*. London: John Murray.
Festing, S. (1986a). 'Rare Flowers and Fantastic Breeds: The 2nd Duchess of Portland and her Circle—I', *Country Life* (12 June), 1684–6.
Festing, S. (1986b). 'Grace without Triviality: The 2nd Duchess of Portland and her Circle—II', *Country Life* (19 June), 1772–4.
Festing, S. (1987). 'Animal Crackers, Menageries and Aviaries in the 18th Century', *Country Life* (June 1987), 124–5.
Festing, S. (1988). 'Menageries and the Landscape Garden', *Journal of Garden History*, 8/4: 104–7.
Fisher, M. H. (2004). *Counterflows to Colonialism: Indian Travellers and Settlers in Britain, 1600–1857*. Delhi: Permanent Black.
Fitzmaurice, Edmund (1912). *Life of William, Earl of Shelburne*. London: Macmillan.
Flinders, Matthew (1814). *A Voyage to Terra Australis: Undertaken for the Purpose of Completing the Discovery of that Vast Country*. London: G. & W. Nichol.
Foister, S. (2006). *Holbein in England*. London: Tate Gallery.
Fontes da Costa, P. (2009). 'Secrecy, Ostentation, and the Illustration of Exotic Animals in Sixteenth-Century Portugal', *Annals of Science*, 66/1: 59–82.
Foote, Jesse (1794). *The Life of John Hunter*. London.
Forster, George (1778). *A Letter to the Earl of Sandwich*. London.
Foster, W. (1926). *John Company*. London: Lane.
Fothergill, John (1781). *Fothergill, John (1712–1780): A Complete Collection of the Medical and Philosophical Works*, ed. John Elliot. London: John Walker.
Fox, G. T. (1827). *Synopsis of the Newcastle Museum*. Newcastle.
Frost, T. (1874). *The Old Showmen and Old London Fairs*. London: Tinsley Brothers.
Garner, T. (1800a). *A Brief Description of the Principal Foreign Animals and Birds, now Exhibiting at the Grand Menagerie, over Exeter-'Change, the Property of Mr Gilbert Pidcock*. [Octavo edn]. London: T. Burton.
Garner, T. (1800b). *A Brief Description of the Principal Foreign Animals and Birds, the Property of Mr Gilbert Pidcock, now Exhibiting at the Grand Menagerie, over Exeter-'Change, Strand: Chiefly Extracted from the Works of Buffon and Goldsmith: Embellished with Elegant Engravings: With Several Additions and Anecdotes*. By T. Garner, Printer, York. Second [Quarto] edn. London: T. Burton.
Gascoigne, John (1994). *Joseph Banks and the English Enlightenment: Useful Knowledge and Polite Culture*. Cambridge: Cambridge University Press.
George Stubbs 1724–1806 (1984). London: Tate Gallery.
George, Wilma (1980). 'Sources and Background to Discoveries of New Animals in the Sixteenth and Seventeenth Centuries', *History of Science*, 18: 79–104.
George, Wilma (1985). 'Alive or Dead: Zoological Collections in the Seventeenth Century', in Oliver Impey and Arthur MacGregor (eds), *The Origins of Museums: The Cabinet of Curiosities in Sixteenth- and Seventeenth-Century Europe*. Oxford: Clarendon Press, 179–87.
Gesner, C. (1554). *Historiae Animalium lib. II de Quadripedibus oviparis* (1st edn). Tiguri: Excudebat C. Froschoverus.
Gesner, C. (1555). *Historiae Animalium lib. III de Quadripedibus qui est de avium natura* (1st edn). Tiguri: Apud C. Fruschoverum.

Gesner, C. (1560). *Historiae Animalium lib. 1 de Quadripedibus viviparis* (second printing of 1st edn). Tiguri: Apud C. Fruschoverum.
Gesner, C. (1602). *Historiae Animalium lib. 1 de Quadripedibus viviparis* (2nd edn). Francofurti: In Bibliopolio Cambierano.
Gibson, William (1841). *Rambles in Europe, in 1839: With Sketches of Prominent Surgeons, Physicians*. Philadelphia: Lea & Blanchard.
Gilpin, W. (1789). *Observations Relative Chiefly to Picturesque Beauty, made in ... 1776, on Several Parts of Great Britain*, ii. London: Blamire.
Gleig, George Robert (1841). *Memoirs of the Life of the Right Hon. Warren Hastings, First Governor-General of Bengal*. London: Bentley.
Glendinning, C. (2013). *Raffles and the Golden Opportunity*. London: Profile Books.
Goldsmith, Oliver (1774). *An History of the Earth, and Animated Nature*. London.
Goodman, G. (1839). *The Court of King James I*. London: Richard Bentley.
Goodrich, S. G. (1845). *Illustrative Anecdotes of the Animal Kingdom*. New York: J. Allen.
Gorgas, Michael (1997). 'Animal Trade between India and Western Eurasia in the Sixteenth Century—the Role of the Fuggers in Animal Trading', in K. S. Mathew (ed.), *Indo-Portuguese Trade and the Fuggers of Germany*. New Delhi: Manohar, 195–225.
Granger, W. G., and Caulfield, J. (1804). *The New Wonderful Museum and Extraordinary Magazine: Being a Complete Repository of... Rarities Of Nature*, ii. London: Hogg & Co.
Grant, T. (2002). 'Polar Performances', *Times Literary Supplement*, 14 June 2002, 14–15.
Green, David (1951). *Blenheim Palace*. London: Country Life.
Green, M. A. E. (1909). *Elizabeth Electress Palatine and Queen of Bohemia*. London: Methuen.
Grier, S. C. (1905). *The Letters of Warren Hastings to his Wife*. Edinburgh & London: Blackwood.
Griffith, Edward (1821). *General and Particular Descriptions of the Vertebrated Animals*. London: Baldwin, Craddock & Joy.
Griffith, Edward, et al. (1827). *The Animal Kingdom Arranged after its Conformity with its Organization. The Class Mammalia*, ii–v. London: Whittaker.
Grigson, Caroline (2015). 'New Information on Indian Rhinoceroses (*Rhinoceros unicornis*) in Britain in the Mid-Eighteenth Century', *Archives of Natural History*, 42/1: 76–84.
Grigson, Caroline, Groves, C., Kitchener, A. C., and Rolfe, W. D. I. (2008).'Stubbs's "Drill and albino hamadryas baboon" in Conjectural Historical Context—a Possible Correction', *Archives of Natural History*, 35: 174–5.
Grigson, Geoffrey (1972). *Shapes and Creatures*. London: Charles & Adam Black.
Hahn, D. (2003). *The Tower Menagerie*. London: Simon & Schuster.
Hamilton-Dyer, S. (2009). 'Animal Bones', in P. Dury and R. Simpson (eds), *Hill Hall: A Singular House Devised by a Tudor Intellectual*. London: Society of Antiquaries, 345–51.
Hamilton-Smith, C. (1846). *The Naturalists Library (edited by Sir William Jardine), vol. XIX. Mammalia, vol. V, Dogs, vol. II*. London: Chatto & Windus.
Hancocks, David (2003). *A Different Nature: The Paradoxical World of Zoos and their Uncertain Future*. Berkeley and Los Angeles: University of California Press.

Harris, Walter (1699). *Description of the King's Royal Palace and Gardens at Loo*. London.
Harrison, G. B. (1946). *A Last Jacobean Journal*. London: Routledge.
Harvey, Robert (1998). *Clive. The Life and Death of a British Emperor*. London: Hodder & Stoughton.
Harvey, William (1653). *Anatomical Exercitations, Concerning the Generation Of Living Creatures to which are Added Particular Discourses, of Births, and of Conceptions, &c.* London: Octavian Pulleyn.
Haydon, Benjamin Robert (1950). *The Autobiography and Journals of Benjamin Robert Haydon (1786–1846)*, ed. Michael Elwin. London: MacDonald.
Hayes, William (1775). *A Natural History of British Birds, &c. with their Portraits, Accurately Drawn, and Beautifully Coloured from Nature, by Mr Hayes*. London: S. Hooper.
Hayes, William (1794–9). '26 Coloured Etchings and 5 Watercolour Drawings of the Birds in the Collection at Osterley Park'. Natural History Museum Library, Zoology, 88 ff H.
Haynes, R. (1989). *Robert Cecil Earl of Salisbury, 1563–1612*. London: Peter Owen.
Heron, R. (1850). *Notes*. Grantham.
Hervey, G. F. (1968). *The Goldfish*. London: Faber & Faber.
Hervieux de Chanteloup, J. C. (1709). *Traité des serins de canarie*. Paris.
Hickey, W. (n.d.). *Memoirs of William Hickey*, ed. Alfred Spencer. 7th edn. London: Hurst & Blackett.
Hill, C. P. (1988). *Who's Who in Stuart Britain*. London: Shepeard-Walwyn.
Hinde, W. (1981). *Castlereagh*. London: Collins.
Hoeniger, F. D., and Hoeniger, J. F. M. (1969a). *The Development of Natural History in Tudor England*. [Charlottesville, VA]: University Press of Virginia.
Hoeniger, F. D., and Hoeniger, J. F. M. (1969b). *The Growth of Natural History in Stuart England from Gerard to the Royal Society*. Charlottesville, VA: University Press of Virginia.
Holder, C. F. (1886). *The Ivory King: A Popular History of the Elephant and its Allies*. New York: Charles Schribner's sons.
Holinshed, R. (1587). *The First and Second Volumes of Chronicles*, i. London.
Home, Everard (1794). 'A Short Account of the Life of the Author', in *A Treatise on the Blood, Inflammation and Gun-shot Wounds, by the Late John Hunter*, ed. Everard Home. London: George Nichol, pp. xiii–lxvii.
Home, Everard (1795). 'Some Observations on the Mode of Generation of the Kanguroo, with a Particular Description of the Organs themselves', *Philosophical Transactions of the Royal Society*, 85: 221–38.
Home, Everard (1800). 'The Croonian Lecture: On the Structure and Uses of the Membrana Tympani of the Ear', *Philosophical Transactions of the Royal Society*, 90: 1–21.
Home, Everard (1808). 'An Account of Some Peculiarities in the Anatomical Structure of the Wombat, with Observations on the Female Organs of Generation (by Bell). Everard Home', *Philosophical Transactions of the Royal Society of London*, 98: 304–12.
Home, Everard (1823). 'On the Difference of Structure between the Human Membrana Tympani and that of the Elephant', *Philosophical Transactions of the Royal Society*, 113: 23–6.

Home, Everard (1829). *Lectures on Comparative Anatomy*, v. London: Longman, Rees, Orme, Brown, & Green.

Home, Everard (1830). 'A Report on the Stomach of the Zariffa', *Philosophical Transactions of the Royal Society*, 120: 85–6.

Hone, W. (1826). *The Every-Day Book and Table Book*. London: William Tegg.

Hone, W. (1832). *The Year Book of Daily Recreation and Information*. London: Thomas Tegg.

Hone, W. (1837). *The Every-Day Book and Table-Book*. London: Thomas Tegg.

Hooke R. (1935). *The Diary of Robert Hooke, Transcribed from the Original*, ed. H. W. Robinson and W. Adams. London: Taylor & Francis.

Horsfield, Thomas (1825). 'Description of the *Rimau-dahan* of the Inhabitants of Sumatra: A New Species of *Felis* Discovered in the Forest of Bencoolen by T. Stamford Raffles', *Zoological Journal*, 1: 542–54.

Horsfield, Thomas (1826). 'Description of the *Helarctos euryspilus*: Exhibiting in the Bear from the Island of Borneo the Type of a Subgenus of *Ursus*', *Zoological Journal*, 2: 221–34.

Hosking, G. L. (1952). *The Life and Times of Edward Alleyn . . . Founder of the College of God's Gift at Dulwich*. London: Jonathan Cape.

Howes, Edmund (1615). *Annales, or Generall Chronicle of England. Begun First by Maister John Stow. And after him Continued vnto the Ende of this Present Yeere 1614, by E. Howes*. London: Thomas Adams.

Howes, Edmund (1631). *Annales, or a General Chronicle of England. Begun by John Stowe: Continued and Augmented with Matters Foraigne and Domestique, Ancient and Moderne, unto the End of this Present Yeare, 1631*. London: Richardi Meighen.

Hughson, David (1805). *London: Being an Accurate History and Description of the British Metropolis and its Neighbourhood*, ii. London: Stratford.

Hugo, T. (1866). *The Bewick Collector*. London: Lovell Reade & Co.

Huish, R. (1829). *Delineations of the Most Distinguished Wild Animals in the Various Menageries of this Country*. London: Thomas Kelly.

Huish, R. (1830a). *The Wonders of the Animal Kingdom; Exhibiting Delineations of the Most Distinguished Wild Animals, in the Menageries of this Country*. London: Thomas Kelly.

Huish, R. (1830b). *Memoirs of George the Fourth*. London: T. Kelly.

Humphrey, Ozias, and Mayer, Joseph (2005). *A Memoir of George Stubbs*. London: Pallas Athene.

Hunter, John (1787). 'Observations Tending to Show that the Wolf, Jackal, and Dog Are All of the Same Species', *Philosophical Transactions of the Royal Society*, 77: 253–66.

Hunter, John (1861). *Essays and Observations on Natural History, Anatomy, Physiology, Psychology, and Geology*, ed. Richard Owen. London: John Van Voorst.

Hunter, William (1771). 'An Account of the Nyl-Ghau, an Indian Animal, not hitherto Described', *Philosophical Transactions of the Royal Society*, 61: 170–81.

Hunter, William (2008). *The Correspondence of Dr William Hunter*, ed. C. Helen Brock. London: Pickering & Chatto.

Hussey, C. (1970). 'Purley Hall, Berkshire', *Country Life*, 5 February, 310–13.

Hyde, M. (1977). *The Thrales of Streatham Park*. Cambridge, MA, and London: Harvard University Press.

Impey, E. J. A. (1963). *About the Impeys*. Worcester: Ebenezer Baylis.
Israel J. I. (1990). *Empires and Entrepôts: The Dutch, the Spanish Monarchy, and the Jews, 1585–1713*. London: Hambledon Press.
Ito, Takashi (2014). *London Zoo and the Victorians, 1828–1859*. Woodbridge: Boydell Press.
Jackson, Christine, E. (1985). *Bird Etchings, the Illustrators and their Books, 1655–1855*. London: Cornell University Press.
Jackson, Christine, E. (1994). *Bird Painting, the 18th Century*. Woodbridge: Antique Collectors Club.
Jackson, Christine, E. (1998). *Sarah Stone: Natural Curiosities from the New World*. London: Merrell Holberton and the Natural History Museum.
Jacques, D. (1822). *A Visit to Goodwood, the Seat of the Duke of Richmond*. Chichester.
Jacques, David, and van der Horst, Arend Jan (1988). *The Gardens of William and Mary*. London. Christopher Helm.
Jacques-Laurent Agasse 1767–1849 (1989). London: Tate Gallery.
Jardine, W. (1833). *The Naturalist's Library (Edited by Sir William Jardine), Volume II: Mammalia Volume I, The Natural History of Monkeys*. Edinburgh: W. H. Lizars; London: Longman, Rees, Orme, Brown, Green & Longman.
Jardine, W. (1834). *The Naturalist's Library (Edited by Sir William Jardine), Volume XVI: Mammalia Volume II, The Felinae*. London: Chatto & Windus.
Jardine, W. (1848). *The Naturalist's Library (Edited by Sir William Jardine), Volume XXVI: Mammalia, Monkeys*. London: Chatto & Windus.
Jenkins, Susan (2007). *Portrait of a Patron: The Patronage and Collecting of James Brydges, 1st Duke of Chandos*. Aldershot: Ashgate.
Johnson, Joan (1984). *Princely Chandos: James Brydges 1674–1744*. Gloucester: Alan Sutton.
Keating, M. B. St. L. (1816). *Travels to Morocco (through France and Spain...) an Account of the British Embassy to the Court of Morocco, under the Late George Payne, Esq. Consul General*. London.
Keeling, C. H. (1984). *Where the Lion Trod: A Study of Zoological Gardens*. Shalford: Clam Publications.
Keeling, C. H. (1989). *Where the Zebu Grazed: A Further Study and Discussion on Forgotten Animal Collections, how they Were Run, and the People who Ran them*. Shalford: Clam Publications.
Keeling, C. H. (1991a). *In the Beginning*. Shalford: Clam Publications.
Keeling, C. H. (1991b). *Where the Elephant Walked*. Shalford: Clam Publications.
Keeling, C. H. (1993). *Where the Macaw Preened*. Shalford: Clam Publications.
Keeling, C. H. (1995). *Where the Penguin Plunged*. Shalford: Clam Publications.
Keeling, C. H. (2000). *The Marvel by the Mersey*. Shalford: Clam Publications.
Keeling, C. H. (2001). 'The Zoological Gardens of Great Britain', in V. N. Kisling (ed.), *Zoo and Aquarium History: From Ancient Animal Collections to Zoological Gardens*. Boca Raton, FL: CRC Press, 49–74, 369–71.
Kelly, Thomas (1829). *The Wonders of the Animal Kingdom Exhibiting Delineations of the Most Distinguished Wild Animals in the Various Menageries of this Country*. London: Thomas Kelly.
Kent, J. (1896). *Records and Reminiscences of Goodwood and the Dukes of Richmond*. London.
Kerry, Earl of (1922). 'King's Bowood Park [No. III]', *Wiltshire Archaeological and Natural History Magazine*, 42: 18–38.

Kidd's New Guide to the 'Lions' of London; or, The Stranger's Directory (1832). London: William Kidd.

Kinch, M. P. (1986). 'The Meteoric Career of William Young, Jr (1742–1785), Pennsylvania Botanist to the Queen', *Pennsylvania Magazine of History and Biography*, 11/3: 359–88.

Knight, C. (1829–40). *The Menageries: Quadrupeds, Described and Drawn from Living Subjects*. 3 vols. London: Charles Knight.

Knox, Tim (2002). 'The Artificial Grotto in Britain', *Magazine Antiques*, 61/6: 100–7.

Kotar, S. L., and Gessler, J. L. (2011). *The Rise of the American Circus, 1716–1899*. Jefferson, NC: McFarlane & Co.

Lai Yu-chih (2013). 'Images, Knowledge and Empire: Depicting Cassowaries in the Qing Court', *Transcultural Studies*, 1: 7–100 (esp. 32–40).

Laird, Mark (1999). *The Flowering of the Landscape Garden: English Pleasure Grounds, 1720–1800*. Philadelphia, PA: University of Pennsylvania Press.

Laird, Mark, and Weisberg-Roberts, Alicia (2009). *Mrs Delany and her Circle*. New Haven, CT: Yale Center for British Art; London: Sir John Soane's Museum.

Lambert, A. B. (1807). 'Description of a New Species of Macropus (*M. elegans*), from New Holland [from a Living Specimen in the Collection at Exeter Change', *Transactions of the Linnean Society*, 8: 318–19.

Lambourne, L. S. (1965). 'A Giraffe for George IV', *Country Life*, 2 December, 1498–1502.

Lambton, L. (1985). *Beastly Buildings: The National Trust Book of Architecture for Animals*. London: Jonathan Cape.

Lansdowne, Marquis of (1928). *The Petty–Southwell Correspondence 1676–1687*. London.

Larwood, J. (1881). *The Story of London Parks*. New edn. London: Chatto.

Laskey, John (1813). *A General Account of the Hunterian Museum, Glasgow*. Glasgow: John Smith.

Latham, John (1781). *A General Synopsis of Birds*, i. London.

Latham, John (1785). *A General Synopsis of Birds*, iii, pt 2. London.

Latham, John (1787). *Supplement to the General Synopsis of Birds*. London.

Latham, John (1790). *Index Ornithologicus*. London.

Latham, John (1801). *Supplement II to the General Synopsis of Birds*. London.

Latham, John (1823). *A General History of Birds*. Winchester.

Laufer, B. (1928). *The Giraffe in the History of Art*. Chicago: Field Museum of Natural History.

Lawrence, John (1819). *Practical Treatise on Breeding, Rearing, and Fattening All Kinds of Domestic Poultry, Pheasants, Pigeons, and Rabbits*. London: Sherwood, Sherwood, Neely, and Jones.

Lawson, Charles (1895). *The Private Life of Warren Hastings, First Governor-General of India*. London: Swann Sonnenschein.

Le Fanu, W. R. (1931). 'John Hunter's Buffaloes', *British Medical Journal*, 26 September, 574.

Lee, Mrs R. [Sarah Bowditch] (1833). *Memoirs of Baron Cuvier*. London: Longman, Rees, Orme, Brown Green & Longman.

Leigh, Ione (1951). *Castlereagh*. London: Collins.

Leigh Thomas, H. (1801). 'An Anatomical Description of a Male Rhinoceros', *Philosophical Transactions of the Royal Society*, 91: 145–52.

Lennie, Campbell (1976). *Landseer: The Victorian Paragon*. London: Hamilton.
Lennox, Charles Gordon, 5th Duke of Richmond (1839). *Catalogue raisonné of the Pictures in the Gallery of the Duke of Richmond*. London: Smith, Elder & Co.
Lennox, Charles, Earl of March (1911). *A Duke and his Friends*. London: Hutchinson.
Lennox-Boyd, C., Dixon, R., and Clayton, T. (1989). *George Stubbs the Complete Engraved Works*. London: Stipple Publishing.
Lever, C. (1977). *The Naturalized Animals of the British Isles*. London: Hutchinson.
Lever, C. (1992). *They Dined on Eland: The Story of the Acclimatisation Societies*. London: Quiller Press.
Lewis, Lesley (1998). *Thomas More Family Group Portraits after Holbein*. Leominster: Gracewing.
Lister, Sam (2011). '[The Aviary, Osterley Park]', *The Times*, 4 April 2011.
Llanover, Lady (1861) (ed.). *The Autobiography and Correspondence of Mary Granville, Mrs Delany*. Series 1. London: Richard Bentley.
Llanover, Lady (1862) (ed.). *The Autobiography and Correspondence of Mary Granville, Mrs Delany*. Series 2. London: Richard Bentley.
Loisel, Gustave (1912). *Histoire des meínageries de l'antiquiteí à nos jours*. Paris.
Longstaffe-Gowan, Todd (2005). *The Gardens and Parks at Hampton Court Palace*. London: Frances Lincoln.
Loudon, J. C. (1831). 'The Surrey Zoological Gardens', *Gardener's Magazine*, 7: 692–3.
Loudon, J. C. ('Observator') (1832). 'A Visit to the Surrey Zoological Gardens', *Magazine of Natural History*, 5: 401–4.
Loudon, J. C. (1846). *An Encyclopædia of Cottage, Farm, and Villa Architecture and Furniture*. London: Longman, Brown, Green, & Longman.
Luttrell, N. (1857). *A Brief Historical Relation of State Affairs from September 1678 to April 1714*. Oxford: Oxford University Press.
Lysons, Daniel (1792). *The Environs of London*, ii. London.
Macaulay, T. B., Lord (1841). 'Warren Hastings', *Edinburgh Review*, 74: 160–255.
Macaulay, T. B., Lord (1864). *The History of England from the Accession of James the Second*. London: Longman, Green, Longman, Roberts, & Green.
McCann, T. (1994). '"Much troubled with very rude company ...": The 2nd Duke of Richmond's Menagerie at Goodwood', *Sussex Archaeological Collection*, 132: 143–9.
MacCarthy, Fiona (1989). *Eric Gill*. London: Faber & Faber.
McEvedy, C. (1961). *The Penguin Atlas of Mediaeval History*. Harmondsworth: Penguin.
MacGregor, Arthur (1983). *Tradescant's Rarities: Essays on the Foundation of the Ashmolean Museum, 1683, with a Catalogue of the Surviving Early Collections*. Oxford: Clarendon Press.
MacGregor, Arthur (1994). 'The Life, Character and Career of Sir Hans Sloane', in Arthur MacGregor (ed.), *Hans Sloane, Collector, Scientist, Antiquary*. London: British Museum Press, 11–44.
MacGregor, Arthur (1995). 'The Natural History Correspondence of Sir Hans Sloane', *Archives of Natural History*, 22: 79–90.
MacGregor, Arthur (2012). *Animal Encounters: Human and Animal Interaction in Britain from the Norman Conquest to World War One*. London: Reaktion Books.

McKendrick, M. (1968). *Ferdinand and Isabella*. London: Cassell.
Macky, John (1722). *A Journey through England, in Familiar Letters from a Gentleman here to his Friend Abroad*. London.
McLean, Antonia (1972). *Humanism and the Rise of Science in Tudor England*. London: Heineman.
Macray, W. D. (1894). *A Register of the Members of St Mary Magdalen College, Oxford, from the Foundation of the College*. London.
Maitland, W. (1775). *The History of London from its Foundation to the Present Time*. London.
Malcolm, J. (1836). *The Life of Robert, Lord Clive*. London: John Murray.
Malcolm, J. P. (1811). *Anecdotes of the Manners and Customs of London from the Roman Invasion to the Year 1700*, i. 2nd edn. London: Longman, Hurst, Rees, Orme & Brown.
Marshall, Alice (1970). *Catalogue of the Anatomical Preparations of Dr William Hunter in the Museum of the Anatomy Department*. University of Glasgow.
Mason, A. Stuart (1992). *George Edwards: The Bedell and his Birds*. London: Royal College of Physicians.
Masseti, M. (2002). 'The Ring-Necked Parakeet, Psittacula krameri Scopoli, 1769, in the Aegean Region', in Marco Masseti (ed.), *Island of Deer: Natural History of the Fallow Deer of Rhodes and of the Vertebrates of the Dodecanese (Greece)*. City of Rhodes: Environmental Organization, 85–8.
Miles, A. E. W., and Grigson, C. (1990). *Colyer's Variations and Diseases of the Teeth of Animals*. Cambridge: Cambridge University Press.
Miller, D. P. (1983). 'Between Hostile Camps: Sir Humphry Davy's Presidency of the Royal Society of London, 1820–1827', *British Journal for the History of Science*, 16/1: 1–47.
Miller, Philip (1732). *The Gardeners Dictionary: Containing the Methods of Cultivating and Improving the Kitchen, Fruit and Flower Garden*. Dublin.
Mitchell, M. D. (2013). '"Legitimate Commerce" in the Eighteenth Century: The Royal African Company of England under the Duke of Chandos, 1720–1726', *Enterprise & Society*, 544–78.
Moiser, C. (2005). 'Gilbert Pidcock's Two Headed Cow', *Journal Bartlett Society*, 16: 12–15.
Monconys, B. (1665). *Journal des Voyages de Monsieur de Monconys*. Lyons.
Moore, Wendy (2005). *The Knife Man*. London: Bantam Press.
Moorhouse, Geoffrey (1971). *Calcutta, the City Revealed*. New Delhi: Penguin Books India.
Morales, E. (1994). 'The Guinea Pig in the Andean Economy: From Household Animal to Market Economy', *Latin American Research Review*, 29/3: 129–42.
Morales, E. (1995). *The Guinea Pig: Healing, Food and Ritual in the Andes*. Tucson: University of Arizona Press.
Moriarty, Colm (2012). <http://irisharchaeology.ie/2012/05/the-death-of-an-elephant-dublin-1681/> (accessed 4 June 2012).
Morley, H. (1880). *Memoirs of Bartholomew Fair—a Verbatim Reprint of the Original Edition*. London: Warne.
Mountague, William (1696). *The Delights of Holland: Or, A Three Months Travel about that and the Other Provinces*. London: John Sturton and A. Bosvile.
Mullen, A. (1682). *An Anatomical Account of the Elephant Accidentally Burnt in Dublin, on Fryday, June 17. in the Year 1681: Together with a Relation of New Anatomical Observations in the Eyes of Animals*. London.

Müller-Haye, B. (1984). 'Guinea-Pig or Cuy', in I. L. Mason (ed.), *Evolution of Domesticated Animals*. London: Longman, 252–7.

Murray, John (1826). *Experimental Researches on the Light and Luminous Matter of the Glow-Worm, the Luminosity of the Sea, the Phenomena of the Chameleon, the Ascent of the Spider into the Atmosphere, and the Torpidity of the Tortoise, etc.* Glasgow: W. R. McPhan.

Nelson, P. D. (2000). *General Sir Guy Carleton, Lord Dorchester: Soldier-Statesman of Early British Canada*. London: Associated University Presses.

Nicholls, H. (2013). *There's Life in this Old Bird Yet* <http://www.theguardian.com/science/blog/2013/jun/28/dead-dodo-old-bird> (accessed 2 December 2013).

Nichols, F. M. (1918). *The Epistles of Erasmus*. London: Longmans Green & Co.

Nichols, John (1828). *The Progresses, Processions, and Magnificent Festivities, of King James the First*. London: J. B. Nichols.

Nichols, John (1977). *The Progresses and Public Processions of Queen Elizabeth*. New York: Kraus Reprint Corporation.

Nichols, R. S. (2003). *English Pleasure Gardens*. Jaffrey, NH: David R. Godine.

Norrington, Ruth (1983). *In the Shadow of a Saint: Lady Alice More*. Waddesdon: Kylin Press.

Northumberland, Elizabeth, Duchess of (1775). *A Short Tour Made in the Year One Thousand Seven Hundred and Seventy One*. London.

Nygren, E. J., and Pressly, N. L. (1977). 'The Catalogue', in J. H. Plumb (ed.), *The Pursuit of Happiness: A View of Life in Georgian England*. New Haven, CT: Yale Center for British Art, 29–67.

O'Regan, H. J. (2002). 'The Archaeology of Zoos', *British Archaeology*, 68: 12–19.

Ollard, R. (1994.) *Cromwell's Earl: A Life of Edward Montagu, 1st Earl of Sandwich*. London: HarperCollins.

Oman, C. (1951). *Henrietta Maria*. London: Hodder & Stoughton.

Oman, C. (1964). *Elizabeth of Bohemia*. London: Hodder & Stoughton.

Ormond, Richard (1981). *Sir Edwin Landseer*. London: Thames & Hudson.

Ormrod, David (1973). *The Dutch in London*. London: HMSO.

Ormrod, David (2003). *The Rise of Commercial Empires: England and the Netherlands in the Age of Mercantilism, 1650–1770*. Cambridge: Cambridge University Press.

Ovenell, R. F. (1992). 'The Tradescant Dodo', *Archives of Natural History*, 19/2: 145–52.

Parish, Judith (1991). 'Walcot Hall: A Short History', in *Walcot Hall, a Brief History*. Privately printed, 13–28.

Parnell, G. (1999). *The Royal Menagerie at the Tower of London*. Leeds: Royal Armouries Museum.

Parsons, J. (1743). 'A Letter from Dr Parsons to Martin Folkes, Esq; President of the Royal Society, Containing the Natural History of the Rhinoceros', *Philosophical Transactions of the Royal Society*, 42: 523–40 and tables II, III.

Parsons, J. (1745). 'An Account of a Quadruped brought from Bengal, and now to be Seen in London', *Philosophical Transactions of the Royal Society*, 43: 465–7 and figure.

Parsons, J. (1760). 'Some Account of the Animal Sent from the East Indies, by General Clive, to his Royal Highness the Duke of Cumberland which is now in the Tower of London', *Philosophical Transactions of the Royal Society*, 51: 648–52 and figure.

Pasmore, Stephen (1976–8). 'John Hunter in Kensington', *Transactions of the Hunterian Society*, vol. 35–36 (unpaginated).
Paterson, W. (1789). *A Narrative of Four Journeys into the Country of the Hottentots and Caffraria*. London: J. Johnson.
Pattacini, L. (1998). 'André Mollet, Royal Gardener in St James's Park, London', *Garden History: The Journal of the Garden History Society*, 26/1: 3–18.
Pearson, John (1983). *Stags and Serpents: The Story of the House of Cavendish and Dukes of Devonshire*. London: Macmillan.
Peck, L. L. (2005). *Consuming Splendor: Society and Culture in Seventeenth-Century England*. Cambridge: Cambridge University Press.
Pennant, Thomas (1771). *Synopsis of Quadrupeds*. Chester.
Pennant, Thomas (1777). *British Zoology. Volume IV. Crustacea. Mollusca. Testacea*. London.
Pennant, Thomas (1781). *History of Quadrupeds*. London.
Pennant, Thomas (1790). *Of London*. London: Robt Faulder.
Pennant, Thomas (1793). *History of Quadrupeds*. 3rd edn. London.
Pennant, Thomas (1798). *Outlines of the Globe: The View of Hindoostan*. London.
Pennant, Thomas (1812). *British Zoology*. London.
Pepys, Samuel (1893–9). *The Diary of Samuel Pepys*, ed. Henry B. Wheatley. London: G. Bell.
Pepys, Samuel (1985). *The Diary of Samuel Pepys: A New and Complete Transcription*, ed. R. Latham and W. Matthews. London: Bell & Hyman.
Phillip, Arthur (1789). *The Voyage of Governor Phillip to Botany Bay; with an Account of the Establishment of the Colonies of Port Jackson and Norfolk Island*. London: John Stockdale.
Picard, Liza (1998). *Restoration London*. London: Weidenfeld & Nicolson.
Picard, Liza (2000). *Dr Johnson's London*. London: Weidenfeld & Nicolson.
Pidcock, Gilbert (1778). *The History, and Anatomical Description of a Cassowar, from the Isle of Java, in the East-Indies: The Greatest Rarity now in Europe*. Bury: W. Green.
Pieters, F. (1998). *Wonderen der Natuur in de Menagerie van Blauw Jan te Amsterdam, zoals gezien door Jan Velten rond 1700*. ETV/Artis Library of Amsterdam.
Pigière, F., Van Neer, W., Ansieau, C., and Denis, M. (2012). 'New Archaeozoological Evidence for the Introduction of the Guinea Pig to Europe', *Journal of Archaeological Science*, 39: 1020–4.
Plot, Robert (1677). *The Natural History of Oxfordshire, Being an Essay toward the Natural History of England*. Oxford: The Theatre.
Plumb, C. (2010a). 'Exotic Animals in Eighteenth Century Britain'. Ph.D. thesis, University of Manchester.
Plumb, C. (2010b). '"In fact, one cannot see it without laughing": The Spectacle of the Kangaroo in London, 1770–1830', *Museum History Journal*, 3/1: 7–32.
Plumb, C. (2011). '"The Queen's Ass": The Cultural Life of Queen Charlotte's Zebra in Georgian Britain', in Samuel Alberti (ed.), *The Afterlife of Animals: A Museum Menagerie*. Charlottesville, VA: University of Virginia Press, 17–36.
Polito, S. (1803). *Description and Natural History of S. Polito's Collection of Living Beasts and Birds*. Edinburgh.
Pontoppidan, Erich (1755). *The Natural History of Norway*. London.
Potter, Jennifer (2006). *Strange Blooms: The Curious Lives and Adventures of the John Tradescants*. London: Atlantic Books.

Power, E. E. (1964). *Medieval English Nunneries*. New York: Biblo & Tannen.
Pückler-Muskau, H. (1832). *Tour in Germany, Holland and England, in the Years 1826, 1827 & 1828*, trans. Sarah Austin. London: Effingham Wilson.
Raat, A. J. P. (2010). *Gideon Loten (1710–1789): A Personal History of a... Dutch Virtuoso*. Hilversum: Verloren.
Raffles, Sir Thomas Stamford (1822). 'Descriptive Catalogue of a Zoological Collection, Made on Account of the Honourable East India Company, in the Island of Sumatra and its Vicinity, under the Direction of Sir Thomas Stamford Raffles, Lieutenant-Governor of Fort Marlborough', *Transactions of the Linnean Society of London*, 13: 239–74.
Raffles, Sophia (1835). *Memoir of the Life and Public Services of Sir Thomas Stamford Raffles, FRS*. London: James Duncan.
Raikes, Thomas (1858). *A Portion of the Journal Kept by T. Raikes, Esq., from 1831 to 1847*. London.
Ramsbottom, J. (1938). 'Old Essex Gardens and their Gardeners', *Essex Naturalist*, 26: 73–87.
Ray, John (1678). *The Ornithology of Francis Willughby: Translated into English, and Enlarged with Many Additions*. London.
Reichenbach, Herman (2002). 'Lost Menageries: Why and how Zoos Disappear (Part 1)', *International Zoo News*, 49/3 (no. 316).
Renault, G. (1959). *The Caravels of Christ*. London: Allen & Unwin.
Richardson, J. (2000). *The Annals of London*. London: Cassell.
Rieke-Müller, A, and Dittrich, L. (1999). *Unterwegs mit wilden Tieren Wandermenagerien zwischen Belehrung und Kommerz 1750–1850*. Marburg and Lahn: Basilisken-Presse.
Ritvo, H. (1990). *The Animal Estate: The English and Other Creatures in the Victorian Age*. London: Penguin.
Roberts, Jane (1997). *Royal Landscape: The Gardens and Parks of Windsor*. London: Yale University Press.
Robinson, R. (1887). *Thomas Bewick his Life and Times*. Newcastle: R. Robinson.
Rodenhurst, W. B. (1802). *A Description of Hawkstone, the Seat of Sir R. Hill, Bart MP*. London.
Rolfe, W. D. I. (1983a). 'William Hunter (1718–83) on Irish "Elk" and Stubbs' Moose', *Archives of Natural History*, 11: 263–90.
Rolfe, W. D. I. (1983b). 'A Stubbs Drawing Recognized', *Burlington Magazine*, 125: 738–41.
Rolfe, W. D. I., and Grigson, C. (2006). 'Stubbs's Drill and Albino Hamadryas Baboon in Conjectural Historical Context', *Archives of Natural History*, 32: 18–41.
Rookmaaker. L. C. (1973). 'Rhinoceroses in Europe', *Bijdragen tot de Dierkunde*, 43/1: 39–63.
Rookmaaker, L. C. (1989). *The Zoological Exploration of Southern Africa 1650–1790*. Rotterdam: Balkema.
Rookmaaker, L. C. (1992). 'J. N. S. Allamand's Additions (1769–1781) to the Nouvelle Edition of Buffon's Histoire Naturelle published in Holland', *Bijdragen tot de Dierkunde, Amsterdam*, 61/3: 131–62.
Rookmaaker. L. C. (1998). *The Rhinoceros in Captivity*. The Hague: SPB Academic.
Rush, R. (1833). *A Residence at the Court of London*. Philadelphia: King & Biddle.

Rush, R. (1845). *A Residence at the Court of London 1819–1825*. Philadelphia: Lee & Blanchard.
Rybot, Doris (1972). *It Began before Noah*. London: Michael Joseph.
Rye, W. B. (1865). *England as Seen by Foreigners in the Days of Elizabeth and James the First*. London: J. R. Smith.
Saussure, Cesar de (1902). *A Foreign View of England in the Reigns of George I and George II*. London: John Murray.
Scherren, H. (1901). *A Short History of the Zoological Society of London*. London: William Clowes.
Scherren, H. (1905). *The Zoological Society of London: A Sketch of its Foundation and Development and the Story of its Farm, Museum, Gardens, Menagerie and Library*. London: Cassell & Co.
Schupbach, William (1986). 'Illustrations from the Wellcome Institute Library: Earl's Court House from John Hunter to Robert Gardiner Hill', *Medical History*, 30: 351–6.
Scott, Walter (1998). *The Journal of Sir Walter Scott*. Edinburgh: Canongate Books.
Seward, Anna (1811). *Letters: Written between the Years 1784 and 1807*. Edinburgh and London.
Sharpe, John (1832). *Popular Zoology, Comprising Memoirs and Anecdotes of the Quadrupeds, Birds and Reptiles, in the Zoological Society's Menagerie*. London.
Shaw, George (1796). *Cimelia physica: Figures of Rare and Curious Quadrupeds, Birds, &c*. London.
Sherwin, O. (1958). 'A Man with a Tail—Lord Monboddo', *Journal of the History of Medicine and Allied Science*, 13: 435– 68.
Simond, L. (1815). *Journal of a Tour and Residence in Great Britain: During the Years 1810 and 1811*, ii. Edinburgh: Archibald Constable; London: Longman, Hurst, Rees, Orm & Brown.
Simons, John (2012). *The Tiger that Swallowed the Boy: Exotic Animals in Victorian England*. Farringdon: Libri Publishing.
Sloane, Hans (1725). *A Voyage to the Islands Madera, Barbados, Nieves, S. Christophers and Jamaica: With the Natural History of the Herbs and Trees, Four-Footed Beasts, Fishes, Birds, Insects, Reptiles, &c. of The Last of Those Islands*, ii. London.
Sloane, Hans (1728). 'An Account of Elephants Teeth and Bones Found under Ground', *Philosophical Transactions of the Royal Society*, 35: 457–71.
Sloane, Kim (2003). *Enlightenment, Discovering the World in the Eighteenth Century*. London: British Museum Press.
Smith, T. (1806). *The Naturalist's Cabinet, Containing Interesting Sketches of Animal History*. London: James Cundee.
Smollett, Tobias (1771). *The Expedition of Humphry Clinker*. London.
Spano, S., and Truffi, G. (1986). 'Il Parrocchetto dal collare (*Psittacula krameri*) al stato libero in Europa, con particolare referimento alle presenze in Italia', *Rivista italiana di ornithologia*, 56: 231–9.
Speaight, R. (1966). *The Life of Eric Gill*. London: Methuen & Co.
Spear, Percival (1975). *Master of Bengal: Clive and his India*. London: Thames & Hudson.
Stewart, Edith Helen Vane Tempest, Marchioness of Londonderry (1958). *Frances Anne: The Life and Times of Frances Anne, Marchioness of Londonderry, and her Husband Charles, Third Marquess of Londonderry*. London: Macmillan.

Strong, R. C. (1998). *The Renaissance Garden in England*. London: Thames & Hudson.

Strong, R. C. (2000). *Henry Prince of Wales and England's Lost Renaissance*. London: Pimlico.

Strype, J. (1720). *A Survey of the Cities of London and Westminster*. London: A. Churchill.

Stukeley, William (1723). 'An Essay towards the Anatomy of the Elephant, from one Dissected at Fort St George Oct 1715, and Another at London Oct 1720', in W. Stukely, *Of the spleen, its description and history*. London.

Surtees, Virginia (1990) (ed.). *A Second Self: The Letters of Harriet Granville 1810–1845*. Salisbury: Michael Russell.

Switzer, Stephen (1718). *Ichnographia Rustica, or, the Nobleman, Gentleman and Gardener's Recreation*. London.

Symonds, H. (1912). 'A Jacobean Zoological Garden', *Home Counties Magazine*, 14: 309–12.

Symonds, H. (1988). 'A Jacobean Zoological Garden in St James's', *London and Middlesex Archaeological Society Transactions*, 31: 133–41.

Tattersfield, Nigel (2011). *Thomas Bewick: The Complete Illustrative Works*. London: British Library.

Taylor, Basil (1970). *George Stubbs's Painting of a Cheetah with Two Indians: Art at Auction 1969–70*. London: Sotheby's.

Taylor, Basil (1971). *Stubbs*. London: Phaidon.

Taylor, John (1826). *The Life, Death and Dissection of the Largest Elephant ever Known in this Country . . . Destroyed at Exeter 'Change*. London.

Temple, R., and Ansley, L. M. (1936) (eds). *The Travels of Peter Mundy in Europe and Asia, 1608–1667*, v. Cambridge: Cambridge University Press.

Thayer, Stuart (2005). *American Circus Anthology, Essays of the Early Years* <http://www.circushistory.org/Thayer/Thayer2a.htm> (accessed 17 October 2011).

Thompson, R. W. (1934). *Wild Animal Man: A Biography of Reuben Castang*. London: Duckworth.

Thomson, D. C. (1882). *The Life and Works of Thomas Bewick*. London: The Art Journal Office.

Tillyard, S. (1995). *Aristocrats, Caroline, Emily, Louisa and Sarah Lennox 1740–1832*. London: Vintage.

Topsell, Edward (1607). *The Historie of Foure-Footed Beastes . . . Collected out of all the Volumes of C. Gesner*. London: Printed by William Iaggard.

Topsell, Edward (1972). *The Fowles of Heauen; or, History of Birdes*, ed. Thomas P. Harrison and F. David Hoeniger. Austin, TX: University of Texas.

Toynbee, P. (1927–8) (ed.). Horace Walpole's Journals of Visits to Country Seats (1763)', *Sixteenth Volume of the Walpole Society*, 10–80.

Tradescant, J. (1656). *Museum Tradescantianum: Or, A Collection of Rarities Preserved at South Lambeth, near London*. London.

Trotter, L. J. (1878). *Warren Hastings: A Biography*. London: W. H. Allen.

Turner, Samuel (1800). *An Account of an Embassy to the Court of the Teshoo Lama, in Tibet: Containing a Narrative of a Journey through Bootan, and Part of Tibet*. London.

Tyson, E. (1680). *Phocaena, or the Anatomy of a Porpess, Dissected at Gresham Colledge: With a Praeliminary Discourse Concerning Anatomy, and a Natural History of Animals*. London: Benj. Tooke.

Tyson, E. (1699). *Orang-outang, sive, Homo sylvestris, or, The Anatomy of a Pygmie Compared with that of a Monkey, an Ape, and a Man*. London: Thomas Bennet & Daniel Brown.

Uglow, J. (2004). *A Little History of British Gardening*. London: Chatto & Windus.

Uglow, J. (2006). *Nature's Engraver: A Life of Thomas Bewick*. London: Faber & Faber.

Ullah, Ansar Ahmed, and Eversley, John (2010). *Bengalis in London's East End*. London: Swadhinata Trust.

Urquhart, Diane (2007). *The Ladies of Londonderry: Women And Political Patronage*. London: Tauris.

Van der Stighelen, K., and James, S. E. (2000). '"Haec habeant longos gaudia tanta dies": New Discoveries Concerning the Portrait of the Family of William Brooke, 10th Lord Cobham at Longleat House', *Dutch Crossing: A Journal of Low Countries Studies*, 23/1: 66–101.

Verey, D. (1979). *The Buildings of England. Gloucestershire I, the Cotswolds*. 2nd edn. Harmondsworth: Penguin.

Verney, F. P. (1892). *Memoirs of the Verney Family*, i and ii. *During the Civil War*. London: Longmans, Green & Co.

Verney, M. M. (1899). *Memoirs of the Verne Family*, iv. *From the Restoration to the Revolution 1660–1696*. London: Longmans, Green & Co.

Vevers, Gwynne (1976). *London's Zoo: An Anthology to Celebrate 150 Years of the Zoological Society of London, with its Zoos at Regent's Park in London and Whipsnade in Bedfordshire*. London: Bodley Head.

Vigors, N., and Broderip, W. J. (1829). *Guide to the Gardens of the Zoological Society March 1829*. London: Richard Taylor.

Walford, E. (1885). *Greater London*. London: Cassell.

Walker, T. E. C. (1968). 'The Clives at Claremont', *Surrey Archaeological Collections*, 65: 91–6.

Walpole, H. (1937–83). *The Yale Edition of Horace Walpole's Correspondence*, ed. W. S. Lewis. New Haven, CT: Yale University Press.

Warne, Richard (1802). *Excursions from Bath*. London: G. G. & J. Robinson.

Warwick, J. E. (1836). *Description and History, with Anecdotes, of the Giraffes, (Camelopardis giraffa, Gmel.) now Exhibiting at the Surrey Zoological Gardens*. London: J. King.

Waterhouse, G. R. (1838). *Catalogue of the Mammalia Preserved in the Museum of the Zoological Society of London*. London: Zoological Society of London.

Watson, J. N. P. (1978). '"Going to See the Lions": The Tower Menagerie', *Country Life* (16 November), 164: 1637–8.

Weinreb, B., and Hibbert, C. (1995). *The London Encyclopaedia*. 2nd edn. London: Macmillan.

Weinstein, R. (1980). 'Some Menagerie Accounts of James I', *Transactions of the London and Middlesex Archaeological Society*, 31: 133–41.

Welch, M. A. (1972). 'Francis Willoughby, F.RS (1635–1672)', *Journal of the Society for the Bibliography of Natural History*, 6/2: 71–85.

Wilkinson, B. J. (2011). *Carrion Dreams: A Chronicle of the Human–Vulture Relationship* <http://archive.org/details/CarrionDreams2.0AChronicleOfTheHuman-vultureRelationship> (accessed 20 July 2012).

Williams, Clare (1933). *Sophie in England, a Translation of the Passages on England in the Journal of a Journey through Holland and England (1788)*. London: Jonathan Cape.

Wing, E. (1977). 'Animal Domestication in the Andes', in C. A. Reed (ed.), *Origins of Agriculture*. The Hague: Mouton, 837–59.
Wood, J. C. (1973). *A Dictionary of British Animal Painters*. Leigh-on-Sea: F. Lewis.
Wood, William (1807). *Zoography: or, the Beauties of Nature Displayed*. London: Cadell & Davies.
Wood Jones, F. (1949). 'John Hunter and his Museum', *Annals of the Royal College of Surgeons of England*, 4: 337–41.
Wood Jones, F. (1951). 'John Hunter's Unwritten Book', *Lancet*, 258: 778–80.
Worlidge, J. (1677). *Systema Horti-culturae: Or, the Art of Gardening*. London.
Wortley Montagu, M. (1906). *Letters from the Right Honourable Lady Mary Wortley Montagu 1709 to 1762*. London: Dent.
Yalden, D. W., and Albarella, U. (2009). *The History of British Birds*. Oxford: Oxford University Press.
Yapp, W. B. (1981). *Birds in Medieval Manuscripts*. London: British Library.
Yorke, M. (1981). *Eric Gill: Man of Flesh and Spirit*. London: Constable.
Zagorodnaya, I. (2006). 'English Diplomats at the Court of the Tzars', in Olga Dmitrieva and Natalya Abramova (eds), *Britannia and Muscovy, English Silver at the Court of the Tsars*. New Haven, CT: Yale University Press, 176–95.

PICTURE CREDITS

The Art Archive: 5.5; The Author: 5.2, 5.3, 5.4, **Plate** 2, **Plate** 6; Biodiversity Heritage Library. Digitized by Smithsonian Libraries (www.biodiversitylibrary.org): 5.6; Bridgeman Art: 6.1, **Plate** 14; British Library: 2.2, 4.1, 4.2, 4.4, 4.8, 4.9, 4.12, 4.13, **Plate** 12; British Museum: 3.1, 3.2; British Newspaper Archive: 4.3; The Master and Fellows of Corpus Christi College, Cambridge: 1.1; Glasgow University Library: 2.1; Hunterian Museum, Glasgow: **Plate** 7; Manchester Art Gallery: **Plate** 8; National Trust: **Plate** 1; Public Domain: 4.6, **Plate** 3; Royal Collection Trust / ©Her Majesty Queen Elizabeth II 2014: **Plate** 11; Royal College of Surgeons of England: 4.10, 4.11; University of Houston: 1.2; University Library of Göttingen: **Plate** 4; University Library of Wisconsin: 3.3; Victoria and Albert Museum: 4.5, **Plate** 9, **Plate** 13; Wellcome Images: 4.7, **Plate** 5; Yale Centre for British Art: 5.1, **Plate** 10

INDEX OF ANIMALS

GENERAL INDEX